Gene Regulation by Steroid Hormones

Edited by
A.K. Roy and J.H. Clark

With 145 Figures

Springer-Verlag
New York   Heidelberg   Berlin

ARUN K. ROY
Professor of Biological Sciences, Oakland University, Rochester, Michigan 48063, USA

JAMES H. CLARK
Professor of Cell Biology, Baylor College of Medicine, Houston, Texas 77030, USA

Cover Picture: Electronmicrograph of a hybrid molecule between ovalbumin gene and ovalbumin mRNA. The intervening sequences are shown as loops. Courtesy of Dr. Eugene Lai, Baylor College of Medicine.

Library of Congress Cataloging in Publication Data
Conference on Molecular Mechanism of Steroid Hormone Action, Oakland University, 1978.
  Gene regulation by steroid hormones.
  Bibliography: p.
  Includes index.
  1. Steroid hormones—Physiological effect—Congresses. 2. Genetic regulation—Congresses. 3. Gene expression—Congresses. I. Roy, Arun K. II. Clark, James H. III. Title.
QP572.S7C66   1978   591.1'927   79-25018

Preface

Within the last two decades endocrinological research has taken a definite turn toward biochemistry and molecular biology. This has resulted in a new discipline called "molecular endocrinology." Studies on the mechanism of hormone action have continued to make headlines with fundamental discoveries in receptor action and gene regulation. Recently the insect endocrinologists have also begun to explore the molecular mechanism of steroid hormone action taking advantage of the vast number of *Drosophila* mutants, the library of *Drosophila* gene, and several well-characterized cell-lines. The availability of the recombinant DNA technology has provided a truly revolutionary tool in the hands of the molecular endocrinologists. "Gene Regulation by Steroid Hormones" is compiled and presented in this frontier spirit, and we hope that this volume will serve not only the active investigators in the field but will also be very useful to students and researchers with a general interest in regulatory biology.

The book is an offshoot of the Conference on Molecular Mechanism of Steroid Hormone Action held at the Meadow Brook Mansion of Oakland University in the fall of 1978. We wish to acknowledge the financial assistance from the National Science Foundation and Oakland University. The conferees will never forget the warmest hospitality of Dr. LOWELL EKLUND and his staff at the Meadow Brook center and we also wish to express personal gratitude to many of our students and colleagues for helping us to make the conference a great success.

Spring 1980
ARUN K. ROY
JAMES H. CLARK

Table of Contents

List of Contributors

C.M. ALVAREZ, Department of Biology, University of California, Los Angeles, CA 90024, USA

C.W. BARDIN, Center for Biomedical Research, The Population Council, The Rockefeller University, New York, NY 10021, USA

J.D. BAXTER, Howard Hughes Medical Institute Laboratories, Metabolic Research Unit and the Department of Medicine, Biochemistry and Biophysics, University of California, San Francisco, CA 94143, USA

C. BECKERS, FB 10 Zoologie, Technische Hochschule, D-6100 Darmstadt, Fed. Rep. of Germany

T.R. BROWN, Division of Pediatric Endocrinology, Department of Pediatrics, Johns Hopkins Hospital, Baltimore, MD 21205, USA

L.P. BULLOCK, Departments of Medicine and Comparative Medicine, The Milton S. Hershey Medical Center, The Pennsylvania State University, Hershey, PA 17033, USA

F.M. BUTTERWORTH, Department of Biological Sciences, Oakland University, Rochester, MI 48063, USA

B. CHATTERJEE, Department of Biological Sciences, Oakland University, Rochester, MI 48063, USA

L. CHERBAS, Cellular and Developmental Biology, The Biological Laboratories, Harvard University, Cambridge, MA 02138, USA

P. CHERBAS, Cellular and Developmental Biology, The Biological Laboratories, Harvard University, Cambridge, MA 02138, USA

J.H. CLARK, Department of Cell Biology, Baylor College of Medicine, Houston, TX 77030, USA

R.G. DEELEY, National Cancer Institute, National Institutes of Health, Bethesda, MD 20205, USA

G. DEMETRI, Stanford University, Palo Alto, CA 94302, USA

R. DENNIS, Institut für Physiologische Chemie, Philipps Universität, D-3550 Marburg(Lahn), Fed. Rep. of Germany

A.K. DESHPANDE, Center for Biomedical Research, The Population Council, The Rockefeller University, New York, NY 10021, USA

A. DUGAICZYK, Department of Cell Biology, Baylor College of Medicine, Houston, TX 77030, USA

T.E. EESSALU, Department of Biological Sciences, Oakland University, Rochester, MI 48063, USA

H. ERIKSSON, Department of Cell Biology, Baylor College of Medicine, Houston, TX 77030, USA

A. GAREN, Department of Molecular Biophysics and Biochemistry, Yale University, New Haven, CT 06520, USA

T.D. GELEHRTER, Department of Human Genetics, University of Michigan Medical School, Ann Arbor, MI 48109, USA

R.F. GOLDBERGER, National Cancer Institute, National Institutes of Health, Bethesda, MD 20205, USA

D.K. GRANNER, Department of Medicine, University of Iowa Medical School, Iowa City, IW 52240, USA

W. GUYETTE, Department of Cell Biology, Baylor College of Medicine, Houston, TX 77030, USA

G. HAGER, National Cancer Institute, NIH, Division of Cancer Cause and Prevention, Laboratory of Tumor Virus Genetics, Bethesda, MD 20205, USA

J. W. HARDIN, Department of Cell Biology, Baylor College of Medicine, Houston, TX 77030, USA

J.M. HARMON, National Cancer Institute, NIH, Division of Cancer Biology and Diagnosis, Laboratory of Biochemistry, Bethesda, MD 20205, USA

R.A. HIIPAKKA, Ben May Laboratory for Cancer Research, University of Chicago, Chicago, IL 60637, USA

L.K. JOHNSON, Howard Hughes Medical Institute Laboratories, Metabolic Research Unit and the Department of Medicine, Biochemistry and Biophysics, University of California, San Francisco, CA 94143, USA

J.-A. LEPESANT, Centre National de la Recherche Scientifique, Institut de Recherche en Biologie Moleculaire, University Paris VII, Paris, France

S. LIAO, Ben May Laboratory for Cancer Research, University of Chicago, Chicago, IL 60637, USA

R.M. LOOR, Ben May Laboratory for Cancer Research, University of Chicago, Chicago, IL 60637, USA

M. MANTEUFFEL-CYMBOROWSKA, Polish Academy of Sciences, Nencki Institute of Experiment Medicine, Warsaw, Poland

B. MARKAVERICH, Department of Cell Biology, Baylor College of Medicine, Houston, TX 77030, USA

P. MAROY, Institute of Genetics, Hungarian Academy of Sciences, Szeged, Hungary

M.J. MATUSIK, Department of Cell Biology, Baylor College of Medicine, Houston, TX 77030, USA

A.R. MEANS, Department of Cell Biology, Baylor College of Medicine, Houston, TX 77030, USA

V.K. MOUDGIL, Department of Biological Sciences, Oakland University, Rochester, MI 48063, USA

G.C. MUELLER, Department of Oncology, McArdle Laboratory, University of Wisconsin, Madison, WI 53706, USA

H. NISHIGORI, Department of Molecular Medicine, Mayo Clinic, Rochester, MN 55901, USA

S.K. NORDEEN, Howard Hughes Medical Institute Laboratories, Metabolic Research Unit and the Department of Medicine, Biochemistry and Biophysics, University of California, San Francisco, CA 94143, USA

M.R. NORMAN, Department of Chemical Pathology, King's College Hospital Medical School, London, England

B. W. O'MALLEY, Department of Cell Biology, Baylor College of Medicine, Houston, TX 77030, USA

J.D. O'CONNOR, Department of Biology, University of California, Los Angeles, CA 90024, USA

D.A. RICHARDS, Department of Cell Biology, Baylor College of Medicine, Houston, TX 77030, USA

J.L. ROBERTS, Howard Hughes Medical Institute Laboratories, Metabolic Research Unit and the Department of Medicine, Biochemistry and Biophysics, University of California, San Francisco, CA 94143, USA

J.M. ROSEN, Department of Cell Biology, Baylor College of Medicine, Houston, TX 77030, USA

A.K. ROY, Department of Biological Sciences, Oakland University, Rochester, MI 48063, USA

B.A. Sage, Department of Biology, University of California, Los Angeles, CA 90024, USA

C. Savakis, Cellular and Developmental Biology, The Biological Laboratories, Harvard University, Cambridge, MA 02138, USA

T.J. Schmidt, National Cancer Institute, NIH, Division of Cancer Biology and Diagnosis, Laboratory of Biochemistry, Bethesda, MD 20205, USA

W.T. Schrader, Department of Cell Biology, Baylor College of Medicine, Houston, TX 77030, USA

Y. Seleznev, Department of Cell Biology, Baylor College of Medicine, Houston, TX 77030, USA

J.P. Stein, Department of Cell Biology, Baylor College of Medicine, Houston, TX 77030, USA

E.B. Thompson, National Cancer Institute, NIH, Division of Cancer Biology and Diagnosis, Laboratory of Biochemistry, Bethesda, MD 20205, USA

D.O. Toft, Department of Molecular Medicine, Mayo Clinic, Rochester, MN 55901, USA

S. Tsai, Department of Cell Biology, Baylor College of Medicine, Houston, TX 77030, USA

S. Upchurch, Department of Cell Biology, Baylor College of Medicine, Houston, TX 77030, USA

W.V. Vedeckis, Department of Cell Biology, Baylor College of Medicine, Houston, TX 77030, USA

A. Venetianer, Institute of Genetics, Biological Research Center, Hungarian Academy of Sciences, Szeged, Hungary

C.M. Williams, Cellular and Developmental Biology, The Biological Laboratories, Harvard University, Cambridge, MA 02138, USA

H.G. Williams-Ashman, Ben May Laboratory for Cancer Research, Department of Biochemistry, and Department of Pharmacological and Physiological Sciences, University of Chicago, Chicago, IL 60637, USA

S.L.C. Woo, Department of Cell Biology, Baylor College of Medicine, Houston, TX 77030, USA

C.D. Yonge, Department of Surgery, Yale Medical School, New Haven, CT 06520, USA

Estrogens and Progestins

Chapter 1
Steroid Hormone Action: Comments on the Nature of the Problem

GERALD C. MUELLER

The mechanism of action of steroid hormones has been the subject of inquiry for more than 25 years. Despite a massive effort, the collection of many data, and highly significant progress, the goal—a description of steroid action in molecular terms—has remained elusive. Why? What are the barriers to achieving this understanding? Are there any particular areas of underdevelopment?

In the quest for understanding there have been several breakthroughs. The first of these was the demonstration that a large measure of the response to steroid hormones resulted from and depended on an acceleration of gene expression. The initial evidence for this conclusion resided in the ability of inhibitors of RNA and protein synthesis to block many of the early effects of estradiol in the rat uterus. This concept now has been validated on many fronts, with nearly all steroid hormones, using the more sophisticated approaches of molecular biology to demonstrate the newly synthesized RNA and protein molecules. We now know with certainty that a large fraction of the response to steroid hormones results from effects of the hormone on the gene expression mechanisms. Whether these effects are direct or indirect is still unanswered. Especially puzzling has been the observation that the effects may be either positive or negative with respect to the production or suppression of different messenger RNAs and proteins. These divergent genetic responses may even occur within the same cell, a situation further complicated by the experience that the spectrum of genes affected, and their relative balance, may change with the duration of the hormonal treatment. In addition, there is the observation that the treatment of cells that are preparing for or are engaging in replication may broaden the range of responses further, with certain cells emerging from a specific induction sequence with an entirely different phenotype. The hormonally induced differentiation may even include a change in the cells' responsiveness to the hormones that precipitated their change. Finally, there are certain acute responses that do not depend at all on the induced synthesis of RNA and protein, but involve the activation of preexisting catalytic components or membrane states to achieve a response.

In attempting to explain the range of responses it is difficult to contemplate a mechanism of hormone action that does not encompass a means for addressing multiple targets. Also implicit in this view is the existence of a recognition mechanism that is shared by the multiple targets and is used by the hormone receptor or other mediator to direct the hormone action to these targets. In my opinion the immediate targets are proteins, and the recognition factor is likely to operate between the hormone receptor and these proteins. Only proteins contain the necessary topological information and conformational responsiveness to fit the diversity of the hormone responses.

To date relatively little effort has been directed toward the search for a common group or principle of operation that might function in the recognition process. This is unfortunate, since such a search might prove equally valuable to the current rush to characterize the affected genes and their transcripts. It appears that a good place to look for a recognition factor might be the chromosomal proteins that envelop the responsive genes. In particular, we should inquire into the nature of any hormone-induced changes in these protein associations.

Are there any common proteins? Are there any common prosthetic groups? Are any of these proteins structurally modified in response to the hormone treatment? Do any of the proteins cooperate in an enzymatically mediated relaxation of the chromatin structure in these regions—especially a relaxation that exposes promoter sites?

The other side of the mechanism of the steroid hormone action problem is the hormone-recognizing component, the receptor. Clearly, the demonstration that target tissues contain proteins that bind the hormone with high specificity constitutes a major breakthrough in the conceptualization of this area of biology. The finding that these receptor proteins undergo steroid-induced conformational changes at normal body temperatures further identifies them as molecular machines with the potential for inflicting structural change on any associating molecule that shares its topography. In this connection it is of interest that the receptor proteins are seldom found as free and independent molecules at physiological salt concentrations. Instead, they exist in molecular associations with other proteins. Very little effort has gone into identifying these associating proteins, despite the observation that they vary in size, representation in different tissues, and their electrostatic properties in different ionic environments.

In my opinion, the proteins in these receptor complexes have been disregarded far too long. In particular, we need to establish their identity. We then can ask whether the interaction of these proteins with the receptor makes them more or less susceptible to the various posttranslational modifying reactions that influence protein structure. It seems reasonable to anticipate that there will be a recognition group or principle shared by different members and that accounts for any of the specificity in the associations with steroid receptor complexes. In particular, the role of such an interaction in the assembly of multienzyme systems, whose activity varies with the hormonal state or the assembly of multiprotein complexes around hormonally responsive genes, should be investigated. It is anticipated that knowledge of these interactions will provide the basis for an explanation of the subtle balance in the positive and negative effects of the hormone on the expression of the different genes that characterize the responses to a single hormone in different cells.

None of the above comments should be construed as depreciating the value of ongoing studies aimed at the characterization of the receptor as a molecule in its own right. Clearly this knowledge is essential for an understanding of hormone action. The low concentration of receptors in tissues and their high propensity to aggregate have seriously hindered these studies. Perhaps we need to employ the approaches of modern molecular biology to obtain a reiteration of the genes for receptors in a productive cell. The receptor molecules are needed both for use as reagents in probing receptor interactions with other cellular components and as substrates for receptor characterization studies.

Finally, progress in our understanding of steroid action, in all probability, will continue to depend on the acquisition of more insight into the structure and function of chromatin, and particularly, into the mechanisms leading to the revision or recapitulation of specific chromatin complexes during nuclear replication. Whereas hormone–receptor complexes can mediate their actions in membranes and other extranuclear settings, it is clearly their potential for action in the genetic scene during replication that provides a molecular basis for determining the biological destiny of cells. As has been pointed out previously, the replication of chromatin depends on the availability of soluble proteins from the cytoplasm and their involvement in the formation of transient molecular complexes with newly replicated DNA. In accordance with the propensity of steroid receptors to interact with other cell proteins, it is proposed that steroid hormone receptors may direct the differentition of cells by interacting with proteins that are critical in the replication and assembly of new chromatin.

In conclusion, it has been the goal of the above discussion to stimulate students of steroid hormone action to inquire imaginatively into the nature of the interactions of steroid hormone receptors with other cell proteins since the range of permutations and combinations offered would appear to provide a basis for spanning the diversity of genetic and nongenetic effects that are mediated by the rather simple steroid molecules.

Chapter 2

The Structure and Regulation of the Natural Chicken Ovomucoid Gene

Joseph P. Stein, Savio L.C. Woo, Achilles Dugaiczyk, Sophia Tsai, Anthony R. Means, and Bert W. O'Malley

I. Introduction

The molecular mechanism by which steroid hormones regulate specific gene expression has been an area of acute interest during the past several years. One particularly attractive model system for studying this hormonal regulation has been the hen oviduct (O'Malley et al. 1969). A number of laboratories, in addition to our own, have utilized this model system for investigations of eucaryotic molecular biology (Oka and Schimke 1969; Palmiter and Schimke 1973; Palmiter et al. 1976; Cox 1977; Hynes et al. 1977; Garapin et al. 1978b; Mandel et al. 1978). Administration of estrogen to the newborn chick stimulates oviduct growth and differentiation and results in the appearance of a number of new specific intracellular proteins (O'Malley et al. 1969; Hynes et al. 1977; Palmiter 1973; Chan et al. 1973; Harris et al. 1973, 1975; Sullivan et al. 1973; O'Malley and Means 1974). The synthesis of one of these proteins, ovalbumin, has been studied extensively. Ovalbumin mRNA has been purified (Rosen et al. 1975), and a full-length dsDNA copy synthesized (Monahan et al. 1976b) and cloned in a bacterial plasmid (McReynolds et al. 1977). More recently, ovalbumin genomic DNA sequences have been isolated from restriction enzyme digests of hen DNA and cloned (Woo et al. 1978). The other three major proteins under estrogenic control in the oviduct tubular gland cell, ovomucoid, conalbumin and lysozyme, have been less extensively studied (Palmiter 1972; Hynes et al. 1977). In order to study the mechanisms by which steroid hormones coordinately regulate the expression of different genes in one tissue, we attempted to isolate a second estrogen-induced gene from the chick oviduct. We now wish to describe the purification of ovomucoid mRNA, the synthesis and cloning of a dsDNA from this ovomucoid mRNA, and the use of this ovomucoid structural gene clone to isolate and amplify the natural ovomucoid gene from chick DNA, and to determine the sequence organization of the native ovomucoid gene.

II. Cloning of the Ovomucoid Structural Gene Sequence

A. Ovomucoid mRNA Purification

Sequential chromatography of a total hen oviduct RNA extract on oligo (dT)-cellulose, Sepharose 4B, which substantially separated the ovalbumin mRNA activity from ovomucoid and lysozyme mRNA activities, and a second oligo (dT)-cellulose column, yielded a mRNA preparation that consisted primarily of two different size classes of RNA (Stein et al. 1978). To separate these two size classes, the mRNA preparation was denatured and applied to 12.2-M linear 5–20% sucrose gradients. Although the RNA banded in a rather broad peak, judicious collection of fractions from the leading edge of the peak resulted in the removal of the bulk of the smaller contaminating mRNA. The final purified mRNA (as recovered from the sucrose gradient) migrated as a single band on a 2.5% agarose slab gel in 0.025 M sodium citrate, pH 3.5, containing 6 M urea. The size of the purified mRNA (mRNA$_{om}$) was estimated to be 800 nucleotides.

The enrichment of mRNA$_{om}$ relative to other mRNA species during the purification was assessed by translation of the RNA preparations in a cell-free translation system derived from wheat germ (Rosen et al. 1975; Stein et al. 1978). Ovomucoid mRNA activity was determined by specific immuno-precipitation of ovomucoid peptides using a partially purified goat antiovomucoid immunoglobin G. The measurement of radioactivity incorporated into trichloroacetic acid (TCA)-insoluble material was used as an indication of total mRNA activity.

The purification of ovomucoid mRNA as followed by translation is shown in Table 1. The combination of purification steps listed resulted in a 270-fold increase in the specific activity of ovomucoid mRNA compared with the total extract. The percentage of immunoprecipitable ovomucoid in the wheat germ postribosomal supernatant fraction compared with the total peptides synthesized increased during purification from 3 to 52%.

B. Synthesis and Cloning of a Complementary DNA Copy

Since it was clear that mRNA$_{om}$ was not a pure species and conventional methods of RNA purification had been exhausted, we decided to complete the purification and isolate the coding ovomucoid DNA sequence at the same time by molecular cloning. To this end dsDNA was prepared from the partially purified mRNA$_{om}$ and the hairpin loop cut with S_1 nuclease as described previously (Monahan et al. 1976b; McReynolds et al. 1977; Stein et al. 1978; Woo et al. 1977). To obtain only full-length dsDNAs for restriction analysis and cloning, a 2% preparative agarose gel was run and a 0.6-cm band containing dsDNA of about 750–900 base pairs was excised. The DNA was recovered from the agarose gel by diffusion (Sharp et al. 1974).

For molecular cloning experiments, the dsDNA was "tailed" with [α-^{32}P]dATP to an average length of 51 dAs per 3' terminus by terminal transferase. Chimeric plasmids were formed by annealing the [^{32}P]dsDNA to

Table 1. Purification of Ovomucoid mRNA

	Specific activity[a] (cpm/μg)	Purification[b] (fold)	Ovomucoid synthesized[c] Total protein synthesized (%)
Total extract	100	1	3
dT-Cellulose bound	1 900	19	12
Sepharose peak	8 000	80	24
dT-Cellulose bound	10 500	105	35
Sucrose gradient peak	27 000	270	52

[a] Defined as the total amount of ovomucoid polypeptides synthesized in the wheat germ translation assay in response to 1 μg of the various RNA preparations.
[b] Purification of mRNA$_{om}$ over the total RNA in the preparation.
[c] Determined as the ratio of immunoprecipitated ovomucoid in the wheat germ translation assay to the total Cl$_3$CCOOH-precipitated polypeptides.

*Pst*I-cut pBR322. Ligation of the hybrid molecule as well as repair of any gaps that might occur owing to the difference in length of the poly(dA) and poly(dT) regions occurred in vivo after transformation (Chang and Cohen 1977). This hybrid molecule was used to transform *E. coli* strain X1776. Transformed colonies containing chick DNA inserts were selected by in situ hybridization with [^{32}P]cDNA synthesized from mRNA$_{om}$ (Grunstein and Hogness 1975). In all, 15 colonies gave a positive signal. However, since the mRNA$_{om}$ from which the probe was synthesized was not homogeneous, the probe should represent many sequences. Thus, it was still necessary to identify a clone unambiguously as containing an ovomucoid insert.

C. Identification of an Ovomucoid Structural Gene Clone

To select a clone containing an ovomucoid DNA insert, an assay based on mRNA hybridization and translation was developed. Recombinant plasmid DNA was isolated from several positive clones, as described by Katz et al (1977). Plasmid DNA (100 μg) was then digested with 30 units of the restriction endonuclease Hha I for 16 h at 37°C, denatured by heating at 100°C for 10 min and quick-cooled. Each sample was diluted into 6 mM Tris-HCl, pH 8.0, 25 mM MgCl$_2$ in 4 × SSC to a final concentration of 10 μg/ml and passed twice through a Millipore filter at 4°C. The filters were then baked at 70°C in a vacuum oven to fix the DNA. Either pure ovalbumin mRNA (mRNA$_{ov}$) or the enriched mRNA$_{om}$ was then hybridized to the DNA filters for 12 h at 42°C in 600 μl of 50% formamide buffer. An equal amount of poly(dA) was added to prevent hybridization of the poly(A) tract of noncomplementary mRNAs to the linkers present in the filter-bound plasmid DNA. After removing the buffer and washing the filters several times to remove nonspecifically adsorbed RNA, any hybridized RNA was eluted by denaturing in 0.01 *M* Tris–HCl, pH 7.4, 3 m*M* EDTA for 2 min at 90°C. Carrier tRNA was added (10 μg/ml) and the RNA precipitated in ethanol. The recovered hybridizable RNA was translated in the wheat germ system, and the

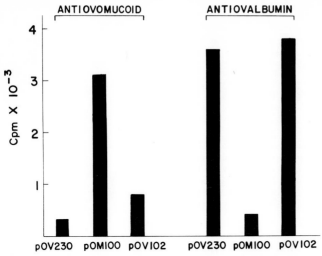

Fig. 1. Hybridization–translation assays of cloned DNAs. The radioactive peptides precipitated with antiovomucoid are shown for all three plasmid DNAs on the left, and the precipitates with antiovalbumin on the right. No correction factor for non-specific trapping of label was applied to the antibody precipitations.

amounts of ovalbumin- and ovomucoid-specific products were compared (Fig. 1). A clone that contains the full-length ovalbumin structural gene sequence, pOV230, was used to test the validity of the assay. The RNA that hybridized to this plasmid DNA was translated into ovalbumin and not ovomucoid. This indicated that filter-bound recombinant plasmid DNA containing a known structural gene DNA sequence could selectively hybridize to a complementary mRNA species, and that this RNA could be removed from the filter and translated into antibody-precipitable peptides in the wheat germ system. The other two plasmids tested in Fig. 1 both resulted from the transformation with $dsDNA_{om}$ reported here. Translation of the mRNA that hybridized to the pOM100 DNA filter yielded immunoprecipitable ovomucoid. The clone from which pOM100 DNA was derived must therefore contain an inserted ovomucoid DNA sequence. The other plasmid tested in Fig. 1, pOV102, hybridized only to $mRNA_{ov}$. The fact that this plasmid contained an ovalbumin DNA sequence was not surprising since about 48% of the dsDNA used in the transformation represented nonovomucoid sequences. Since the mRNA used to synthesize the dsDNA was considerably shorter than native ovalbumin mRNA, the mRNA preparation must contain fragments of about 800 base pairs derived from the 3′ end of ovalbumin mRNA.

D. Confirmation of an Ovomucoid Insert by DNA Sequencing

In order to confirm the identity of the cloned DNA of pOM100 as ovomucoid cDNA a partial DNA sequence was determined. Recombinant plasmid pOM100 was cleaved with *Pst*I and digested with snake venom phosphodiesterase (McReynolds et al. 1978). This treatment enabled labeling of the 5′

Fig. 2. Autoradiograph of the nucleotide sequence derived from *Pst*I fragment P1 of pOM100. Plasmid DNA was cleaved with *Pst*I, and the 5' ends were dephosphorylated and labeled with [γ-³²P]dATP. Fragment P1 was isolated and cleaved with *Hinf*I. The sequence shown corresponds to the coding (mRNA) strand near the left *Pst*I site toward the *Hinf*I site in P1 (see map, Fig. 3). Band doubling in the C + T and C lanes resulted in some anomalous bands. This may result from incomplete β elimination.

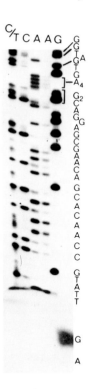

termini with bacterial alkaline phosphatase and T4 polynucleotide kinase (Maxam and Gilbert 1977). The 200-base pair *Pst*I fragment (designated P1) was isolated in a 12% polyacrylamide slab gel, then incubated with *Hinf*I, which cleaves P1 into two fragments of 116 and 81 base pairs in length, respectively. The DNA sequence of fragment P1 of pOM100, shown in Fig. 2, codes for portions of two glycosylated ovomucoid peptides whose amino acid sequence was determined by Beeley (1976). The matched sequence included a 13 amino acid peptide:

[72]SER TYR ALA ASN THR THR SER GLU ASP GLY LYS VAL MET[84]
5'AGT TAT GCC AAC ACG ACA AGC GAG GAC GGA AAA GTG ATG3'

This sequence match is well beyond random probability and proved that the inserted DNA of pOM100 contained ovomucoid sequences.

E. A Partial Restriction Map of pOM100

A restriction map of the pOM100 clone is shown in Fig. 3. The cloned DNA contains 646 base pairs of ovomucoid DNA, including a 138-base pair sequence of noncoding DNA at the 3' end. The cloned DNA is not a complete copy of the mRNA, but rather is missing about 120 nucleotides at the 5' end. The cutting sites for the restriction endonucleases shown in Fig. 3 were determined from the nucleotide sequence; many have been confirmed through restriction endonuclease digestions. The two *Pst*I sites, which de-

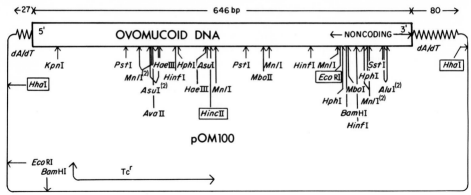

Fig. 3. A partial restriction map of pOM100.

fine the P1 fragment used for the sequence analysis described above, are located at 130 and 335 base pairs, respectively, from the 5′ end. These are the only *Pst*I sites in plasmid pOM100. *Eco*R1 and *Bam*H1 each cut the cloned DNA only once, at sites in the 3′ noncoding sequence separated only by 12 base pairs. *Hinc*II also cuts only once, in the middle of the cloned DNA. *Hha*I does not cut the cloned ovomucoid sequence, but cuts the plasmid DNA at two sites close to either end of the insert. This enzyme therefore was used to cut out the entire ovomucoid DNA sequence for the preparation of hybridization probes.

III. Regulation of Expression of the Ovomucoid Gene by Estrogen

A. Preparation of a Pure Hybridization Probe

A specific probe for ovomucoid mRNA sequences was prepared from pOM100, as illustrated diagrammatically in Fig. 4. The pOM100 was digested with the restriction endonuclease *Hha*I. The *Hha*I fragment containing the ovomucoid-specific sequences and flanking plasmid sequences then was purified from the remaining plasmid fragments by agarose gel electrophoresis (Lai et al. 1978). The purified fragment was labeled with [^{3}H]dCTP by nick translation to a specific activity of 6×10^6 cpm/μg. A single-stranded ovomucoid probe then was prepared by hybridizing these labeled fragments with a 100-fold excess of ovomucoid mRNA to a R_0t of 2.5×10^{-2}. Twenty micrograms each of poly(U) and poly(A) was added to avoid self-reannealing of the *Hha*I fragment caused by the poly(A) and poly(T) linker. Under these conditions, the hybridization reaction was complete and essentially no DNA–DNA reassociation occurred (C_0t of 2.5×10^{-4}). At the end of the incubation period, the reaction mixture was treated with S$_1$ nuclease to destroy the flanking plasmid sequences and anti-coding sequences which remained single-stranded. The hybridization with excess ovomucoid mRNA and S$_1$ nuclease digestion were repeated once

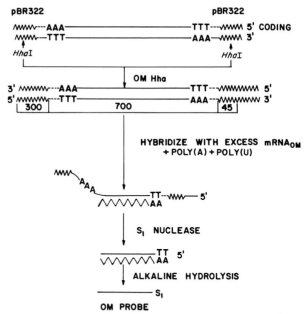

Fig. 4. Schematic representation of the preparation of pure hybridization probe for transcripts of [³H]DNA$_{om}$. Fragments corresponding to DNA$_{om}$ were labeled with [³H]dCTP by nick translation to a specific activity of 6×10^6 cpm/μg. Pure single-stranded probes were obtained as described in the text.

more to ensure complete removal of the plasmid and anticoding sequences. Finally, the single-stranded DNA probe ([³H]DNA$_{om}$) was obtained by alkaline hydrolysis of the messenger RNA.

B. Ovomucoid is a Single Copy Gene

In order to determine the copy number of the ovomucoid gene in the chick genome, the kinetics of hybridization of the single-stranded [³H]DNA$_{om}$ probe to an excess of chick oviduct and liver DNA was determined (Fig. 5). The probe reacted similarly to both chick oviduct and liver DNA, with an apparent $C_0t_{1/2}$ value of 1.5×10^3. This agrees well with the value predicted for a single copy gene in the haploid chick genome (Harris et al. 1973; Monahan et al. 1976a; Harrison et al. 1974; Bishop and Rosbach 1973). Therefore, the synthesis of ovomucoid mRNA resembles that observed for other proteins that are synthesized in large amounts (ovalbumin, globin, silk fibroin, etc.) and is accomplished by preferential transcription rather than by gene amplification (Bishop and Rosbach 1973; Suzuki et al. 1972).

C. The Effect of Estrogen on Ovomucoid mRNA Accumulation

The [³H]DNA$_{om}$ probe was also used to quantitate the induction of ovomucoid mRNA during secondary estrogen stimulation (Tsai et al. 1978). As shown in Table 2, column 2, nuclear RNA isolated from chick oviducts stimulated chronically with estrogen reacted rapidly with the [³H]DNA$_{om}$

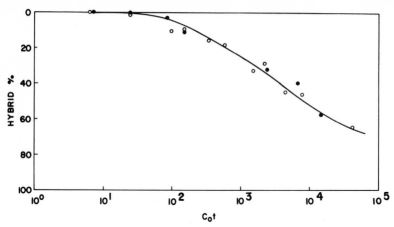

Fig. 5. Hybridization of ^{3}H-labeled ovomucoid probe to excess chick DNA. Chick DNA (1.3 mg) isolated from oviducts (○) or from liver (●) was hybridized to 0.06 ng of $[^3H]DNA_{om}$ in 200 μl.

probe. At an equivalent R_0t value of 50, the hybridization reaction was essentially complete. The apparent $R_0t_{1/2}$ for the reaction is 1.7. This indicates that 0.1% of the RNA sequences in stimulated oviduct nuclei are ovomucoid sequences. (Table 2, column 3). Nuclear RNA prepared from chicks first stimulated and then withdrawn from estrogen reacted 1000 times more slowly ($R_0t_{1/2}$ of 1.6×10^3) than nuclear RNA isolated from stimulated chicks. Thus, removal of estrogen resulted in a marked decrease in the concentration of ovomucoid sequences.

The number of molecules of ovomucoid RNA sequences in a given RNA preparation can be estimated by comparing the $R_0t_{1/2}$ value of the hybridization curve to that obtained for purified $mRNA_{om}$. Since the $mRNA_{om}$ is not pure, the $R_0t_{1/2}$ of pure $mRNA_{om}$ to $[^3H]DNA_{om}$ probe can only be estimated using $mRNA_{ov}$ as a standard. Ovomucoid mRNA is estimated to be 800 nucleotides in length with a molecular weight of 297 000 (Stein et al. 1978). Thus, pure $mRNA_{om}$ is calculated to react with the probe at an equivalent $R_0t_{1/2}$ of 2×10^{-3} under our present experimental conditions (Table 2). Since the RNA/DNA ratio in the isolated nuclei is 0.25, about 1600 molecules of $mRNA_{om}$ are calculated to be present in the cell nucleus of estrogen-stimulated oviducts (Table 2, column 6). Withdrawal of estrogen from chicks results in a marked decrease of $mRNA_{om}$ to less than one molecule per cell nucleus.

The marked diminution of ovomucoid sequences in the RNA preparation upon withdrawal of estrogen made it interesting to study the temporal effect of hormone on the accumulation of $mRNA_{om}$ during acute stimulation. Chicks withdrawn from hormone for 14 days were subsequently given a single injection of estrogen and sacrificed at the time intervals indicated. For the 48 h time point, one additional injection was given 24 h prior to sacrifice. Nuclear RNA then was isolated and hybridized to the $[^3H]DNA_{om}$

Table 2. Concentration of mRNA$_{om}$ in Total Nuclear RNA during Secondary Stimulation

Tissue	$R_0t_{1/2}$ [a]	Fraction of mRNA$_{om}$ [b]	RNA/DNA	Tubular gland cells / Total cells	Molecules of mRNA$_{om}$ [c] / Cell nucleus	Molecules of mRNA$_{om}$ / Tubular gland cell nucleus
Oviduct$_{DES}$	1.7	1.18×10^{-3}	0.25	0.8	1550	1940
Oviduct$_W$	1.6×10^3	1.25×10^{-6}	0.07	0.15	0.5	3
Oviduct$_W$ + 2h$_{DES}$	4.25×10^2	4.70×10^{-6}	0.09	0.15	2	12
Oviduct$_W$ + 4h$_{DES}$	2.65×10^2	7.55×10^{-6}	0.15	0.16	6	38
Oviduct$_W$ + 8h$_{DES}$	1.65×10^2	1.21×10^{-5}	0.15	0.20	10	50
Oviduct$_W$ + 16h$_{DES}$	6.6×10	3.0×10^{-5}	0.15	0.20	24	120
Oviduct$_W$ + 48h$_{DES}$	1.2×10	1.67×10^{-4}	0.25	0.35	220	620

[a] The $R_0t_{1/2}$ of pure mRNA$_{om}$ is calculated from the following equation: $(R_0t_{1/2})_{om} = [(N_{om}/\sqrt{N_{om}})/(N_{ov}/\sqrt{N_{ov}})](R_0t_{1/2})_{ov} = 2 \times 10^{-3}$; $N_{om} = 900$; $N_{ov} = 1930$; $(R_0t_{1/2})_{ov} = 3 \times 10^{-3}$.

[b] Fraction mRNA$_{om}$ = $R_0t_{1/2}$ pure ovomucoid message/$R_0t_{1/2}$ given RNA.

[c] Molecules of RNA$_{om}$/cell nucleus = fraction of mRNA$_{om}$ (RNA/DNA)(2.6×10^{-12} g DNA)(6.02×10^{23} molecules)/(2.97×10^5 g mRNA$_{om}$)].

probe.   Within 2 h after readministration of estrogen to withdrawn chicks, the rate of hybridization increased fourfold ($R_0t_{1/2}$ = 4.3 × 10^2; Table 2, column 2).   It gradually increased to tenfold after readministration of estrogen for 8 h ($R_0t_{1/2}$ = 1.7 × 10^2).   By 16 h, the rate of hybridization still appeared to increase at a similar rate ($R_0t_{1/2}$ = 6.6 × 10).   However, after a second injection of hormone, the $R_0t_{1/2}$ at 48 h further decreased to 1.2 × 10^{-1}. This rate of hybridization was only sevenfold slower than that calculated for nuclear RNA from oviducts of chronically stimulated chicks.   Thus, within 48 h, the concentration of ovomucoid RNA sequences per cell nucleus increased almost 200-fold.

Since the proportion of tubular gland cells in the chick oviduct during secondary stimulation has been determined (Oka and Schimke 1969; Yu et al. 1971), one can estimate the number of molecules of ovomucoid RNA per tubular gland cell nucleus.   As shown in Table 2, column 7, the number of ovomucoid RNA molecules in chronically stimulated chick oviducts is about 1900.   Upon withdrawal of hormone for 14 days, the number of ovomucoid RNA molecules decreased to 3 per tubular gland cell nucleus.   Four hours after readministration of hormone, 38 molecules per cell were found.   The number of molecules increased further to 600 per cell with a second injection of hormone, which is 200 times higher than that observed for withdrawn chicks.

D. Ovomucoid mRNA Transcription in Isolated Nuclei

The transcription of ovomucoid sequences in isolated nuclei from several tissues and hormonal states was also investigated.   The [³H]RNA was synthesized in nuclei isolated from hormone-stimulated chick oviducts, withdrawn oviducts, or chick liver or spleen and then assayed for the presence of $mRNA_{om}$ sequences by hybridization to filters containing pOM100 DNA. As shown in Table 3, radiolabeled ovomucoid mRNA sequences were found in the in vitro transcripts from chronically stimulated chick oviducts but very little if any were observed in withdrawn oviduct, liver or spleen.   Thus, transcription of the ovomucoid gene sequences is relatively tissue specific, confirming the results of the isolated nuclear RNA hybridization studies presented previously.

Of the total RNA synthesized in nuclei isolated from stimulated oviducts, 0.018% corresponds to $mRNA_{om}$ sequences.   This value is approximately 200-fold greater than would be expected if random transcription of the haploid chick genome occurred.   Since only 10% of the DNA is expressed in oviduct cells (Liarakos et al. 1973), ovomucoid sequences are actually transcribed at a frequency 20-fold greater than random transcription of the available sequences.   It is thus clear that the ovomucoid gene is transcribed preferentially in stimulated oviduct nuclei.   Of interest also is the observation that 0.1% of the RNA synthesized corresponds to $mRNA_{ov}$ sequences.   This suggests that the ovomucoid gene is transcribed at a rate 5.5 times slower than the ovalbumin gene on a mass basis.   The amount of $mRNA_{om}$ (nucleo-

Table 3. In Vitro Transcription of $mRNA_{om}$ and $mRNA_{ov}$ from Isolated Nuclei

Filter	[³H]RNA[a]	Competitor	[³H]RNA hybridized (cpm)	Hybridizable gene sequences (cpm)	mRNA in total RNA (%)
pOM100	Oviduct$_{DES}$	—	232	1077	0.018
		mRNA$_{om}$	78		
		Yeast RNA	248	1287	0.021
E. coli DNA	Oviduct$_{DES}$	—	96	N.D.	—
		mRNA$_{om}$	46		
Rat liver DNA	Oviduct$_{DES}$	—	84	N.D.	—
		mRNA$_{om}$	98		
pHb1001	Oviduct$_{DES}$	—	86	N.D.	—
		mRNA$_{om}$	68		
pOM100	Oviduct$_W$	—	74	N.D.	—
		mRNA$_{om}$	86		
pOM100	Liver	—	106	N.D.	—
		mRNA$_{om}$	90		
pOM100	Spleen	—	76	N.D.	—
		mRNA$_{om}$	80		
pOM100	Oviduct$_{DES}$	—	36	N.D.	—
	+ α-amanitin	mRNA$_{om}$	38		
pOV230	Oviduct$_{DES}$	—	1130	6300	0.106
		mRNA$_{ov}$	166		

[a] Input [³H]RNA:oviduct$_{DES}$, 5.9×10^6 cpm, oviduct$_W$, 4.4×10^6 cpm; liver, 4.9×10^6 cpm; oviduct$_{DES}$ + α-amanitin, 1.6×10^6 cpm.

tides) present in the nuclear RNA extracted from oviducts of chronically stimulated chicks is 0.11%, whereas that for ovalbumin sequences is 0.4% (Roop et al. 1978). The fourfold difference in the relative amount of these two nuclear mRNA sequences in vivo is very similar to the relative rate of transcription of the two genes in isolated nuclei. Therefore, transcription rates might contribute significantly to differential expression of the two genes. This by no means rules out potential contributions of posttranscriptional factors as the differences in concentration of these two proteins in the cytoplasm is greater than the difference in concentration with respect to their mRNAs in nuclei.

IV. Restriction Mapping of the Ovomucoid Gene in Chick DNA Digests

Does the Ovomucoid Gene Contain Intervening Sequences?

Recently, it was shown that the structural sequences of the ovalbumin gene in native chick DNA are not contiguous (Lai et al 1978, 1979; Dugaiczyk et al. 1978). Rather, multiple noncoding DNA sequences (intervening sequences) are interspersed within the structural ovalbumin gene. To determine if the ovomucoid structural gene is similarly split, limited restriction mapping studies of hen liver DNA were undertaken.

HindIII HpaII HincII TaqI PvuII

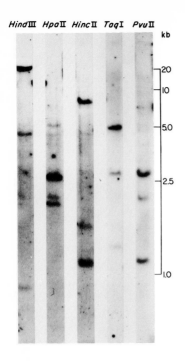

Fig. 6. Radioautogram of an agarose gel containing hen liver DNA after digestion with various restriction endonucleases and Southern hybridization. In this experiment, 12.5 μg of hen liver DNA was digested by the respective restriction endonuclease listed above each lane in the radioautogram. Electrophoresis was carried out in 1.5% agarose gels and DNA was transferred to nitrocellulose filters. Hybridization was carried out with ^{32}P-labeled OM_T.

Hen liver DNA was digested with various restriction endonucleases, and the restricted DNA fragments were separated by agarose gel electrophoresis and transferred onto a nitrocellulose filter by the method of Southern (Southern 1975). The filter was treated according to the procedure of Denhardt (1966) and those DNA fragments on the filter containing structural ovomucoid sequences were identified by hybridization to a ^{32}P-labeled ovomucoid probe (OM_T) followed by radioautography (Laskey and Mills 1977). The radioautograms of five of these digests are shown in Fig. 6. Only one of the endonucleases used, *Hinc*II, cuts the ovomucoid structural gene clone. It thus was expected to produce two hybridizable fragments. Instead, three were observed. The other four enzymes displayed here do not cut the cloned ovomucoid DNA, and thus should have resulted in only one hybridizable band. However, *Hind*III, *Taq*I, and *Pvu*II resulted in three hybridizable fragments, and *Hpa*II digestion yielded four. Therefore, noncoding sequences that contain target sites for these enzymes must be interspersed within the ovomucoid natural gene.

V. Cloning of the Natural Ovomucoid Gene

Although limited restriction mapping of the ovomucoid gene in digests of hen liver DNA did substantiate the existence of intervening sequences within this gene, the number and location of these sequences could not be accurately determined. In order to define precisely the sequence organization of this gene within genomic chick DNA, it was necessary to isolate and clone natural gene fragments. It was evident from total hen DNA restriction mapping experiments that the structural ovomucoid gene sequences were contained within two *Eco*RI fragments of 15 and 7 kilobases (kb), respectively, and that most of the structural gene sequences, as well as the 5′ region, were contained in the larger fragment. Therefore, this 15-kb fragment was enriched from total *Eco*RI-digested chick DNA by preparative agarose gel electrophoresis and employed for cloning of the ovomucoid gene.

λ Phage Charon 4A contains three *Eco*RI cleavage sites and yields four DNA fragments upon digestion with *Eco*RI. The two upper phage DNA arms were separated from the middle *Eco*RI DNA fragments by preparative agarose gel electrophoresis. The isolated phage DNA arms were allowed to ligate in the presence of the DNA fraction enriched for ovomucoid DNA, using T4 DNA ligase. The ligation mixture was employed for cloning in a P3 facility by in vitro packaging according to the method of Sternberg et al. (1977) as modified by Faber, Kiefer, and Blattner (personal communication, 1977). Approximately 20 000 phage plaques were generated in this experiment. These phages were screened for the recombinant phage containing the ovomucoid gene by hybridization with the OM_T probe using an amplification procedure as described previously (Woo et al. 1978). Three phage plaques gave positive signals on radioautograms in this test. These phages were cultured individually and their DNA prepared. Digestion of DNA from any of these phages with *Eco*RI produced two DNA fragments corresponding to the arms of Charon 4A and an additional 15-kb DNA fragment, which hybridized with the ovomucoid structural sequence OM_T upon Southern analysis. This 15-kb DNA fragment of the natural ovomucoid gene was subsequently recloned in pBR322 in *E. coli* X1776 and was designated pOM15.

VI. Restriction Mapping of the Cloned 15-kb Ovomucoid DNA

To generate ovomucoid-specific probes for restriction mapping of the cloned natural gene, the ovomucoid probe OM_T was cut with the restriction endonuclease *Hinc*II. This enzyme cuts the ovomucoid structural gene sequence in the middle, creating a 630-base pair (bp) probe specific for the 5′ half of the ovomucoid gene (OM_L), and a 410-bp probe specific for the 3′ half of the gene (OM_R). The 15-kb ovomucoid DNA (OM15) was then excised from pBR322 DNA by *Eco*RI digestion of the recombinant plasmid pOM15, followed by preparative agarose gel electrophoresis. The excised ovomucoid DNA was digested further with a variety of other restriction endonucleases, and those DNA fragments containing the left and right halves of the structural ovomucoid gene sequence were identified by agarose gel electrophoresis, followed by Southern hybridization with the nick-translated OM_L and OM_R probes, respectively. Some of the data are summarized in Table 4. These data were employed to construct a preliminary map of the natural ovomucoid gene as shown in Fig. 7.

Table 4. Sizes of Ovomucoid Sequence-Containing Chicken DNA Fragments after Restriction Endonuclease Digestion

| | Size after digestion (kb) | | | | | | Probe used |
	*Eco*RI	*Bam*HI	*Hind*III	*Bg*/II	*Hinc*II	*Taq*I	
	15.0	8.0	18.0	8.4	6.7	4.5	OM$_L$
			4.5		1.1		
			0.8				
+*Eco*RI	15.0	8.0	4.5	5.8	6.7	4.5	OM$_L$
			3.9		1.1		
			0.8				
+*Hinc*II	6.7	6.3	3.7	4.1	6.7	4.5	OM$_L$
	1.1	1.1	1.1	1.1	1.1	1.1	
			0.8				
	15.0	8.0	18.0	8.4	6.0	2.9	OM$_R$
	7.0	1.2					
+*Eco*RI	15.0	8.0	6.0	5.9	5.4	2.5	OM$_R$
	7.0	1.2	3.9	2.6	0.6	0.4	
+*Hinc*II	5.4	1.2	6.0	3.2	6.0	2.9	OM$_R$
	0.6	0.6					

Restriction enzyme cutting sites that exist within the structural gene sequence are shown above the schematic drawing of the ovomucoid gene. Six such noncontiguous structural sequences, each containing at least one restriction enzyme site identified in the structural gene clone pOM100, were identified by restriction mapping analysis. Because these sites were separated by distances greater than the corresponding distances in the structural gene clone, the presence of intervening sequences between each of these structural sequences was indicated. The five intervening sequences shown on the 3' side of the gene were identified in this manner. The other two intervening sequences at the 5' end were identified by electronmicroscopic mapping of the ovomucoid natural gene, as described below.

These results therefore substantiate the preliminary restriction studies

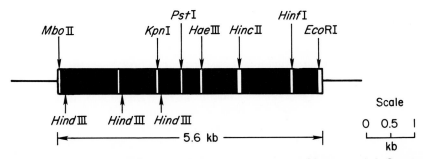

Fig. 7. Structural organization of the natural ovomucoid gene. (□) Ovomucoid structural sequences; (■) intervening sequences; (——) flanking DNA sequences. Various restriction endonuclease sites on the ovomucoid gene are indicated by the arrows. Those above the schematic drawing indicate restriction sites present in the structural ovomucoid gene and those below are sites within the intervening sequences.

carried out on total DNA digests of hen liver. The structural ovomucoid gene sequences are not contiguous and are apparently separated into several portions by multiple intervening sequences within the chick genome. The existence of five intervening sequences as established by limited restriction analyses of the cloned OM15 DNA should be a minimum estimation of the number of intervening sequences.

VII. Electronmicroscopic Mapping of the Ovomucoid Gene

Although intervening sequences within structural genes can be detected by restriction enzyme mapping, a complete map requires other methods independent of the chance occurrence of restriction sites. Short structural gene regions may be missed owing to the relative instability of the hybrids formed during hybridization, resulting in weak signals upon Southern filter analysis. In an attempt to more fully characterize the structure of the ovomucoid gene, hybrid molecules formed between ovomucoid mRNA and the cloned 15-kb ovomucoid DNA were examined by electronmicroscopy.

The DNA was thermally denatured and incubated with ovomucoid mRNA under conditions that permit only RNA–DNA hybridization but not DNA–DNA reassociation. Thus, homologous regions between the mRNA and the cloned DNA would appear as double-stranded regions. These DNA sequences nonhomologous with the mRNA would appear as single-stranded loops. Such a hybrid molecule formed between ovomucoid mRNA and the cloned OM15 DNA is shown in Fig. 8. A total of seven intervening DNA loops of various sizes are present in this molecule (labeled A–F in the direction 5'–3'). All of the loop structures occurred at one DNA terminus and occupied about one-third of the total length of the OM15 DNA molecule. Two additional intervening sequences at the 5' end of the gene were detected in these studies, but not by restriction mapping analysis. Although restriction mapping is more quantitative in that fragment sizes and position can be correlated with known distances in the structural sequence, in the absence of favorably placed restriction sites some intervening sequences may be missed.

Electronmicroscopy permits more direct analysis of intervening sequences. Hence, electronmicroscopy analysis has detected two additional intervening sequences; yet the overall structure is in good qualitative agreement with the limited restriction map. Thus, there are at least seven intervening sequences in the ovomucoid gene. Since the ovomucoid structural gene is only about 750 nucleotide pairs in length, the frequence of occurrence of these intervening sequences is striking. At one intervening sequence per 107 nucleotide pairs of structural gene sequence, it represents the highest frequency of intervening sequences observed among all eucaryotic genes examined to date (Garapin et al. 1978a,b; Mandel et al. 1978; Woo et al. 1978; Lai et al. 1979a; Weinstock et al. 1978; Tilghman et al. 1977; Tiemeier et al. 1978; Maniatis et al. 1978; Tonegawa et al. 1977, 1978; Brack and Tonegawa 1977; Seidman et al. 1978a,b).

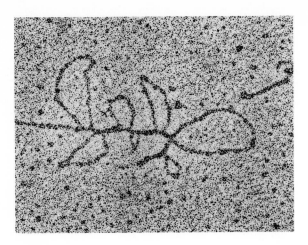

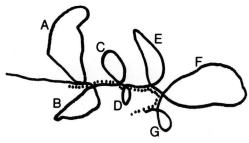

Fig. 8. Electronmicrograph and line drawing of a hybrid molecule formed between the ovomucoid gene (OM15) and ovomucoid mRNA. Hybridization was carried out using 10 μg/ml of OM15 DNA and 20 μg/ml of ovomucoid mRNA in 70% deionized formamide containing 100 mM Tris–HCl, pH 7.6, 10 mM Na$_2$EDTA and 150 mM NaCl. The mixture was heated at 80°C for 5 min to denature the DNA and hybridization was carried out at 43°C for 2 h.   (——) OM15; (---) mRNA$_{om}$.

We now have the entire ovomucoid gene, along with several kilobases of flanking DNA on the 5′ and 3′ ends, in one clone isolated from a library of oviduct DNA fragments, supplied to us by Thomas Maniatis and Richard Axel. Fragments of this clone are now being labeled and used for hybridization to in vivo nuclear RNA to further delineate the extent of the gene that is transcribed. This clone is now being sequenced, and in the near future comparative sequence data on the intron–exon boundaries should be available.

VIII. Ovomucoid mRNA Precursors in Chick Oviduct Nuclei

A. Identification of High Molecular Weight Species of Ovomucoid RNA in Chick Oviduct Nuclei

Recently several high molecular weight species of ovalbumin sequence-containing RNA were identified in chick oviduct nuclei (Roop et al. 1978). These species presumably represent a large nuclear precursor RNA in different stages of processing.   Since we have demonstrated that the ovomucoid gene contains multiple intervening sequences, there also might exist in oviduct nuclei species of ovomucoid RNA larger than mature $mRNA_{om}$.   To search for these species, nuclear RNA from oviducts of estrogen-stimulated chicks was subjected to agarose gel electrophoresis in the presence of the denaturing agent, methylmercury hydroxide (10mM).   Following transfer of the RNA from the gel onto diazobenzyloxymethyl (DBM) paper by the method of Alwine et al. (1977), hybridization to a ^{32}P-labeled ovomucoid cDNA probe, and radioautography, multiple discrete bands containing ovomucoid structural sequences were observed (Fig. 9, lane A). The band with the greatest mobility comigrated with mature $mRNA_{om}$ (Fig. 9, lane B). Five major bands were significantly larger than ovomucoid mRNA (800 nucleotides).   In addition, two minor bands were detected in certain nuclear RNA preparations.   As shown in Table 5, the sizes of the seven species of high molecular weight ovomucoid RNA, which were labeled a–g, ranged from 1.5 to 5 times the size of ovomucoid mRNA.   The mobilities of the prominent ovomucoid RNAs (bands a, b, d, e, g) were reproducibly observed in all five preparations of oviduct nuclear RNA that were analyzed.

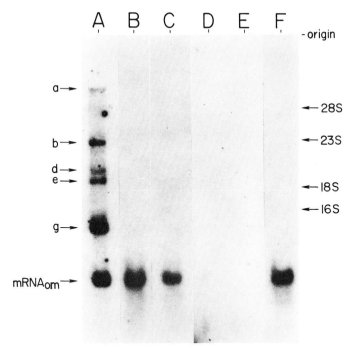

Fig. 9. Identification of ovomucoid mRNA sequences in high molecular weight nuclear RNA.

A oviduct nuclear RNA, 20 μg (high molecular weight ovomucoid RNA species "c" and "f" were not detected in this preparation). **B** Ovomucoid mRNA purified to approximately 40%, 15 ng. **C** Oviduct cytoplasmic RNA, 20 μg. **D** Spleen nuclear RNA, 20 μg. **E** liver nuclear RNA, 20 μg. **F** RNA isolated after mixing liver nuclei with oviduct cytoplasm, 20 μg.   The position of 28S, 23S, 18S, and 16S ribosomal RNA markers is indicated.

Table 5. Sizes of the Ovomucoid Sequence-Containing RNA from Chick Oviduct Nuclei[a]

RNA$_{om}$ species	Relative mobility (to mRNA$_{om}$)	Length (nucleotides)
a	0.25	5400 ± 120
b	0.45	3100 ± 60
c	0.48	2750 ± 60
d	0.55	2500 ± 60
e	0.59	2300 ± 50
f	0.65	2000 ± 30
g	0.76	1700 ± 75
mRNA$_{om}$	1.0	1100 ± 110

[a] Relative mobility was calculated as the average of 5 individual experiments. Length was calculated by comparison with the electrophoretic mobility of RNA and DNA standards. RNA standards were chicken 27S rRNA (4530 nucleotides), chicken 18S rRNA (2076 nucleotides), *Escherichia coli* 23S rRNA (3129 nucleotides) and *E. coli* 16S rRNA (1550 nucleotides). DNA standards were obtained by digestion of λ-DNA with *Eco*RI (21 746, 7524, 5920, 5523, 4793, and 3380 nucleotides) or with *Hind*III (23 222, 9730, 6460, 4492, 2301, 2000, and 682 nucleotides). In the conditions of electrophoresis (in the presence of methylmercury hydroxide), DNA becomes single stranded, and its mobility is similar to that of RNA. Error limits indicate the standard deviation.

Based upon its intensity, band g, which has 1700 nucleotides, appears to be the most abundant species of high molecular weight ovomucoid RNA. Based upon the radioactivity associated with each band, as determined by liquid scintillation counting, we estimate that 20–45% of the nuclear ovomucoid sequences reside in the forms higher in molecular weight than mRNA$_{om}$.

High molecular weight species of RNA that contain ovomucoid sequences were not found in oviduct cytoplasm, where only mature mRNA$_{om}$ was detected (Fig. 9, lane C). In addition, the presence of nuclear high molecular weight ovomucoid RNA, as well as that of mature mRNA$_{om}$, was tissue specific. No ovomucoid bands were detected in the nuclear RNA from spleen or liver that was isolated from estrogen-stimulated chicks (Fig. 9, lanes D, E). Thus, high molecular weight ovomucoid RNA, as well as mRNA$_{om}$, is present in significant amounts only in those tissues that actively synthesize egg white proteins.

Although the conditions employed for electrophoresis were rigorously denaturing (Bailey and Davidson 1976; Alwine et al. 1977), an attempt was made to generate high molecular weight species of ovomucoid RNA by mixing liver nuclei and oviduct cytoplasm (both prepared from chicks stimulated with diethylsilbestrol, DES) and then extracting the RNA from the mixture. However, mRNA$_{om}$ was the only species detected (Fig. 9, lane F). Thus, as was observed previously for ovalbumin RNA, the high molecular weight species of ovomucoid RNA do not appear to result from aggregation of mRNA$_{om}$ with itself or other RNA molecules.

B. Identification of Ovomucoid Intervening Sequences in High Molecular Weight Nuclear RNA

In order to confirm the hypothesis that these large nuclear ovomucoid RNA species represent mRNA precursors in different stages of processing, it was necessary to identify ovomucoid intervening sequences in these species. Therefore, a hybridization probe was prepared by nick translation of the OM15 clone, which contains all of the ovomucoid structural and intervening sequence regions (see Fig. 7). When this probe was hybridized to nuclear RNA from chronically stimulated chick oviducts, all of the ovomucoid nuclear RNA bands, except band f, that hybridized to the ovomucoid cDNA probe were again detected (Fig. 10, lane −mRNA). In addition, two bands (located between bands a and b) that were extremely faint after hybridization to the ovomucoid cDNA probe were clearly detected by the natural ovomucoid gene probe. In order to detect only those bands that contained intervening sequences, it was necessary to hybridize in the presence of excess unlabeled ovomucoid mRNA to compete out the structural sequences in the natural gene. Therefore, ovomucoid mRNA (20 μg, approximately 20% pure) was preincubated with the ^{32}P-labeled probe in 1.0 M of hybridization buffer at 42°C for 1 h. The mixture was diluted to 5.0 M with hybridization buffer and then hybridized to the DBM paper containing the bound nuclear RNA. Under these hybridization conditions, detection of the band corresponding to mature $mRNA_{om}$ was eliminated (Fig. 10, lane +mRNA). In contrast, all of the bands of high molecular weight ovomucoid RNA were still detected. This indicates that all of the species of ovomucoid RNA larger than mature $mRNA_{om}$ contain sequences complementary to the intervening sequences of the ovomucoid gene.

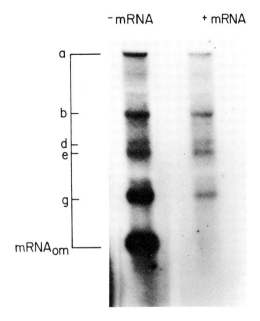

Fig. 10. Identification of ovomucoid intervening sequences in high molecular weight ovomucoid nuclear RNA. Oviduct nuclear RNA from chronically stimulated chicks, 20 μg, was electrophoresed, transferred to DBM paper (2 × 14 cm) and allowed to hybridize to the ^{32}P-labeled natural ovomucoid gene probe (10^7 cpm, specific activity 6 × 10^8 cpm/μg) in the absence (lane −mRNA) or presence of unlabeled ovomucoid mRNA as a competitor (lane +mRNA). The radioautography of both samples was for the same length of time (about 1 week).

C. Estrogen-Induced Accumulation of High Molecular Weight
Species of Ovomucoid RNA

The accumulation in the oviduct nucleus of species of ovomucoid RNA larger than mature mRNA$_{om}$ was dependent upon estrogen. From 20 μg of oviduct nuclear RNA isolated from chicks that had been withdrawn from DES treatment for 14 days, neither high molecular weight ovomucoid RNA nor mRNA$_{om}$ was detected (Fig. 11, left, lane W). Thus, in agreement with our studies described above using R_0t analysis, the amount of ovomucoid RNA sequences in withdrawn oviduct nuclei is very low. When 100 μg of withdrawn oviduct nuclear RNA was analyzed, a faint band of mRNA$_{om}$ was observed (Fig. 11, right, lane W). Thus, the few ovomucoid RNA sequences that remain after hormone withdrawal appear to be present mainly in the form of mature mRNA$_{om}$.

As soon as one hour after administration of DES to hormone-withdrawn chicks, both mature mRNA$_{om}$ and high molecular weight ovomucoid RNA species have accumulated. Analysis of 20 μg of nuclear RNA revealed the accumulation of the most abundant high molecular weight species, band g

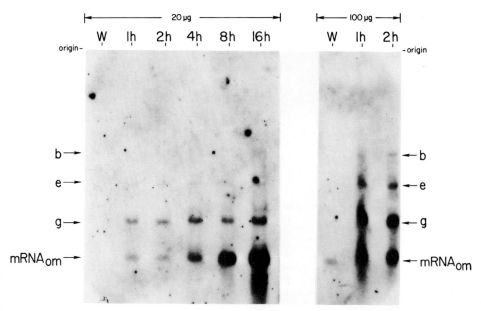

Fig. 11. Induction of high molecular weight ovomucoid RNA by secondary estrogen stimulation. Oviduct nuclear RNA was isolated from chicks that had received daily subcutaneous injections of 2.5 mg of DES for 14 days and subsequently withdrawn from hormone for 14 days (W). In addition, hormone withdrawn chicks were injected with a single dose of 2.5 mg of DES and the oviduct nuclear RNA was isolated after 1, 2, 4, 8, and 16 h. The RNA was electrophoresed, transferred to DBM paper (15 × 15 cm) and hybridized to the [^{32}P]-labeled ovomucoid probe (3.3 × 10^7 cpm, spec. act. 8 × 10^8 cpm/μg).

Left, 20 μg of nuclear RNA was analyzed; *right,* 100 μg of nuclear RNA was analyzed.

(Fig. 11, left), while analysis of 100 μg of RNA revealed the accumulation of the three most predominant species (b, e, and g) larger than $mRNA_{om}$ (Fig. 11, right). Species a, c, d, and f were not detected, probably because these are the least abundant of the high molecular weight ovomucoid RNA molecules.

After short periods (1–2 h) of secondary estrogen stimulation, the relative intensity of the ovomucoid RNA bands (Fig. 11, right) was similar to that observed after chronic estrogen stimulation (Fig. 9, lane A). The bands corresponding to species g and $mRNA_{om}$ were high and of roughly equal intensity, whereas that of species e was of lesser intensity and that of species b was lesser yet. After longer periods (greater than 4 h) of secondary stimulation, this pattern shifted. The concentration of $mRNA_{om}$ increased at each time point so that 16 h after the DES injection a very intense $mRNA_{om}$ band was observed. In contrast, the concentration of the most abundant high molecular weight ovomucoid RNA species, band g, increased only slightly during this period and no significant increases in the concentration of ovomucoid species e were observed. Thus, 4–16 h after DES injection, mature $mRNA_{om}$ was clearly more abundant than any of the high molecular weight ovomucoid RNA species. Taken together, these results suggest that ovomucoid mRNA may be synthesized by the processing of a large primary gene transcript and that the multiple bands of ovomucoid RNA larger than $mRNA_{om}$ may represent transcripts at different stages of processing.

Attempts to isolate and study the high molecular weight nuclear RNA precursors of ovomucoid mRNA are underway. We already have established (unpublished observations) that the poly(A) tail at the 3' end of the mRNA is added very early, perhaps at the same time as transcription of the primary gene product itself, as recently determined for Adeno 2 viral mRNAs (Nevins and Darnell 1978). Further studies of those molecules should provide much information about the synthesis and processing of ovomucoid mRNA.

Acknowledgment. The research in this contribution was supported by NIH grant HD-8188 (B.W.O.), American Cancer Society Research Grant BC-101 and Robert A. Welch Foundation Grant Q-611 (A.R.M.) and American Cancer Society Fellowship PF-1211 (J.P.S.). S.L.C.W. is an Associate Investigator of the Howard Hughes Medical Institute.

References

Alwine JC, Demp DJ, Stark GR (1977) Proc Natl Acad Sci U.S.A. 74: 5350–5354
Bailey JM, Davidson N (1976) Anal Biochem 70: 75–85
Beeley AG (1976) Biochem J 159: 335–345
Bishop JO, Rosbach M (1973) Nature (London) New Biol 241: 204–207
Brack C, Tonegawa S (1977) Proc Natl Acad Sci USA 74: 5652–5656
Chan L, Means AR, O'Malley BW (1973) Proc Natl Acad Sci USA 70: 1870–1874
Chang S, Cohen SN (1977) Proc Natl Acad Sci USA 74: 4811–4815
Cox RF (1977) Biochemistry 16: 3433–3443
Denhardt D (1966) Biochem Biophys Res Commun 23: 641–646

Dugaiczyk A, Woo SLC, Lai EC, Mace ML Jr., McReynolds LA, O'Malley BW (1978) Nature 274: 328–333

Garapin AC, Cami B, Roskam W, Kourilsky P, LePennec JP, Perrin F, Gerlinger P, Cochet M, Chambon P (1978a) Cell 14: 629–639

Garapin AC, LePennec JP, Roskam W, Perrin F, Cami B, Krust A, Breathnach R, Chambon P, Kouvilsky P (1978b) Nature 273: 349–354

Grunstein M, Hogness DS (1975) Proc Natl Acad Sci USA 72: 3961–3965

Harris SE, Means AR, Mitchell WM, O'Malley BW (1973) Proc Natl Acad Sci USA 70: 3776–3780

Harris SE, Rosen JM, Means AR, O'Malley BW (1975) Biochemistry 14: 2072–2080

Harrison PR, Birnie GD, Hell A, Humphries S, Young BD, Paul J (1974) J Mol Biol 84: 539–544

Hynes NE, Groner B, Sippel AE, Njuyen-Huu, MC, Schutz G (1977) Cell 11: 923–932

Katz L, Williams PH, Sato S, Leavitt RW, Melinski DR (1977) Biochemistry 16: 1677–1683

Lai EC, Woo SLC, Dugaiczyk A, Catterall JF, O'Malley BW (1978) Proc Natl Acad Sci USA 75: 2205–2209

Lai EC, Woo SLC, Dugaiczyk A, O'Malley BW (1979b) Cell 16: 201–211

Lai EC, Stein JP, Catterall JF, Woo SLC, Mace ML, Means AR, O'Malley BW (1979a) Cell *18*: 829–842

Laskey RA, Mills AD (1977) FEBS Lett. 82: 314–316

Liarakos CD, Rosen JM, O'Malley BW (1973) Biochemistry 12: 2309–2816

Mandel JL, Breathnach R, Gerlinger P, LeMeur M, Gannon F, Chambon P (1978) Cell 14: 641–653

Maniatis T, Hardison RC, Lacy E, Lauer J, O'Connell C, Quon D, Sim GK, Efstratiatis A (1978) Cell 15: 687–701

Maxam A, Gilbert W (1977) Proc Natl Acad Sci USA 74: 560–564

McReynolds LA, Catterall JF, O'Malley BW (1977) Gene 2: 217–231

McReynolds LA, O'Malley BW, Nisbett A, Fothergill JE, Fields S, Givol D, Robertson M, Brownlee GG (1978) Nature (London) 273: 723–728

Monahan JJ, Harris SE, O'Malley BW (1976a) J Biol Chem 251: 3738–3748

Monahan JJ, McReynolds LA, O'Malley BW (1976b) J Biol Chem 251: 7355–7362

Nevins JR, Darnell JE, Jr. (1978) Cell 15: 1477–1493

Oka T, Schimke RT (1969) J Cell Biol 43: 123–137

O'Malley BW, McGuire WL, Kohler PO, Korenman SG (1969) Recent Progr Horm Res 25: 105–160

O'Malley BW, Means AR (1974) Science 183: 610–620

Palmiter RD (1972) J Biol Chem 247: 6450–6461

Palmiter RD (1973) J Biol Chem 248: 8260–8270

Palmiter RD, Schimke RT (1973) J Biol Chem 248: 1502–1512

Palmiter RD, Moore PB, Mulvihill ER (1976) Cell 8: 557–572

Roop DR, Nordstrom JL, Tsai SY, Tsai M-J, O'Malley BW (1978) Cell 15: 671–685

Rosen JM, Woo SLC, Holder JW, Means AR, O'Malley BW (1975) Biochemistry 14: 69–78

Seidman JG, Edgell MH, Leder P (1978a) Nature 271: 582–585

Seidman JG, Leder A, Edgell MH, Polsky F, Tilghman SM, Tiemeier DC, Leder P. (1978b) Proc Natl Acad Sci USA 75: 3881–3885

Sharp PA, Gallimore PM, Flint ST (1974) Cold Spring Harbor Symp. Quant. Biol. 39: 457–474

Southern EM (1975) J Mol Biol 98: 503–518

Stein JP, Catterall JF, Woo SLC, Means AR, O'Malley BW (1978) Biochemistry 17: 5763–5772

Sternberg N, Tiemeier D, Enquist L (1977) Gene 1: 255–280

Sullivan D, Palacios R, Staunezer J, Taylor JM, Taras AJ, Kiely ML, Summer NM, Bishop JM, Schimke RT (1973) J Biol Chem 248: 7530–7539

Suzuki Y, Gage LP, Brown DD (1972) J Mol Biol 70: 637–649

Tiemeier D, Tilghman S, Polsky FE, Seidman JC, Leder A, Edgell MH, Leder P (1978) Cell 14: 237–245

Tilghman SM, Tiemeier DC, Polsky F, Edgell MH, Seidman JG, Leder A, Enquist L, Norman B, Leder P. (1977) Proc Natl Acad Sci USA 74: 4406–4410

Tonegawa A, Brack C, Hozumi N, Schuller R (1977) Proc Natl Acad Sci USA 74: 3518–3522

Tonegawa S, Maxam AM, Tizard R, Bernard O, Gilbert W (1978) Proc Natl Acad Sci USA 75: 1485–1489

Tsai SY, Roop DR, Tsai M-J, Stein JP, Means AR, O'Malley BW (1978) Biochemistry 17: 5773–5780

Weinstock R, Sweet R, Weiss M, Cedar H, Axel R (1978) Proc Natl Acad Sci USA 75: 1299–1303

Woo SLC, Chandra T, Means AR, O'Malley BW (1977) Biochemistry 16: 5670–5676

Woo SLC, Dugaiczyk A, Tsai M-J, Lai EC, Catterall JR, O'Malley BW (1978) Proc Natl Acad Sci USA 75: 3688–3692

Yu JYL, Campbell LD, Marquardt RR (1971) Can J Biochem 49: 348

Discussion of the Paper Presented by B.W. O'Malley

MUELLER: Do you think that the receptor mechanism could influence gene expression via some interaction with intervening sequences?

O'MALLEY: In the past year we have really concentrated our efforts on gene structure and search for the primary transcript. We have done few direct experiments on receptor interactions in any in vitro transcription systems. Dr. Schrader is going to mention some initial binding studies with receptor and cloned gene fragments. I would prefer not to get into this at the present time, other than to say that we have no evidence that there is a specific binding site in inducible genes for receptor.

GAREN: I would like to know whether the full transcript that contains the intervening sequences could be translated in vitro, for example, in the reticulocyte cell-free system.

O'MALLEY: We have not actually tried that, but would predict, no. I don't see how the ribosomes could read through those intervening sequences. When you look at the intervening sequences, which we have sequenced, it just doesn't resemble a sequence that could code for proteins. It has long runs of polymers, and has many stop signals for protein synthesis.

GOLDBERGER: Have you tried to do an acute withdrawal and look at the half-life of the message?

O'MALLEY: Since a number of other laboratories have done this study, we have not carried out such experiments. We can only say that if you withdraw hormone and pulse label mRNA, by 60 h there is no synthesis.

SIDDIQUI: Do you think that the processing enzyme could be controlled by hormones?

O'MALLEY: I can't say definitely at the present time. However, our studies suggest that at time zero, processing enzyme is not rate limiting to process the initial precursor transcripts in that these can be processed to mature message immediately. I think this is only logical since the same processing enzyme is probably used for this reaction in all cells.

EDELMAN: Bert, I want to clarify one point, and to question whether it is the primary transcriptional regulation alone that dictates hormone action, which is increasingly troubling to many people in the field. You have one result, in which you pulsed one animal given DES, and then measure the first hybridizable product and found that the first hybridizable product is mature messenger. Then subsequently you recovered more high molecular weight RNA, and you interpreted that as indicating that there is a full complement of the enzymes. Then what happened was that as soon as you turned on the gene, you see the final product first; then you see the intermediates. Is that right?

O'MALLEY: Yes.

EDELMAN: O.K., now there is a derivation in the literature, in which you do a simple linear precursor–product mathematical analysis. It turns out that, if you do that, in order to get an increase in the final product, you must first get an increase in the precursor. In other words, when you start at the beginning of the chain and then run the thing through, the mathematics of the precursor–product relationship implies, in fact, that a series of built-up curves occurs. So it would seem to me that the result you showed alone implies that it cannot be a simple production of the primary transcript, followed then by this being the driving force in appearance of all the subsequent components.

O'MALLEY: I can perhaps clarify the issue by saying that in that our assay, we do have synthesis of precursors first as shown by our radioactive pulsing experiments. I have not shown these data today. You also probably build up some large molecular weight precursors in the Northern transfer experiments, but by using mass analysis and filter hybridization techniques the RNA is spread over such a large surface area that you don't see the lower concentrations of precursor well. In other words, there is no block to processing in the absence of hormone. Precursor RNA forms first and is then converted to mature message.

MUELLER: I would like to ask just one more question. Relative to the situation at the present time, it seems as if everything is in favor of processing of precursors. Is it absolutely ruled that the loop-out mechanism for DNA plays any role in this process?

O'MALLEY: No evidence for transcription of genes where intervening sequences are looped out exists at present.

MUELLER: What do you think about what this multiplicity or these domains really mean in terms of the function of the receptor mechanism. Does it mean that we have multiple spots all throughout the DNA, or do you think that these are contiguous, or what is your best guess?

GOLDBERGER: I think probably that there are not only multiple genes, that respond to hormonal stimulation; there are also multiple mechanisms by which that stimulation is effected. Now, Bert O'Malley showed you something really important this morning, which meant to me that we now have to accept that hormones are truly having an effect on gene transcription because that half-lives that have been measured for messenger RNAs are not sufficient to explain the accumulation and disappearance rate. We will begin to accept that there is a transcriptional effect of hormones. In addition, we know that there is an effect on half-lives of messages and I don't want to catalog the whole thing. But it is clear that we have a very complex set of responses of hormone action as well as a complex set of genes that respond.

EDELMAN: I have no intention of discussing this still further. Both I and Gerry Mueller, when we were young, published papers showing the role of hormones in gene transcription. Therefore, in principle that is not the issue here. The problem is the increasing complexity of organization, the increasing number of possible modes through which regulation can occur. The alternatives are not simply synthesis and degradation. The turnover or half-life of the mature messenger is not the issue. There may be a series of rescue steps that have nothing to do with degradation, in other words, in any step in the process that may be isolating one of the products and stabilizing it so that it can be acted upon, and that will not be revealed in the turnover studies. In fact, I think ultimately what will be the final proof of the matter will be in the reconstruction experiments. But, nevertheless, even though the doubt is relatively small that it is the primary transcriptional events that regulate gene expression, we will still have some element of doubt remaining until you can reconstruct the system from primary components and prove your point.

MUELLER: I'm personally impressed that, starting back in 1950, we have been increasing our capabilities in measuring what is happening in terms of a hormone response. It started out with the weighing of the eggs, but we are getting down to where we are just weighing parts of molecules, etc. I'm just wondering here, for instance, with all this resolution, what do the experts here feel is the most likely molecular mechanism by which the regulation could be effected, or if we were to come back to it, what kind of information you feel that we need before we can even answer that question?

GOLDBERGER: I think that we are all aiming toward the same thing, and that is a system in which genes can be transcribed in vitro, and the system can be manipulated, and it will. So far this is not what is happening, as we are trying to probe the system without having it work in isolation. Bert O'Malley has done some really interesting studies moving toward that goal. I feel that this is the approach with which one can get to regulation. If you have a piece of a chromosome or chromosomal equivalent, you can transcribe it with the homologous or even an endogenous polymerase, and you can get it to do faithfully what happens inside a cell, or inside a nucleus, then you have the means for studying what the signals are, what the regulatory proteins are, etc. I think until we have this system, that is, a good in vitro system for studying transcription, we are limited.

O'MALLEY: I agree. Our tack has been to take a relatively narrow approach and to try to decide what "the primary" event is, then describe in precise details the chemistry of this reaction. However, there is no doubt as to the complexities of the cell. There may be many induced changes occurring at other steps in the overall reaction that could range through changes in membrane permeability, changes in net receptor concentration, changes in processing enzyme content, changes in ribosomal and tRNA, and the amount of ribosomes available to effect translation of the message. Our particular approach has been to try to isolate and study the primary response. I would think that in future years we will be able to describe, in detail, not only how genes turn on, but how they are turned off, and finally to understand all the interacting ingredients in a cell that are involved in achieving the final production of a protein.

MUELLER: Does anyone here have any feeling as to what you most need in terms of tools or techniques to reach this next step? I realize that what you would like to have is the cloned material, full genes, and then do the reconstitution type of approach.

ROSEN: Just a comment. I think that it has been pretty difficult, even just using isolated nuclei, to get transcription with fidelity, and there are obviously lots of things that we don't know about transcription so that the chance of being able to take a piece of isolated DNA and reconstitute it is a slim one right now. It is something that we have to work toward, but a lot can be learned by dealing with in vitro systems that involve whole cells as well. I think that there is a fair amount of emphasis now to try to develop cell systems that you can study in the whole cell to look at regulation as well. You can learn a little bit there; you can never learn the precise chemical mechanism that Bert O'Malley wants to learn unless you can dissect and put it back together. But the question is: Do we have the type of technology now, which is what you are asking, to really put the system back together? I think that one of the problems is that there is a high-order structure in chromatin that you may not be able to reconstitute just by taking an isolated, even 10 or 20-kb pieces of DNA and adding proteins back to it. This is going to be a problem, notwithstanding the problems of getting polymerase and factors and other things that you need that you don't even know what they are.

LIAO: It's very convenient if you use the message that you can isolate in large amounts, for example, ovalbumin or casein, and you have shown that the effect can be seen in 1 h, but then you see a very rapid increase in the rate of accumulation. So, even though Bob Goldberger does not like the word "lag phase," the question that I must ask is: Can we exclude the possibility that the hormone is stimulating the synthesis of another protein which is needed for the effect to be seen? Do you really believe that there is a direct effect? I mean direct in the sense that you're synthesizing the message.

O'MALLEY: I couldn't rule out the existence of any intermediates. Our recent data, however, reveal that induction starts within 20–30 min and can reach a maximum rate in about 1 h. This makes complicated intermediates less likely.

THOMPSON: If you are considering a way to increase the rate of accumulation of message, suppose that you are trying to decide whether the increase is due to increased initiation of transcripts, or some sort of accumulation resulting from altered processing. As you have said, nobody really suspected the existence of the processing enzyme until very recently. If one imagines another kind of processing enzyme that we still haven't picked up, which rather destroys the transcript, one could envision a mechanism in which the steroid with receptor would directly interact with this processor to block its action. There are models for steroids interacting with enzymes, in fact, to alter their function; GDH is one, dehydrogenase is another. So this would permit then access of pooled transcripts of the precursor hnRNA to be available for the processing enzyme now we know exists. Could you suggest any possible experiment that could be done which would allow that distinction between that model and the model of increased initiation of transcription via any hormone receptor complex?

O'MALLEY: I'll just say one quick thing. The processing is posttranscriptional, would be a posttranscriptional phenomena, is what you're talking about—right? Oh, you mean that it might be on the template? O.K., if it is posttranscriptional, you can rule it out by looking at the rate of initiation of the gene in vitro before and after hormone in chromatin or nuclei. Both of these studies seemingly very clearly show there is a very inefficient, very poor, very low rate of initiation of transcription of the ovalbumin gene in the nuclei or chromatin in the absence of hormone, and as you return hormone, very quickly you allow this gene to be read out. Now, if you postulate that there was an enzyme bound in the chromatin now, that in the absence of hormone caused immediate degradation of what was made as it was being transcribed, then I don't think that you could rule that out by these experiments, and I don't know of an experiment immediately that could rule that out. I think that it is highly unlikely and depends on how hard you want to hold on to something for which there is no precedent. There is no precedent for an enzyme which works that fast because you can calculate how fast that has to be in turning over the RNA. It has to have the ability to turn it over faster than it is made. That's quite a constriction to put on that enzyme system, and I just would take the position that it probably doesn't exist at the present time. Believe me, everybody is keeping their eyes open for something that would show up that may show a contribution from this type of a fact.

GOLDBERGER: Brad Thompson, although you used the word "processing," I guess what you meant was that the degradation of mature message could fit into what you were talking about. Is that correct? The disappearance of message or its half-life, right?

THOMPSON: I mean the hydrolysis of precursor to mature messages.

GOLDBERGER: I see what you mean.

ROSEN: The only way to definitely answer that is if you could do a rapid pulse chase in mammalian cells. If you could instantaneously pulse for a minute or two and stop, you should be able to look at the turnover or something, if you could do a short enough pulse chase. You can't do that kind of experiment in a mammalian cell with a short enough pulse chase to really look at the turnover. I think that, as Bert O'Malley said, when you are doing a hybridization assay, you don't have to assay intact molecules. Anything bigger than 50 bases is probably going to hybridize in a stable hybrid, and most of these things are endonucleolytic and are not exonucleolytic anyway, so that if you are dealing with endonucleases that are chopping things up to pieces, you'll still pick those pieces up unless you degrade it down to less than 50 nucleotides. The chance of that happening in such a rapid time as a few seconds is highly unlikely, considering the rates of synthesis of these molecules.

MUELLER: Do we have any kind of method of interrupting the processing at this point? It seems to me that what we need for that area of study is a good inhibitor, or a sensitive gene for the processing entity.

ROSEN: I think that what has been primarily done is in systems where they have used enzymes like RNAse-III to process and attack double-stranded regions. You can design inhibitors like double-stranded RNA that will inhibit that enzyme. You can use intercalating agents such as ethidium bromide, or things that will block that. But we really don't know enough about the enzyme involved in the processing, even in terms of its cleavage and ligation, to design an inhibitor yet. People tried lower temperatures in the silk fibroin case to grow the insect at lower temperature, and that didn't seem to work any better. So, the difficulty is that most of the things that block processing may block transcription as well. You need to design a specific inhibitor that will not block transcription and yet block processing.

MUELLER: That very fact may be telling us something about the whole process that is, these two may be tightly coupled.

Discussants: I.S. EDELMAN, A. GAREN, R.F. GOLDBERGER, S. LIAO, G.C. MUELLER, J.M. ROSEN, and E.B. THOMPSON

Chapter 3

The Effect of Estrogen on Gene Expression in Avian Liver

ROBERT F. GOLDBERGER AND ROGER G. DEELEY

I. Introduction: Choosing the System

About four years ago, we turned our attention from regulation of gene expression in prokaryotes and began to search for a system that would tell us something about regulation in eukaryotic cells. We were guided in this search by the following considerations. First, we did not want to work with a system in which cellular differentiation was itself an integral part of the regulatory response. We felt that we could obtain more clear-cut information if we could study a response in a tissue that was already fully differentiated and in which the regulatory response did not require DNA synthesis. We recognized that a hormonally responsive system would be the easiest to manipulate, and for practical reasons we wanted a system in which the hormonal response was of very great magnitude. And as long as were asking, we thought we might as well put in a request that the tissue we study also contain a whole group of genes that are hormonally responsive, so that we would be able to study not only the effect of hormone on a specific gene but also the mechanisms that coordinate the expression of the hormone-responsive domain of the tissue. In addition, we wanted our hypothetical tissue to have a gene expressed at a high level that is not responsive to hormonal stimulation. Such a gene would provide an ideal control for in vitro transcription studies, where one must show that the specificity of hormonal responsiveness reflects that of the tissue in vivo. The system that we finally chose, and that actually does fulfill all these criteria, is the avian liver and its response to estrogen.

It has been known for almost 30 years that the liver is the site of synthesis of avian yolk proteins, and that synthesis of these proteins is normally carried out in the female and can be induced in the male—even in the mature male—by administration of estrogen (Clegg et al. 1951). Historically, egg yolk proteins were divided into two groups: the low- and the high-density lipoproteins. The major egg yolk phosphoproteins are found among the high-density lipoproteins (Bernardi and Cook 1960). They are known as phosvitin (actually there are two different phosvitins) and lipovitellin. It had been shown that induction of these three proteins by estrogen was tightly coordinated (Ramney and Chaikoff 1951), and so it seemed to us that they could serve as a starting point for studies on the estrogen-responsive genes of a highly active, terminally differentiated tissue—the liver—in which the hormonal response must, of necessity, be superimposed on a plethora of functions that the liver fulfills before, during, and after the hormonal response. One of these constitutive hepatic functions is the synthesis of large amounts of serum albumin, the system we chose as our control.

II. Vitellogenin and Its Messenger RNA

When we began to study the avian egg yolk phosphoproteins, it soon became clear that our selected group of estrogen-responsive genes was in fact a group of one; the egg yolk phosphoproteins turned out to be products of cleavage of a single precursor protein, vitellogenin (Deeley et al. 1975). Clearly we would have to look beyond vitellogenin to capture a group of estrogen-responsive genes. In this chapter we describe some of the work we and our colleagues have done with the vitellogenin system, how we have approached the problem of our control system, serum albumin, and how we have begun to isolate cloned sequences representing various parts of the estrogen-responsive domain of avian liver.

A schematic representation of avian vitellogenesis is shown in Fig. 1. Vitellogenin is synthesized in the liver of the female as a huge polypeptide with a molecular weight of 240 000 (Deeley et al. 1975). This polypeptide is subsequently phosphorylated, glycosylated, and associated with lipid. It is carried in the blood as a dimer (with a molecular weight of about 500 000) to the ovary, where it is taken up by the developing oocyte and cleaved specifically to form the egg yolk phosphoproteins, lipovitellin and phosvitin (Bergink and Wallace 1974). The function of vitellogenin has never been elucidated, though most authors state (without any apparent hesitation) that it serves as a phosphate storage protein for the developing embryo or as a metal ion transport protein. Whatever the true function of vitellogenin may be, it is certainly a molecule to be reckoned with. Each egg contains about 1 g of vitellogenin, which represents approximately 150 mg of high-energy phosphate.

Figure 2 summarizes our current model for the vitellogenin polypeptide, derived from studies of the molecular weights and phosphate contents of its component parts, as well as from immunological, peptide, and amino acid analysis data. The molecule consists of one lipovitellin polypeptide at the amino terminal end and two similar (but slightly different) phosvitin polypeptides clustered towards the carboxy terminal end (Gordon et al. 1977). The whole molecule is unusually large, requiring a messenger RNA of at least 6000 nucleotides to encode it. In fact, the enormous length of the vitellogenin mRNA has allowed us to purify it to homogeneity, using size alone as the criterion for separating it from the other mRNAs of avian liver (Dee-

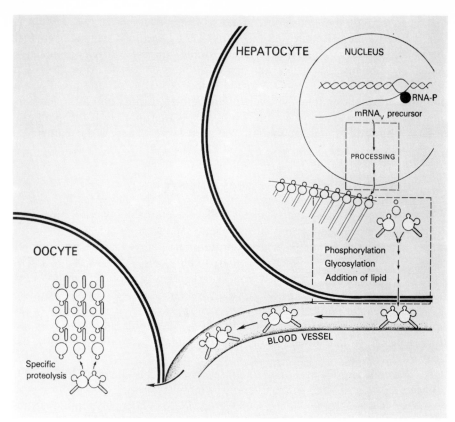

Fig. 1. Avian vitellogenesis. Vitellogenin is synthesized in the hepatocyte on membrane-bound polysomes. After posttranslational modifications, it is immediately secreted into the blood, where it exists as a dimer associated with lipid. It is taken up by the oocyte, possibly by pinocytosis, and cleaved specifically to form the egg yolk phosphoproteins, phosvitin and lipovitellin.

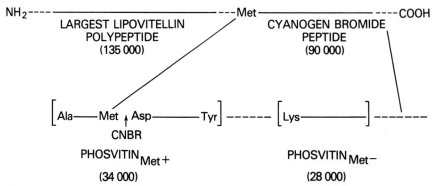

Fig. 2. Model of the avian vitellogenin polypeptide monomer (molecular weight 250 000). The experimental basis for this model is described by Gordon et al. (1977). The dashed lines indicate that the precise positions of the lipovitellin polypeptide and the phosvitin lacking methionine are not presently known.

Fig. 3. Electrophoresis of purified vitellogenin mRNA in a 1.5% agarose gel containing methylmercuric hydroxide (5 m*M*). The gel was stained with ethidium bromide and photographed under ultraviolet light. The figure shows a photograph of the gel to which had been applied 2.1 μg of purified vitellogenin mRNA and 2.5 μg of 18S and 28S ribosomal RNA from chicken liver. (Reprinted from Deeley RG, Gordon JI, Burns ATH, Mullinix KP, Bina-Stein M, Goldberger RF (1977a) J Biol Chem 252: 8310, with permission of the authors and publisher.)

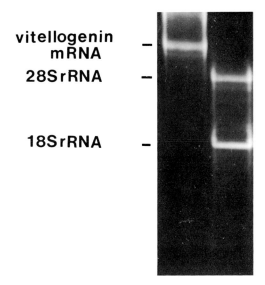

vitellogenin mRNA   –

28SrRNA   –

18SrRNA   –

ley et al. 1977a). A preparation of pure vitellogenin mRNA (along with 18 and 28S ribosomal RNA markers) subjected to electrophoresis in agarose containing methylmercuric hydroxide as the denaturing agent is shown in Fig. 3. Its size, determined by gel electrophoresis, as shown here (but with many RNA standards of known size), and by contour analysis of electron photomicrographs, as shown in Fig. 4, is 2.35 million daltons, or 7000 nucleotides, approximately 600 nucleotides longer than that required to specify the vitellogenin polypeptide (Deeley et al. 1977a).

III. Kinetics of mRNA Accumulation

We were now in a position to synthesize vitellogenin cDNA, which we needed as a probe to examine the kinetics of accumulation of vitellogenin mRNA in vivo. One point of interest was to determine whether vitellogenin mRNA could be detected at all in the liver of the mature rooster; another was to follow up on an old observation that suggested that, in addition to the transient inductive effect of primary stimulation with estrogen, there was a long-term effect reflected in the increased magnitude and rapidity of the vitellogenic response to secondary estrogen stimulation. An old experiment of Goldstein and Hasty (1973) is shown in Fig. 5, in which vitellogenin protein was monitored in rooster serum after stimulation with estrogen. Depending upon what technique is used, one picks up vitellogenin in the circulation of the rooster within a few hours after estrogen stimulation; it reaches a maximum in about three days, and has disappeared about 7 days later. One of the very interesting features of the vitellogenic response is the greater rapidity and greater magnitude of the response when the same animal is later given a second stimulation with estrogen, days or weeks, or even months after the first. This effect, which is shown on the right of Fig. 5, is

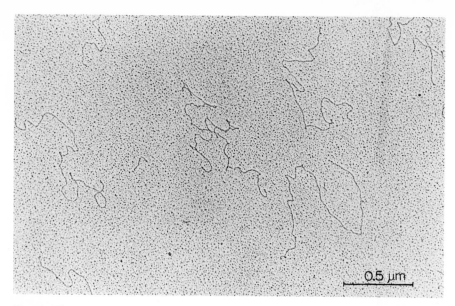

Fig. 4. Electron photomicrograph of purified vitellogenin mRNA. The RNA was spread in formamide (75%, v/v) and urea (6 *M*). Its length was determined, by measurement of many such molecules, to be 2 μm. (Reprinted from Deeley RG, Gordon JI, Bums ATH, Mullinix KP, Bina-Stein M, Goldberger RF (1977a) J Biol Chem 252: 8310, with permission of the authors and publisher.)

reminiscent of the anamnestic response of the immune system, and has been the subject of much speculation. We repeated these experiments, but instead of using an antibody to monitor the accumulation of vitellogenin protein in the serum, we used specific cDNA to monitor the accumulation of vitellogenin mRNA in the liver (Deeley et al. 1977a). With this approach, utilizing nucleic acid hybridization, it is possible to detect as little as a single molecule of vitellogenin mRNA per cell. The results of several experiments on the primary vitellogenic response (Burns et al. 1978) are shown in Fig. 6. In Figure 6A are the results obtained when cDNA was used to monitor the amount of vitellogenin mRNA present before, and at various times after, a single primary stimulation with estrogen. Before stimulation, there are between zero and five molecules of vitellogenin mRNA per cell. Following stimulation, the level of vitellogenin mRNA increases very rapidly, reaching almost 2000 molecules per cell within the first 12 h, and 6000 molecules per cell at the height of the response, 3 days after estrogen treatment. Thereafter, the vitellogenin mRNA disappears from the liver at a rate consistent with that expected for a molecule with a half-life of about 30 h, as shown by the descending broken line. Using this estimate of the half-life of the message, and the observed rate of its accumulation, we calculate a transcription rate of about 340 nucleotides per second per cell. The theoretical accumulation curve for such a transcription rate and half life is shown by the ascending broken line in Fig. 6. It extrapolates to a steady state level of about 7000 molecules per cell, which, interestingly enough, is the steady state level

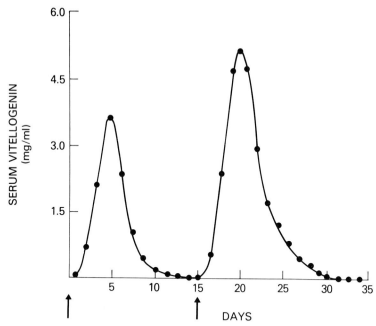

Fig. 5. Accumulation of vitellogenin in the plasma of a rooster after primary and secondary administration of 17β-estradiol. (This figure is redrawn, with permission of the authors, from the data of Goldstein and Hastey, 1973. The data have been recalculated to give the weight of vitellogenin rather than phosvitin, although the original measurements were expressed as micrograms of protein-bound phosphate, a measurement that subsequent work has shown to provide a valid estimate of the vitellogenin content of plasma.)

of vitellogenin mRNA found in the liver of the laying hen. Once again, the levels of vitellogenin mRNA are shown in Fig. 6B, this time determined both by hybridization analysis (the solid line and dots) and by translation of the mRNA in a wheat germ cell-free system (broken line and squares). The results obtained by both methods agree very well, providing assurance that the hybridization analyses really reflect the level of intact translatable vitellogenin mRNA molecules. Figure 6C gives the temporal relationship between the accumulation and disappearance of vitellogenin mRNA (solid circles), the accumulation and disappearance of vitellogenin protein from the plasma (open circles), and the serum levels of 17β-estradiol (dashed line) at various times after injection of the hormone. It is interesting to note that at the time when the level of vitellogenin mRNA begins to fall drastically, the serum level of 17β-estradiol in the rooster is still 10–20 times higher than that found in the laying hen. This level is more than adequate to saturate the high-affinity estrogen binding sites that have been shown to exist in rooster liver nuclei (Gschwendt and Kittstein 1974), so it is rather a puzzle why the level of vitellogenin mRNA should begin to fall so drastically. It may be that the rooster possesses a mechanism for excluding estrogen from the liver cell or for overriding the effect of estrogen.

The rate at which vitellogenin mRNA accumulates during the first 4 h fol-

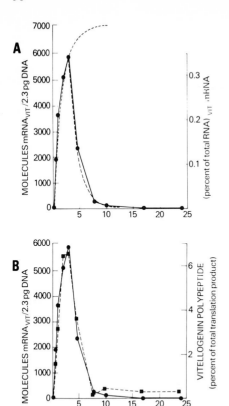

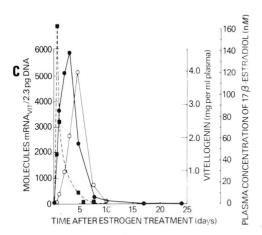

Fig. 6. Kinetics of accumulation and degradation of vitellogenin mRNA following a primary injection of 17β-estradiol. Total RNA was prepared from the liver of cockerels (55 g) at various times after injection of 17β-estradiol. Vitellogenin mRNA was quantified by RNA excess hybridization to vitellogenin cDNA, using purified vitellogenin mRNA as a standard, and also by translation in a wheat germ cell-free system.

A The levels of vitellogenin mRNA (●) determined by R_0t analysis and the theoretical accumulation and decay curves (–––) for a mRNA with a half-life of 29 h. B A comparison of the levels of vitellogenin mRNA determine by R_0t analysis (●) and by translation in a wheat germ cell-free system (■). C The temporal correlation between vitellogenin mRNA levels (●) in the liver and the levels of 17β-estradiol (■) and vitellogenin (○) in plasma. (Reprinted from Burns ATH, Deeley RG, Gordon JI, Udell DS, Mullinix KP, Goldberger RF (1978) Proc Natl Acad Sci USA 75: 1815, with permission of the authors and publisher.)

lowing either a primary or a secondary injection of hormone (Deeley et al. 1977b) is shown in Fig. 7. The curves indicate that vitellogenin mRNA begins to accumulate about 30 min after injection of the hormone following both primary and secondary stimulation. However, there is a six- to sevenfold difference in the rates. Following a primary injection (the lower curve),

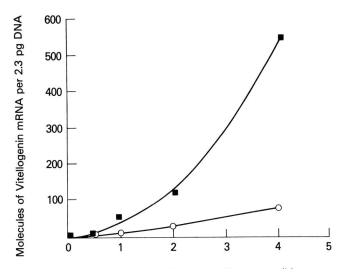

Fig. 7. Initial rates of accumulation of vitellogenin mRNA during primary and secondary stimulation with 17β-estradiol. Cockerels were injected with hormone (20 mg/kg of body weight) and killed at the times indicated. The animals used for the secondary stimulation had received a primary injection of hormone 1 month previously. Total cellular RNA (free of DNA) was extracted from the liver, and hybridization to vitellogenin cDNA under conditions of RNA excess was carried out by the procedures reported by Deeley et al. (1977a). The amount of vitellogenin mRNA in each sample was determined by comparison of the $R_0 t_{1/2}$ value with that obtained using purified vitellogenin mRNA. The number of molecules of vitellogenin mRNA per nuclear equivalent of DNA (2.3 pg) was calculated for each sample from the hybridization data, using the known molecular weight of vitellogenin mRNA (2.35 $\times$ 10^6), the yield of RNA per gram of tissue (5–6 mg/g), and the DNA content of the tissue (2–2.5 mg/g). The data demonstrate that during the first 4 h, the rate of accumulation of vitellogenin mRNA following secondary hormonal stimulation (■) was about 6–7 times higher than following primary stimulation (○). (Reprinted from Deeley RG, Udell DS, Burns ATH, Gordon JI, Goldberger RF (1977b) J Biol Chem 252: 7913, with permission of the authors and publisher.)

the mRNA accumulates linearly at a rate of only 50 nucleotides per second per cell for the first 4 h or so. In contrast, following a secondary injection of hormone (the upper curve), accumulation begins at a high rate—approximately 350 nucleotides per second per cell. A more extended period of the primary response is shown in Fig. 8. Although the rate of accumulation of vitellogenin mRNA begins at a very low rate, it switches to a high rate (about 350 nucleotides per second per cell) after 4 h, so that the primary response is markedly biphasic. In contrast, as mentioned above, during the secondary response accumulation of vitellogenin mRNA starts right away at a high rate (and then increases only slightly—about 1.4-fold). Figure 9 summarizes all of the previous data (Deeley et al. 1977b). During the primary response, shown by the lower curve, vitellogenin mRNA begins to accumulate at a very low rate (which is just barely visible in the figure because of the compressed scale) and then increases greatly after about 4 h, reaching approxi-

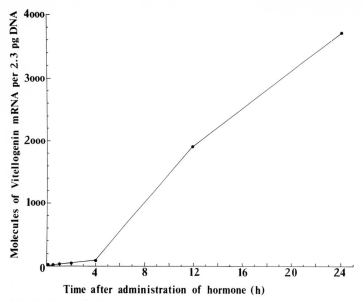

Fig. 8. Biphasic kinetics of accumulation of vitellogenin mRNA during the first 24 h after primary hormonal stimulation. The experiments were carried out and the data analyzed as described in the legend to Fig. 7. The data demonstrate that there was a six- to sevenfold increase in the accumulation rate about 4 h after hormonal stimulation. (Reprinted from Deeley RG, Udell DS, Burns ATH, Gordon JI, Goldberger RF (1977b) J Biol Chem 252: 7913, with permission of the authors and publisher.)

mately 6000 molecules per cell after three days. During the secondary response, shown by the upper curve, accumulation begins at the high rate and reaches approximately 9000 molecules per cell after 3 days. In both cases, the mRNA decays with a half-life of about 30 h.

Thus we see that the anamnestic response of the vitellogenin system, which was mentioned in the beginning of this chapter, in which secondary stimulation with estrogen causes more rapid and more extensive accumulation of vitellogenin protein in rooster plasma, can be explained on the basis of the difference in the rates and extents of accumulation of vitellogenin mRNA in the liver (Deeley et al. 1977b). On the basis of our findings, we have considered the possibility that primary exposure to estrogen not only causes an increased rate of transcription of the previously dormant vitellogenin gene, a phenomenon that lasts for a few days, but also causes a long-term change in the structure of the vitellogenin gene so that this gene is more readily available for induction upon subsequent exposure to estrogen. Unfortunately, although we have been able to repeat the studies of Weintraub and Groudine (1976) on the sensitivity of the expressed versus the nonexpressed hemoglobin gene to digestion with DNase I, we have not yet been successful with such experiments on the vitellogenin gene. For the time being, those experiments have been left in abeyance, and we have turned our attention to other aspects of the vitellogenin problem.

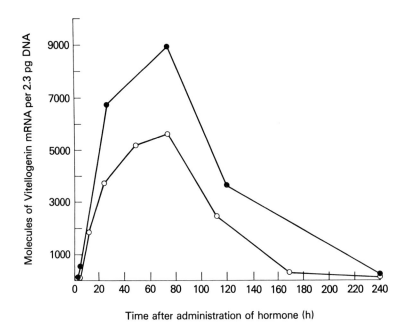

Fig. 9. Rates of accumulation and decay of vitellogenin mRNA after primary and secondary hormonal stimulation. The experiments were carried out and the data analyzed as described in the legend for Fig. 7. The data demonstrate that after primary (○) or secondary (●) stimulation, the mRNA accumulated for a period of 3 days. Its rate of disappearance during the next 7 days was consistent with that expected for a molecule with a half-life of approximately 30 h. The maximum level of vitellogenin mRNA reached following secondary stimulation was 1.5 times higher than that reached following primary stimulation. (Reprinted from Deeley RG, Udell DS, Burns ATH, Gordon JI, Goldberger RF (1977b) J Biol Chem 252: 7913, with permission of the authors and publisher.)

IV. The Natural Vitellogenin Gene

Since we are ultimately interested in studying regulation of the vitellogenin gene in vitro, we were anxious to find out something about the structure and organization of the vitellogenin gene in the chicken genome, and to clone DNA sequences that would provide potential substrates and probes for in vitro transcription studies. To get some idea of the size of the DNA sequence over which the vitellogenin gene is spread, we took a look at the *Eco*RI restriction fragments of chicken genomic DNA, in collaboration with Dr. Phillip Leder. An autoradiograph of an agarose gel in which we ran fractions from an RPC-5 column of the retriction digest is shown in Fig. 10. The probe was vitellogenin cDNA about 1000 nucleotides long and this represents the sequences found at the 3′ end of the vitellogenin mRNA. As shown in the figure, this portion of the vitellogenin structural gene is represented in two fragments which together have a length of about 12 000 base pairs. To identify the sequences encoded in the entire length of the mRNA,

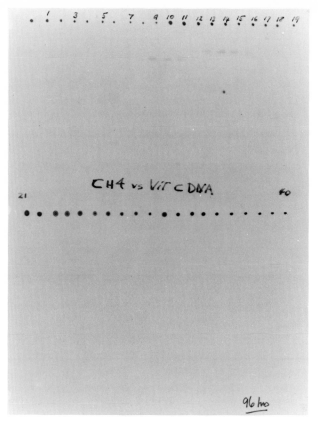

*Fig. 10. Eco*RI fragments of the chicken genome that hybridize with vitellogenin cDNA. An *Eco*RI digest of chicken DNA was fractionated on an RPC-5 column, and the fractions were then subjected to electrophoresis in agarose. The figure shows an autoradiograph of a blot of this gel, probed with radioactive vitellogenin cDNA. See text for details.

we used radioactively labeled fragments of purified vitellogenin mRNA as a probe for similar blots of a genomic *Eco*RI digest. The results are shown in Fig. 11. Here we see a faint reaction with the same two fragments that hybridize to the cDNA, and a much stronger reaction with two other fragments, which would be expected, therefore, to contain the bulk of the structural gene sequences. All the fragments together add up to a size of approximately 20 000 base pairs. Of course, this makes it somewhat difficult to clone the natural vitellogenin gene in its entirety. Therefore, while such studies are in progress, we have also begun to clone double-stranded cDNA made from vitellogenin mRNA.

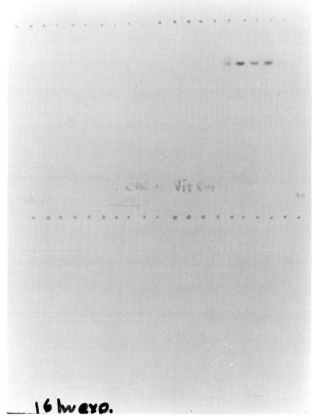

Fig. 11. _Eco_RI fragments of the chicken genome that hybridize with vitellogenin mRNA.   A blot of the same agarose gel used for the experiment in Fig. 10 was probed with radioactively labeled fragments of vitellogenin mRNA.   See text for details.

V. Cloning Double-Stranded Vitellogenin cDNA

The technique we have used for cloning cDNA is summarized in Fig. 12. Here we see the synthesis of double-stranded cDNA, using reverse transcriptase for both strands.   After trimming with S1 nuclease, we used terminal transferase to add tails of 20–30 deoxycytidylic acid residues to both 3′ ends.   Meanwhile, the bacterial plasmid vector, PBR322, was prepared for insertion of this tailed double-stranded cDNA.   First, the circular plasmid DNA was cut with the restriction endonuclease, _Pst_I.   There is a single site for this enzyme located within the ampicillin resistance gene of the plasmid. The nucleotide sequence cleaved by _Pst_I is CTGCAG.   After cleavage, a protruding adenylic acid residue is left as the 3′ terminus, providing an excellent substrate for terminal transferase, which is used in the next step to add 20–30 deoxyguanylic acid residues to both ends of the plasmid DNA. The tailed cDNA and plasmid DNA are allowed to hybridize through their

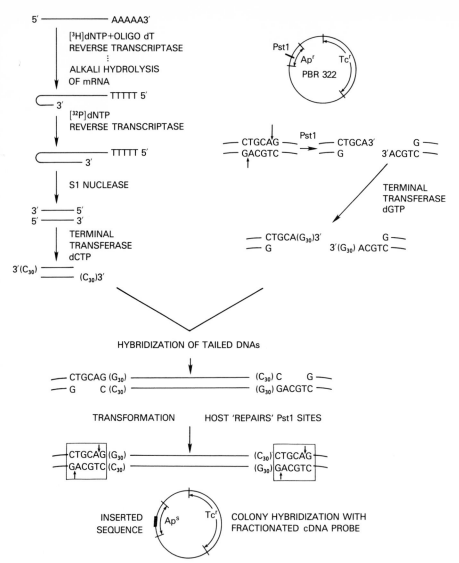

Fig. 12. Overall scheme employed in cloning double-stranded cDNA. (Reprinted from Gordon JI, Burns ATH, Christmann JL, Deeley RG (1978) J Biol Chem 253: 8629, with permission of the authors.)

complementary tails, and the circularized molecules are used to transform the enfeebled *E. coli* host, χ1776 (Bolivar et al. 1977). Note that following transformation the host repairs the *two* *Pst*I sites, allowing the subsequent excision of the DNA with this enzyme. We now have a collection of vitellogenin cDNA clones and are using them in studies that are still too preliminary to discuss here.

VI. The Control System: Serum Albumin

And now we turn to the question of how we have approached our control system, chicken serum albumin. In Fig. 13 what happens to the level of serum albumin mRNA in rooster liver after stimulation with estrogen during the period when vitellogenin mRNA increases 5000- to 10,000-fold is shown. If we express the albumin mRNA concentration as a fraction of the total RNA, we get the lower curve, which shows approximately a twofold decrease in the relative concentration. This results from the fact that the total RNA content of the liver doubles during this period and is responsible for the erroneous notion that has crept into the literature recently that albumin mRNA synthesis decreases slightly upon estrogen stimulation. In fact, as seen by the upper curve, which is a plot of the total albumin mRNA of the liver, there is no significant change during the vitellogenic response. These data were obtained by translation of total liver mRNA in a wheat germ cell-free system. But we get the same results when the level of albumin mRNA is monitored with complementary DNA. To obtain such cDNA, we faced the problem that albumin mRNA is very difficult to purify because it has the same size as many other mRNAs, and so we developed a procedure for cloning a specific sequence from a partially purified preparation of messenger RNA (Gordon et al. 1978).

We started with a fraction of rooster liver that was a by-product of our purification of vitellogenin mRNA, a fraction that contained the bulk of the albumin mRNA, enriched about two-fold. We used this enriched, but still very crude, mRNA as a template for reverse transcriptase to make dou-

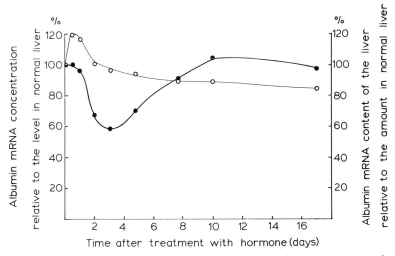

Fig. 13. Levels of albumin mRNA during the primary vitellogenic response. (●) albumin mRNA concentration expressed as a fraction of the total liver RNA; (○) albumin mRNA concentration expressed as the total amount per liver.

ble-stranded cDNA.   We then cloned the whole mixture, knowing that most
of the clones would not contain albumin sequences.   The cloning was accomplished by the G–C tailing technique described above.

Our next step was to identify which of our clones were the right ones.   To
do this, we went back to our original partially purified preparation of mRNA,
and carried out a R_0t analysis, the results of which are shown in Fig. 14.
The right-hand curve shows the results we obtained.   The abundant species
of mRNA (corresponding to albumin) is clearly visible as the first transition,
and constitutes about 30% of the RNA.   We then repeated this reaction but

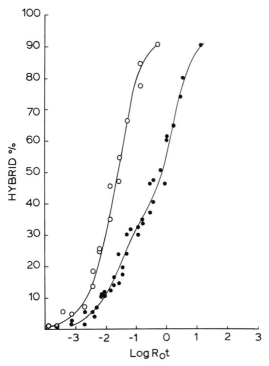

Fig. 14. cDNA complexity analysis of partially purified albumin mRNA and isolation
of abundant class cDNA. [^{3}H]cDNA (0.067 ng), transcribed from partially purified
albumin mRNA was hybridized with 1–100 ng of the mRNA in NaCl (0.6 M), 4-(2-hy-
droxyethyl)-1-piperazineethanesulfonic acid (10 mM, pH 7.0), and Na$_2$EDTA (1 mM)
at 68°C to the R_0t values indicated.   The extent of hybridization was determined by
S1 nuclease digestion of the reaction mixture.   For isolation of abundant class
cDNA, preparative hybridization reactions containing 90–200 ng of [^{3}H]cDNA and
1 μg of partially purified mRNA were incubated to R_0t = 3.98 × 10^{-2} mol s liter^{-1}
(log R_0t, 1.4).   Hybridized abundant class cDNA was isolated by S1 nuclease digestion.   Abundant class cDNA (0.067 ng) was subsequently hybridized "back" to the
partially purified albumin mRNA preparation (5–10 ng).   The hybridization curves
shown are: (●) total cDNA to partially purified mRNA; (○) isolated abundant class
cDNA to partially purified mRNA.   The lines drawn are those theoretically expected
for ideal pseudo-first-order reactions. (Reprinted from Gordon JI, Burns ATH,
Christmann JL, Deeley RG (1978) J Biol Chem 253: 8629, with permission of the authors.)

carried it only to a log R_0t of -1.2, at which point the abundant cDNA was in hybrid form and could be isolated by digesting the rest of the cDNA with S1 nuclease. Now we had a probe that is specific. When we hybridized this probe back to the mRNA, we got the curve shown on the left in Fig. 14. It hybridized with the ideal pseudo-first-order kinetics expected for a single species. We used this probe to identify which of our clones carried albumin DNA. We then confirmed the identity of the selected clones by two other techniques.

First, we used the hybridization arrest procedure (Paterson et al. 1977) to show that the cloned DNA specifically inhibits translation of albumin mRNA in the wheat germ cell-free system. In Fig. 15 fluorographs are shown of denaturing acrylamide gels of the total polypeptides synthesized in the wheat germ cell-free system in response to total rooster liver RNA (left) and an immunoprecipitate of this reaction mixture, using monospecific anti-chicken albumin antibody (right). Having identified the prealbumin polypeptide, we were now ready for the hybridization arrest experiments. Our results are shown in Fig. 16. Each track represents the polypeptides synthesized in the wheat germ cell-free system in response of total rooster liver RNA. The two tracks on the right tell the story. Track 9 shows that when the plasmid DNA from one of our putative albumin clones was hybridized to the RNA before translation, the synthesis of the prealbumin polypeptide was specifically inhibited. Track 10 shows that when the hybrids are formed in

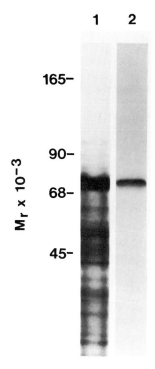

Fig. 15. Cell-free translation of albumin mRNA. Total cellular RNA from rooster liver (20 μg/incubation) was added to a nuclease-treated wheat germ cell-free translation system. After a 2-h incubation in the presence of [^{35}S]methionine, phenylmethylsulfonyl fluoride (50 μg/ml), and benzamidine (1 mM) were added, polysomes were sedimented, and the supernatant, containing released polypeptide chains, reacted with purified monospecific antichicken serum albumin antibody. The immunoprecipitate was washed in detergent, denatured, and electrophoresed in polyacrylamide slab gels (7.5%) containing SDS (0.1%). The figure shows a fluorograph of the gel.
1 total polypeptides synthesized from total rooster liver RNA; **2** polypeptide immunoprecipitated by anti-albumin antibody from the incubation in 1. (Reprinted from Gordon JI, Burns ATH, Christmann JL, Deeley RG (1978) J Biol Chem 253: 8629, with permission of the authors.)

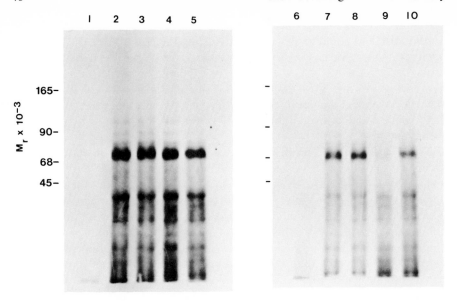

Fig. 16. Clone A26 DNA plasmid reversibly arrests cell-free translation of preproalbumin mRNA.   One microgram of either *Eco*RI-digested PBR322 or *Eco*RI-digested clone A26 plasmid DNA was hybridized to 20 μg of total rooster liver RNA.   Hybridization reactions were split in half, one-half heated at 100°C for 1 min to dissociate any mRNA–DNA hybrids formed, and both aliquots were ethanol-precipitated prior to wheat germ cell-free translation in the presence of [³⁵S]methionine.   The figure shows a fluorograph of ³⁵S-labeled polypeptides synthesized in vitro and separated on a polyacrylamide slab gel (7.5%) containing SDS (0.1%).

1 endogenous translation; **2** RNA alone after incubation under hybridization conditions; **3** heat melt of 2; **4** RNA after incubation with 1.0 μg of PBR322 DNA under hybridization conditions; **5** heat melt of 4; **6** endogenous translation; **7** RNA alone after incubation under hybridization conditions; **8** heat melt of 7; **9** RNA after incubation with 1.0 μg of clone A26 plasmid DNA under hybridization conditions; **10** heat melt of 9.   (Reprinted from Gordon JI, Burns ATH, Christmann JL, Deeley RG (1978) J Biol Chem 253: 8629, with permission of the authors.)

the same way but then destroyed by melting, the ability of albumin mRNA to be translated is restored.   The other pairs of tracks are controls representing hybridization and melt in the absence of any DNA or in the presence of plasmid DNA without any albumin insert.   Such controls are extremely important in hybridization arrest experiments to assure that the conclusion is valid.

As a further check on the identity of the albumin clones, we used recombinant plasmid DNA to probe blots of total rooster liver RNA (fractionated on denaturing agarose gels) by the procedure of Alwine et al. (1977).   The results obtained are shown in Fig. 17, in which the left-hand track is a photograph of the original gel of total RNA stained with ethidium bromide.   Some of the ribosomal RNA was removed so that a band can be visualized corresponding to the abundant albumin mRNA.   The RNA from such a gel was transferred to a sheet of diazotized paper by the technique of Alwine et al. (1977), and probed with nick-translated DNA from a plasmid containing the

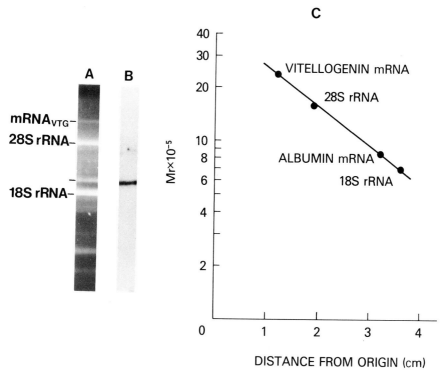

Fig. 17. Sizing of albumin mRNA by hybridization of clone A26 plasmid DNA to total poly(A)+ hen liver RNA. Total poly(A)+ hen liver RNA (3 μg) was electrophoresed in 1.2% agarose gels containing methylmercuric hydroxide (5 m*M*). A portion of the gel was stained with ethidium bromide to visualize RNA bands. The remaining portion was incubated in NaOH (50 m*M*) and 2-mercaptoethanol (5 m*M*) prior to transfer of the RNA to diazobenzyloxymethyl paper. ³²P-labeled *Eco*RI-digested clone A26 DNA was hybridized to the DNA blot.
A 1.2% Agarose gel of hen liver poly(A)+ RNA stained with ethidium bromide and viewed under ultraviolet light. **B** Autoradiograph of blot after hybridization with [³²P]plasmid DNA from clone A26. **C** Sizing of albumin mRNA based on the data presented in **A** and **B**. (Reprinted from Gordon JI, Burns ATH, Christmann JL, Deeley RG (1978) J Biol Chem 253: 8629, with permission of the authors.)

putative albumin insert. As is clear in the figure, the cloned DNA hybridized specifically to the albumin mRNA. This band has a mobility expected for an RNA with a mass of 850 000 daltons, which corresponds to a length of 2600 nucleotides. This length is about 600 nucleotides longer than the minimum size necessary to code for the preproalbumin polypeptide, a feature commonly found in mRNAs that specify secreted proteins.

The reason we have taken the time to describe the details of these experiments is that the methodology is generally applicable. It can be used to obtain clones of specific DNA fragments not only in those cases in which the mRNA is abundant to begin with, but also, and perhaps more important, whenever the concentration of a particular mRNA can be altered by any means, which brings us to the last subject that we wish to discuss here.

VII. The Estrogen-Responsive Domain of Avian Liver

It should be recalled that at the beginning of this chapter we enumerated the features of the system we hoped to find. Well, we have our tissue that is terminally differentiated and highly active metabolically; we have our vitellogenin gene that responds dramatically to estrogen stimulation without requiring DNA synthesis; and we have our control system, serum albumin, which is truly a constitutive function of the liver. But we are still missing specific probes for other genes that respond to estrogen, which we need if we are to be able to study the coordinated response of the liver to stimulation by estrogen.

As a first step to obtaining such probes, it was necessary to define in some way the estrogen-responsive domain of the rooster liver (King et al., 1979). Before presenting detailed data, which are all in the form of R_0t analyses, we would like to show an idealized graph that summarizes the information that can be extracted from such hybridization studies (Fig. 18). What we have plotted here is the rate of formation of RNA–cDNA hybrids as a function of the concentration of RNA (R_0) times the time of incubation (t). In such an experiment, those RNA species that are most abundant in the RNA population hybridize first, whereas those that are least abundant hybridize last. Mainly for convenience, the various RNA species are

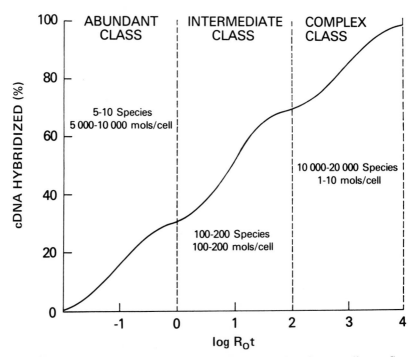

Fig. 18. Idealized R_0t analysis of the complexity of total mRNA of rooster liver. See text for description.

grouped into three abundance classes. Typically, the most abundant group (on the left) represents 5–10 different species of RNA, each present at a concentration of about 5000–10 000 molecules per cell. At the other end of the spectrum, we have the complex group (right), which is composed of 10 000–20 000 different species of RNA, each present at a concentration of only 1–10 molecules per cell. You can see right away from Fig. 18 that one of the disadvantages of using an analysis of this type to estimate changes in the overall complexity of the messenger RNA population of the cell stems from the fact that although the complex class of mRNA may constitutes only 15–30% of the total mRNA by weight, it contains 95–98% of the different mRNA species found in the cell. From a practical standpoint it is also the most difficult region of the hybridization curve from which to obtain accurate data.

There are, however, some ways around this problem. For example, in addition to hybridizing a cDNA back to the template from which it was made, one can also hybridize it to RNA from the tissue in a different state of hormonal response, in which the mRNA population may be different. In this way, we compared the complexities of the mRNA populations of normal and estrogen-stimulated rooster liver by cross-hybridizing the two populations of RNA with their respective cDNAs. For example, as shown in Fig. 19, we have prepared cDNA from normal rooster liver RNA, and then hy-

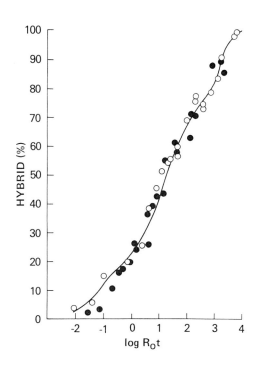

Fig. 19. Hybridization of cDNA prepared from total RNA from the liver of a normal rooster to its homologous RNA and to total RNA from the liver of an estrogen-treated rooster. Total liver cDNA (0.066 ng) was incubated with various amounts of total RNA from the liver of either a normal rooster (●) (0.5–μg) or an estrogen-treated rooster (○) (0.5–350 μg). Incubations were carried out at 68°C in capillaries containing 10 or 20 μl of 0.01 M HEPES, pH 7.0, NaCl (0.6 M), and Na$_2$EDTA (0.02 M). The R_0t values shown are those reached during the experiment and have not been corrected for ribosomal RNA content. The line drawn is a theoretical curve derived solely from the homologous hybridization data. The curve was generated using a Texas Instruments printing calculator programmed to fit the data to 1, 2, or 3 independent pseudo-first-order reactions. Data from the heterologous hybridization reaction were added later for comparison.

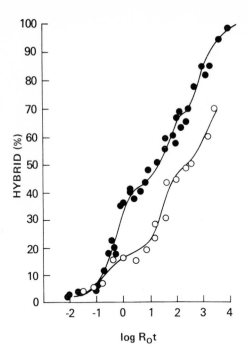

Fig. 20. Hybridization of cDNA prepared by transcription of total RNA from the liver of an estrogen-treated rooster to its homologous RNA, to RNA from a normal rooster and to RNA from hen liver. Details of the hybridization procedure are described in the legend to Fig. 19. (●) Total RNA from the livers of estrogen-treated roosters; (○) total RNA from the livers of normal roosters. The curves shown for the homologous hybridization reactions are curves calculated to fit three abundance classes as described in the legend to Fig. 19. Any curves describing heterologous hybridization reactions were simply drawn through the data.

bridized it both to its template (open circles) and to RNA from the liver of an estrogen-stimulated rooster (solid circles). It is clear that the curves are superimposable, so there are no apparent differences in any of the three conventional abundance classes. We conclude that all those mRNA sequences present before treatment with estrogen are found in essentially the same concentration after treatment. However, when the reciprocal experiment is carried out, a very different result is obtained, as shown in Fig. 20. Here we see the cDNA made from the liver RNA of an estrogen-stimulated rooster hybridized to its template (solid circles) and to normal (unstimulated) rooster liver RNA (open circles). The differences are so great, in fact, that it is impossible to say whether they result from changes in the concentration of mRNA of one abundance class or all abundance classes.

Here we come to a second means for increasing the accuracy of the hybridization analysis. By selecting cDNA corresponding to one particular abundance class of mRNA, we are able to study its hybridization in isolation from the other abundance classes. So, in order to place an upper limit on the total number of sequences that change in concentration upon estrogen stimulation, we prepared cDNA to the most complex class of mRNA, which represents about 15 000 different messages. Using this cDNA in hybridization analyses of normal and estrogen-stimulated rooster liver RNA, as shown in Fig. 21, we found that there is no detectable difference. This would put an upper limit of 1–200 on the number of genes expressed in the presence of estrogen that are not expressed in normal rooster liver, and these sequences must be limited to the abundant or intermediate classes of

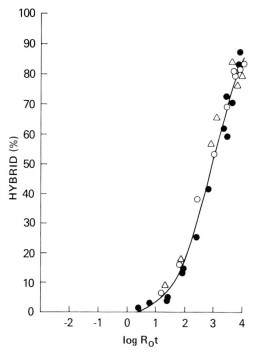

Fig. 21. Hybridization of "complex class" cDNA from the livers of estrogen-treated roosters with total RNA from the livers of normal, estrogen-treated or hormone-withdrawn roosters. Complex class cDNA was isolated as that fraction of total cDNA that had not hybridized with its homologous RNA by a R_0t value of 100 nucleotides s liter^{-1}. Hybridization conditions were as described in the legend to Fig. 19. (●) Total RNA from an estrogen-treated rooster; (○) total RNA from a normal rooster; (△) total RNA from a hormone-withdrawn rooster.

mRNAs (King et al., 1979). Figure 22 shows that this is indeed the case. Here we see the same cross-hybridization experiments, but in this case they were done with cDNA representing the abundant and intermediate classes of mRNA only. There is a point in the hybridization reaction where essentially all of the cDNA has hybridized to its template RNA (from

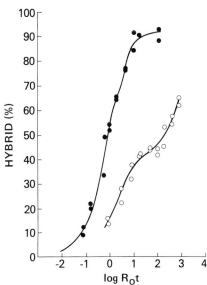

Fig. 22. Hybridization of "abundant and intermediate class" cDNA from the livers of estrogen-treated roosters with total RNA from the livers of normal or estrogen-treated roosters. Abundant and intermediate class cDNA was isolated as that fraction of total cDNA that had hybridized with its homologous RNA by a R_0t of 100 nucleotides s liter^{-1}. Hybridization conditions were as described in the legend to Fig. 19. (●) Total RNA from the livers of estrogen-treated roosters; (○) total RNA from the livers of normal roosters.

estrogen-stimulated rooster, solid circles), while only 45% of it has hybridized to the RNA from a normal rooster liver (open circles). This difference
is so enormous that it became apparent to us that it should be possible to
separate, physically, a group of cDNA sequences complementary to those
mRNA species that change in concentration upon estrogen stimulation, and
that is exactly what we did. First, we decided to concentrate only on abundant sequences. Therefore, we began by isolating cDNA that had been
driven into hybrid form very early in the R_0t curve (by log R_0t 1.0), and the
rest was digested away with S1 nuclease because it remained single
stranded. Of course, we could have decided on any degree of abundance
we wanted. By these techniques we could look at the messages in any given
part of the complexity curve and demand any degree of induction by estrogen. Since we did not know what result to expect initially, we looked only
at the abundant class of messages and demanded a very high induction ratio.

The result of our first experiment is given in Fig. 23, which shows an autoradiograph of a gel containing several different preparations of total liver
RNA, probed with cDNA that is specific for sequences that are induced by
more than two orders of magnitude by estrogen. Track 1 contains normal
rooster liver RNA and, of course, it shows nothing. Track 2 contains poly-
adenylated RNA from hen liver, and tracks 3 and 4 contain two different
amounts of total RNA from the liver of an estrogen-treated rooster. It is
clear that in the last three tracks only two bands appear, representing two
different messenger RNAs. One of them, as expected, is vitellogenin
mRNA, and the other is another message, approximately 800 nucleotides
long, that specifies a protein that we have not yet identified. Using the tech-

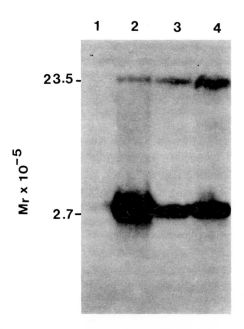

Fig. 23. Hybridization of "estrogen-
inducible" cDNA to a "northern"
blot of liver RNA. Samples of RNA
from the livers of hens and of normal
or estrogen-treated roosters were subjected to electrophoresis in 1.5%
agarose gels containing methylmercuric hydroxide. Following electrophoresis, RNA was transferred from
the gel to diazotized filter paper and
hybridized with "estrogen-inducible"
cDNA (10^7 cpm, s.a. 30 cpm/pg) according to the procedure of Alwine et
al. (1977). RNA samples applied to
the tracks shown on the autoradio-
gram: **1** total RNA from the liver of a
normal rooster (20 μg); **2** poly(A)$^+$
RNA from hen liver (5 μg); **3** and **4**
total RNA from the liver of an estrogen-treated rooster (10 μg and 20 μg,
respectively).

niques described before, we have recently cloned this sequence (King et al., 1979).

At first we were surprised that we ended up with only one sequence in addition to vitellogenin, though it is true that our requirements of abundance and extreme degree of inducibility were rather stringent. But having this sequence in hand, we are intrigued by the fact that it and vitellogenin are the only two major sequences from the estrogen-responsive domain of rooster liver that respond to the same dramatic extent to estrogen stimulation. We hope that at some time in the future we will understand why this is the case and what is the mechanism by which expression of these two genes is so tightly coordinated.

VIII. Summary

We have described the system we are studying—the avian liver and its response to estrogen—and have given our rationale for choosing a fully differentiated tissue in which one can study the response to a hormone against a background of constitutive cellular functions. We have described some of our studies on the vitellogenin system, including measurements of the kinetics of accumulation and disappearance of the messenger RNA following primary and secondary stimulation with estrogen, cloning of vitellogenin cDNA, and studies on the genomic DNA fragments in which this gene is represented. We have also described our work on another gene in avian liver —chicken serum albumin—which we have utilized as a control for our investigation of hormonal responsiveness. Finally, we introduced our studies on the estrogen-responsive domain of avian liver. In the experiments described here, we have placed an upper limit on the diversity of the hormonal response: there are not more than 100–200 genes whose expression is altered by exposure to estrogen. Using hybridization complexity analyses in a novel way, we have begun to dissect the estrogen-responsive domain of avian liver and to isolate, by recombinant DNA techniques, various members of the hierarchy of estrogen-responsive genes. We began by looking through a very narrow window to see only those genes that respond to estrogen by an enormous change in expression, but the methodology is flexible. The window can be widened at will to take a look at an ever increasing number of estrogen-responsive genes—and, of course, the methodology is applicable to other tissue and other hormones. We have stressed the latter approach for this reason: we think that if we are going to understand something about hormonal regulation, we will have to concentrate on more than single gene systems. It is, after all, the complex set of responses elicited by a hormone that sets hormones apart from the operon-specific regulatory signals of prokaryotic cells. We believe that it will be necessary to confront this complexity of hormonal response before we will be able to understand how hormones act.

References

Alwine JC, Kemp DJ, Stark GR (1977) Proc Natl Acad Sci USA 74: 5350

Bergink EW, Wallace RA (1974) J Biol Chem 249: 2897

Bernardi G, Cook WH (1960) Biochim Biophys Acta 44: 86

Bolivar F, Rodriguez RL, Greene PJ, Betlach MC, Heyneker HL, Boyer HW (1977) Gene 2: 95

Burns ATH, Deeley RG, Gordon JI, Udell DS, Mullinix KP, Goldberger RF (1978) Proc Natl Acad Sci USA 75: 1815

Clegg RE, Sanford PE, Hein RE, Andrews AC, Hughes JS, Muetter CD (1951) Science 114: 437

Deeley RG, Mullinix KP, Wetekam W, Kronenberg HM, Meyers M, Eldridge JD, Goldberger RF (1975) J Biol Chem 250: 9060

Deeley RG, Gordon JI, Burns ATH, Mullinix KP, Bina-Stein M, Goldberger RF (1977a) J Biol Chem 252: 8310

Deeley RG, Udell DS, Burns ATH, Gordon JI, Goldberger RF (1977b) J Biol Chem 252: 7913

Goldstein JL, Hasty MA (1973) J Biol Chem 248: 6300

Gordon JI, Deeley RG, Burns ATH, Paterson BM, Christmann JL, Goldberger RF (1977) J Biol Chem 252: 8320

Gordon JI, Burns ATH, Christmann JL, Deeley RG (1978) J Biol Chem 253: 8629

Gschwendt M, Kittstein W (1974) Biochim Biophys Acta 361: 84

King CR, Udell D, Deeley RG, (1979) J Biol Chem 254: 6781

Paterson BM, Roberts BE, Kuff EL (1977) Proc Natl Acad Sci USA 74: 4370

Ramney RE, Chaikoff IL (1951) Biochim Biophys Acta 44: 38

Weintraub H, Groudine M (1976) Science 193: 848

Discussion of the Paper Presented by R.F. Goldberger

SCHRADER: I have a couple of questions.   First, you talked at the very beginning about how you selected the system.   You asked about finding one where no differentiation was involved and what not.   My first question is: To whom did you address these questions, and how was the answer forthcoming?   My next question is, in a less ecclesiastical vein: When you did the Alwine experiments, looking for the albumin message, was that done on total cellular RNA?   Did you ever try it on nuclear RNA, the sort of experiment that Dr. O'Malley was talking about?

GOLDBERGER: It was done on total cellular RNA; we haven't yet examined nuclear RNA.

SCHRADER: When you did the experiments on the complexity analysis that you showed us at the end, was that done on primary stimulation or secondary stimulation?

GOLDBERGER: Those were all secondary or tertiary.   The experiments that I showed here where we only picked up the very highly induced sequences.

SCHRADER: I ought to know this, but I can't remember what the answer is.   Do all the cells or do a preponderance of the cells in the liver make vitellogenin?

GOLDBERGER: You shouldn't apologize for that question, as it is a very good and important question.   It is not known how many and what proportion of the cells in the liver make the vitellogenin.   We suspect strongly that this is at the heart of our problems in trying to repeat the Weintraub experiments on the vitellogenin system.

SCHRADER: If that is so, you made a statement at the beginning in which you were looking for a system in which there was no DNA synthesis required for the response.   In view of the fact that the time course of the primary induction takes a good number of days to get started, is it possible that in fact in the rooster liver, DNA synthesis, or differentiation of the gene, is required in the primary response and that the reason for that lag in induction in the primary response compared to the secondary is merely rescuing more cells to be differentiated for the vitellogenic response?

GOLDBERGER: First, I don't like the word "lag" because we actually start immediately to accumulate the message. You mean the time required to reach the maximum response. I think, probably, that I would keep an open mind on the possibility that you could recruit more and more cells to do this task after you inject estrogen into the rooster. But I think that's extremely unlikely. There have been some whole animal experiments with inhibitors that don't mean much; they are terrible experiments. They have been interpreted to show that DNA synthesis is not required. I don't think that means too much. Now, recently, there are some organ culture systems— Larry Wang has one for *Xenopus* liver—and in those systems, it does seem as though DNA synthesis is not required for the vitellogenin response. But I don't think that it is absolutely settled yet.

O'MALLEY: I wonder if you could summarize again the albumin shut-off data. Exactly what data do you have and what are your feelings relative to what is available in the literature? This is perhaps what I am trying to ask: How do you think hormones can turn genes off? It's a very exciting question and there have been two systems mentioned recently as possible examples of hormonal repression of gene expression. One is the dexamethasone suppression of ACTH mRNA in pituitary cells, and the other you referred to is albumin mRNA and estrogen. It seems that someone must look at the rate of synthesis of these mRNAs.

GOLDBERGER: If you are referring to a recent manuscript by Richard Palmiter, in which he did make the claim that albumin mRNA decreased by 50% or so, I really think that he made an error in calculation and did not take into account the fact that the total RNA content of the liver increases by 50%. But, you know the answers I am giving you are the total steady state level of albumin messenger RNA. I am not telling you synthesis rate, degradation, etc. Those are details that we will have to examine to know whether the gene is in any way perturbed by estrogen. Look at the level of the total amount of albumin mRNA in the liver; it does not change throughout the estrogen response, while other systems change many thousandfold.

MUELLER: In the case of the Friend leukemia cells, the globin messenger expression has a negative effect with corticosteroids, and I think that it is probably one of the better examples. Another question I wanted to ask Bob, since some of your intermediate classes etc. were not affected at all, is this: Does this sort of say to you that the major effect of the regulation has got to be kind of a positive?

GOLDBERGER: No, many mRNAs are hidden in the hybridization complexity analysis. What we are concentrating on is isolating more and more of those messenger RNA species from the abundant and the middle abundant classes that are induced to *higher* concentrations in the estrogen stimulated state. To get at the *missing* messages in estrogen responses is going to be very difficult because, as we have shown, there are none that are diminished to a very great degree.

Discussants: R.F. GOLDBERGER, G.C. MUELLER, B.W. O'MALLEY, and W.T. SCHRADER

Chapter 4

Steroid–Hormone Modulation of Prolactin Action in the Rat Mammary Gland

JEFFREY M. ROSEN, DONALD A. RICHARDS, WILLIAM GUYETTE, AND ROBERT J. MATUSIK

I. Introduction

Hormonal regulation of specific gene expression in target tissues usually is not governed by a single hormone, but rather by the complex interaction of several hormones. Many such multihormonal systems have been described that involve primary regulation by steroid hormones. For example, the induction of α_{2u}-globulin gene expression in rat liver is regulated by androgens, glucocorticoids, and thyroid hormone with possible modulation at the posttranscriptional level by growth hormone (Sippel et al. 1975; Kurtz et al. 1978; Roy et al. Chap. 14, this volume). The synthesis of growth hormone mRNA in pituitary cells is controlled by glucocorticoids and thyroid hormone (Martial et al. 1977a,b). Finally, the induction of egg white protein mRNAs in the hen oviduct is maximally stimulated in the presence of estrogen, progesterone, and testosterone (Palmiter and Haines 1973; Palmiter and Smith 1973). The mammary gland, however, provides a unique model system for studying multiple hormonal interactions because in this instance the primary regulator of gene expression appears to be a peptide hormone, prolactin (Devinoy et al. 1978; Matusik and Rosen, 1978).

The effect of prolactin on milk protein gene expression is modulated by several steroid hormones. One such hormone, hydrocortisone, potentiates prolactin action, while a second steroid, progesterone, antagonizes the effect of the peptide hormone (Matusik and Rosen, 1978). Certain other hormones, for example, thyroxine, also influence milk protein synthesis in the mammary gland (Vonderhaar 1975, 1977). In addition to the control of specialized function, mammary gland growth and differentiation are under the influence of several steroid and peptide hormones. The complex effects of these hormones on milk protein synthesis both in vivo and in vitro have been studied extensively (Topper 1970). However, the mechanisms by which these hormones interact at the molecular level still are not well defined.

The level of a specific mRNA is determined by several factors including the rate of gene transcription, the efficiency of processing of the primary transcript, and the stability of the mature mRNA. Hormonal regulation of mRNA accumulation could therefore occur at any or all of these stages. In order to elucidate the molecular mechanism of prolactin action on casein gene expression, hormonal effects on both the rate of synthesis, as well as on steady state levels of casein mRNA, have been analyzed. In addition pulse-chase studies have been employed to measure the half-life of casein mRNA. These studies were designed to differentiate hormonal effects at the transcriptional level from those at the posttranscriptional level. It cannot be assumed that an increase in the concentration of a given mRNA reflects only the action of a hormone on gene transcription. In fact, many such hormonal responses may involve coordinated responses at both the transcriptional and posttranscriptional levels. This may be especially true in multihormonal systems, in which the level of a specific mRNA is controlled by the interaction of several hormones. In the following chapter data supporting this concept will be presented. Furthermore, an understanding of the precise mechanism of hormone action will require knowledge of the structure of the hormonally responsive genes and their primary transcription products. As the initial step in this undertaking, the cloning of the three major rat casein structural genes and the α-lactalbumin gene are reported in the last section of this chapter. These studies represent the first successful attempt at cloning and analyzing peptide hormone-responsive genes and may yield valuable insight into the mechanisms of peptide hormone action.

II. Hormonal Regulation of Casein Gene Expression During Rat Mammary Development

Development of the mammary gland during pregnancy is characterized by an increased synthesis of rRNA, polysomes, and tRNA (Gaye et al. 1973; Banerjee and Banerjee 1973), including changes in the ratios of specific isoaccepting species of tRNA (Elska et al. 1971). This increase in RNA synthesis is accompanied by the extensive development of endoplasmic reticulum and the appearance of membrane-bound polysomes (Wynn et al. 1976; Oka and Topper 1971; Turkington and Riddle 1970). Furthermore, an increased DNA content (Anderson and Turner 1968) and the proliferation of alveolar cells occurs (Munford, 1963), resulting in a highly sophisticated protein-synthetic machinery capable of secreting several grams of casein per day during lactation (Jenness, 1974). The regulation of casein synthesis and secretion therefore would not be expected to be an "all or none" phenomenon dependent solely upon the induction of casein mRNA, but rather reflects coordinated changes in many components of the protein synthetic and secretory machinery.

Two different methods have been employed in our laboratory in order to determine the levels of casein mRNA present during normal mammary gland development: cell-free translation (Rosen et al. 1975) and molecular hybridization (Rosen and Barker (1976). The results obtained with both of these methods are essentially similar, although the more sensitive hybridization analysis permits the quantitation of casein mRNA sequences, when they are not detectable by cell-free translation. A specific casein $cDNA_{15S}$ probe was hybridized with an excess of RNA isolated from mammary tissue obtained from a 6-month-old virgin animal, at different stages of pregnancy, during lactation, or following regression of the gland after weaning (Rosen and Barker 1976). This extremely sensitive and quantitative cDNA probe was able to detect a limited amount of casein mRNA even in the virgin mammary gland (Rosen et al. 1978). A series of parallel hybridization curves were generated that displayed progressively faster rates of hybridization. The slowest rate was observed with RNA extracted from virgin mammary tissue, while a maximal rate was found using RNA obtained from 8-day lactating tissue. Using these hybridization data and the equivalent $R_0t_{1/2}$ for the highly purified 15S casein mRNA the percentage of casein mRNA in each total RNA extract was determined. As shown in Fig. 1, casein mRNA sequences represented 0.52% of the total cellular RNA in the 8-day lactating tissue, a 19-fold increase over the amount present at 5 days of pregnancy and an overall 300-fold increase relative to the virgin mammary gland. Since the RNA content of the lactating gland is considerably greater than the virgin,

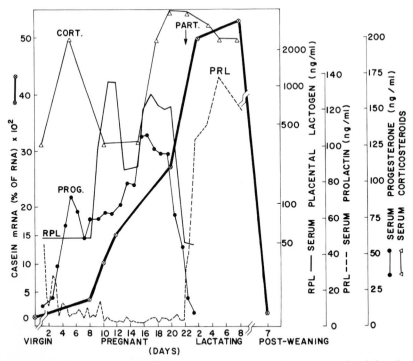

Fig. 1. Changes in casein mRNA levels during normal mammary gland development: correlation with serum hormone levels. Casein mRNA levels were measured by cDNA hybridization (Rosen and Barker 1976; Rosen et al. 1978). Prolactin and placental lactogen levels were obtained from Shiu et al. (1973), progesterone levels were taken from Morishige et al. (1973) and hydrocortisone serum levels were reported by Simpson et al. (1973).

this increase represents almost a 4000-fold increase in the total number of casein mRNA molecules per gram of tissue during this developmental period (Rosen et al. 1978). Following regression of the gland, the level of casein mRNA sequences decreased markedly until casein mRNA comprised only 0.014% of the total RNA.

The amount of 15S casein mRNA sequences can also be expressed as the number of molecules of casein mRNA per alveolar cell (Rosen and Barker 1976). A 12-fold increase in the number of molecules per alveolar cell was observed between 5 days of pregnancy and 8 days of lactation, reaching a maximal level of 79 000 molecules of the 15S casein mRNAs per cell. Thus, in highly specialized tissues, which are producing large amounts of a given protein, it is not unusual to observe mRNA levels as high as 80 000–100 000 specific mRNA molecules per cell.

By comparing the changes in casein mRNA during mammary development with the known changes in several serum hormone levels it may be possible to determine which hormones are the principle regulators of casein mRNA accumulation in vivo. For example, the observed alterations in casein mRNA levels (Fig. 1) may be correlated with the serum levels of prolactin and placental lactogen, which undergo marked changes in the rat during pregnancy, lactation, and after weaning (Shiu et al. 1973); Morishige et al. 1973). Thus, the small amount of casein mRNA activity observed in the early pregnant mammary gland (0–7 days) may result from the increase in serum prolactin to 50 ng/ml that occurs in the rat after coitus, with levels remaining elevated (20–30 ng/ml) for the first three days of pregnancy (Morishige et al. (1973). The additional increase in casein mRNA that occurs at approximately day 8 of pregnancy may then be attributable to the vast increase in rat placental lactogen that occurs at this time, reaching levels as high as 1200 ng/ml by day 12 of pregnancy (Kelly et al. 1976). This conclusion is given additional support by the observation that a 90-fold greater difference in the amount of casein mRNA is present in the rat versus the rabbit during midpregnancy (Kelly et al. 1976). This difference, therefore, may reflect the high concentrations of placental lactogen in the rat, compared to very low levels observe in the rabbit, for example, a maximal level of only 25 ng/m was found at day 30 of pregnancy (Kelly et al. 1976). Thus, placental lactogen may be of primary importance in initiating rat mammary gland development during pregnancy.

While serum placental lactogen levels decrease markedly at parturition, prolactin levels are known to increase to approximately 30 ng/m just prior to parturition. A further increase to levels as high as 300 ng/m is observed within 9–10 h postpartum (Amenomori et al. 1970). This may account for the additional rise in casein mRNA reported during early lactation. Following weaning prolactin levels fall to a basal level of only 10 ng/m, and accordingly, casein mRNA activity is barely detectable. Thus, prolactin and placental lactogen are performng a dual role in the mammary gland; they initiate alveolar differentiation and proliferation, as well as selectively induce casein synthesis (Turkington, 1968).

The effects of these two peptide hormones are also modulated by several steroid hormones. Thus, in the rat the plasma level of progesterone increases as early as day 4 of pregnancy to levels of 80 ng/ml and reaches levels of 120 ng/ml during midpregnancy (Morishige et al. 1973). A dramatic fall in serum progesterone then occurs at parturition, which may be the signal for the onset of lactation. Finally, plasma corticosteroid levels during midpregnancy in the rat have been reported to be comparable to serum progesterone levels approaching levels of 200 ng/ml (Simpson et al. 1973). Thus, the development of secretory capacity may depend upon the relationship between the stimulatory effects of corticosteroids and the inhibitory effect of progesterone during pregnancy (Wynn et al. 1976).

III. Steroid Hormone Modulation of Prolactin-Induced Casein Gene Expression in vivo

As previously mentioned the levels of casein mRNA and casein synthesis observed during midpregnancy may reflect both the positive and negative effects of several peptide and steroid hormones. Thus, two peptide hormones, prolactin and placental lactogen, have been shown to induce casein synthesis (Turkington 1968; Lockwood et al. 1966) and casein mRNA (Devinoy et al. 1978; Matusik and Rosen 1978; Houdebine 1978) both in vivo and in mammary gland organ culture. Furthermore, high levels of progesterone during pregnancy may antagonize the lactogenic effects of these hormones (Rosen et al. 1978); Houdebine, 1976). Finally, while the continued presence of hydrocortisone is not required for prolactin induction of casein mRNA, hydrocortisone does potentiate the action of prolactin. These conclusions are based upon studies performed either in the pseudopregnant rabbit, the ovariectomized midpregnant rat, or in organ cultures derived from midpregnant mammary tissue from both species. The latter in vitro studies will be discussed in more detail in Sec. IV.

Ovariectomy of the midpregnant rat and hormone replacement have been used as a model for studying the regulation of both casein mRNA levels and polysomal casein synthesis in vivo (Rosen et al. 1978). Previous studies by Liu and Davis (1967) demonstrated that ovariectomy of rats approximately midway through pregnancy induced a lactation-like response, characterized by the appearance of a milklike secretion with the immunological properties of casein. This response could be blocked by the administration of progesterone at the time of ovariectomy (Davis et al. 1972). The 12–14-day pseudopregnant rabbit has also been used as an alternative model in which to study the hormonal regulation of casein mRNA and casein synthesis (Gaye et al. 1973); Houdebine 1976). Administration of prolactin usually twice-daily in doses of 12.5 I.U. per injection was accompanied by either twice-daily progesterone injections of 5 mg each or hydrocortisone acetate at 2 mg or 7.5 mg per injection.

Studies using the pseudopregnant rabbit mammary gland demonstrated that progesterone administered either prior to or simultaneously with prolactin will inhibit the induction of casein mRNA, casein synthesis, and subsequent lactogenesis (Houdebine, 1976). This suggests that the continuous presence of progesterone is necessary during pregnancy to exert its inhibitory effect on lactogenesis (Chatterton et al. 1975). Once maximal induction is obtained, the marked inhibitory effect of progesterone is not observed. This observation may have important consequences for understanding the mechanism of action of progesterone in inhibiting lactogenesis. Presumably, a direct competition with lactogenic hormones at the transcriptional level to regulate the synthesis of casein mRNA is unlikely.

The role of progesterone as the principal hormone suppressing lactation during pregnancy was given additional support by the experiments performed in our laboratory using the ovariectomized midpregnant rat (Rosen et al. 1978). Removal of progesterone by ovariectomy resulted in an increased level of casein mRNA, increased polysomal casein synthesis, an increased concentration of intracellular casein, and finally the appearance of a white, milklike secretion. Following ovariectomy, a twofold increase in both casein mRNA activity (determined in a wheat germ translation assay) and casein mRNA sequence concentration (measured using the selective cDNA hybridization probe) was observed. This effect on mRNA accumulation was first observed within 16 h following ovariectomy and was maximal between 24 and 48 h. Progesterone, but not estradiol or hydrocortisone, administered at the time of ovariectomy prevented the increase in casein mRNA (Fig. 2). Administration of progesterone after maximal induction of casein mRNA was obtained, either following ovariectomy or during lactation, was unable to significantly reduce the levels of casein mRNA.

Following ovariectomy of a mid-pregnant rat, a shift in polysome profiles from monosomes, disomes, and trisomes to the larger polysomes containing 7–11 ribosomes found during lactation was also observed (Rosen et al. 1978). Accordingly, a twofold increase in polysomal casein synthesis was detected, reaching a level comparable to that observed in a lactating polysome preparation. Thus, following ovariectomy, a twofold increase in both mRNA activity and polysomal casein synthesis was observed. Progesterone administration at the time of ovariectomy blocked both these responses and increased the amount of small polysomes present. Ovariectomy also resulted in an increased level of intracellular casein as detected by radioimmunoassay and the appearance of a white, milklike fluid (Rosen et al. 1978).

Progesterone appears to regulate casein synthesis and secretion in a pleiotropic fashion. At least three different potential mechanisms of action have been suggested:

1) Progesterone has been reported to counteract the self-regulated increase in prolactin receptors observed in the mammary gland (Djiane and Durand 1977). This may account for the relatively low levels of prolactin receptors detected in the rabbit mammary gland during pregnancy (Djiane et al. 1977), and may result in a reduced ability of prolactin or placental lactogen to induce casein mRNA.

2) Progesterone is also an effective competitor for glucocorticoid receptor

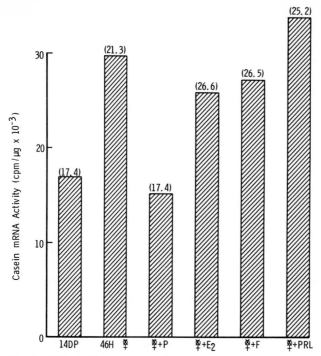

Fig. 2. Effect of hormone administration at the time of ovariectomy on casein mRNA activity. Hormones were administered as described in Rosen, et al. (1978) for 46 h following ovariectomy of a 14-day pregnant rat. The casein mRNA activity determined in the wheat germ assay is shown on the ordinate and the percentage of total mRNA activity is shown in parentheses. The following abbreviations are used. 14 DP (14-day pregnant); 46 h ♀ (46 h following ovariectomy); ♀ + P (ovariectomized rat given progesterone); ♀ + E₂(ovariectomized rat given estradiol benzoate); ♀ + F (ovariectomized rat given hydrocortisone succinate; ♀ + PRL (ovariectomized rat given prolactin).

binding sites in the mammary gland (Shyamala, 1973) and therefore may prevent the glucocorticoid-induced development of the rough endoplasmic reticulum necessary for lactogenesis (Wynn et al. 1976).

3) Finally, a specific, unique progesterone receptor has been demonstrated in mammary carcinomas (Horwitz et al. (1975).

These potential multiple sites of progesterone action may explain the coordinated responses to ovariectomy observed during midpregnancy, that is, an increase in casein mRNA levels, an increase in large polysomes and casein synthesis, and finally increased levels of intracellular casein and the initiation of secretion.

Results of both in vivo and in vitro studies have demonstrated that hydrocortisone alone, that is, in the absence of prolactin, does not stimulate casein mRNA synthesis (Matusik and Rosen 1978; Devinoy and Houdebine, 1977). Both prolactin and glucocorticoids are known to be required for the initiation of milk protein synthesis (Topper 1970). Studies in the pseudo-pregnant rabbit have demonstrated that glucocorticoids can amplify the capacity for prolactin to increase the concentration of casein mRNA, as well as casein synthesis (Devinoy and Houdebine 1977). However, when administered alone, under conditions where the endogenous concentration of prolactin is reduced by prior treatment with an ergot alkaloid, bromocryptine, no effect of the steroid hormone was observed. However, in the presence of endogenous prolactin or a low dose (12.5 I.U) of exogenous hormone a stimulatory effect of hydrocortisone on both casein mRNA levels and casein synthesis was observed (Devinoy and Houdebine 1977). Conversely, when a high dose (100 I.U.) of exogenous prolactin was administered, no further stimulatory effect of hydrocortisone was seen (Devinoy and Houdebine 1977). Thus, glucocorticoids are capable of potentiating the effect of prolactin and increasing the levels of casein mRNA and casein synthesis in the presence of the peptide hormone. This result is supported by the observation that adrenalectomy has been reported to decrease the level of polysomal casein synthesis and casein mRNA in lactating mice (Terry et al. 1977). Furthermore, cortisol treatment at the time of adrenalectomy was able to prevent the marked reduction in casein mRNA activity caused by adrenal ablation. These data do not by themselves, however, support the concept that glucocorticoids alone can affect the rate of transcription of casein mRNA. The synergistic effect of glucocorticoids on casein gene expression will be discussed in more detail in Sec. IV. Studies in mammary gland organ culture are necessary to distinguish between effects of the steroid at the transcriptional and posttranscriptional levels.

IV. Effect of Steroid Hormones on Prolactin Action in Mammary Gland Organ Culture

A. Prolactin-Mediated Casein mRNA Accumulation

Mammary gland organ culture has been utilized as a well-characterized and operable system in which to study steroid hormone modulation of prolactin action (Denivoy et al. 1978; Matusik and Rosen 1978). One advantage of this culture system is that it is performed in a serum-free, chemically defined medium, in which the effective concentration of both peptide and steroid hormones can be precisely controlled. Since no cloned, prolactin-responsive mammary epithelial cell line is presently available for study, organ culture provided a useful alternative for these initial experiments.

Midpregnant mammary gland explants were employed in the following experiments rather than explants derived from virgin tissue in order to study the early effects of prolactin in preexisting, differentiated alveolar cells. In order to reduce the high levels of casein mRNA (Rosen et al. 1978) that existed in the midpregnant rat, explants were exposed for 48 h to a medium containing only insulin and hydrocortisone. Following the first 24 h in culture only 10% of the original amount of casein mRNA remained and after the next 24 h a further decrease to 4% of the original level was observed (Fig. 3). In the continued presence of insulin and hydrocortisone for 72 h, the casein mRNA level decreased to near steady state conditions. Routinely, ovine prolactin was added after the initial 48-h insulin and hydrocortisone time period. In agreement with our previous results (Matusik and Rosen 1978), a rapid induction of casein mRNA was observed following the addition of prolactin. The rate of accumulation appeared to be linear resulting in 1.6-fold increase by 4 h and a sevenfold increase from the insulin-hydrocortisone baseline within 24 h. A small effect of prolactin on casein mRNA levels, that is, 1,3-fold, was also usually observed within one hour (Matusik and Rosen 1978). Similar kinetics of induction have been observed following prolactin addition when the level of casein mRNA sequences were determined by either RNA excess or cDNA excess hybridization. The latter technique was especially useful when only small quantities of RNA were available for hybridization. These results suggested that prolactin had a rapid effect on either the transcription or turvover of casein mRNA.

Using the data shown in Fig. 3 and several other experiments, it was

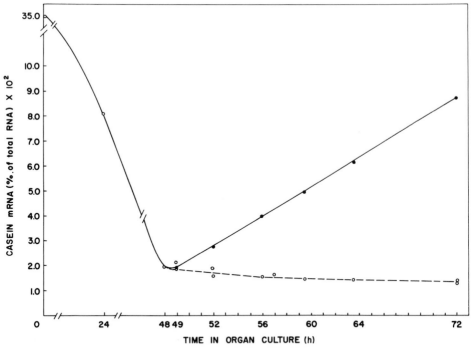

Fig. 3. Analysis of casein mRNA accumulation in the presence and absence of prolactin. Within the first 48 h of culture in the presence of insulin and hydrocortisone (○) the endogenous level of casein mRNA observed in tissue from 15-day pregnant rat (shown at zero time) was decreased by 96%. At this time the incubation was continued in the presence of insulin and hydrocortisone alone (○) or with insulin, hydrocortisone and prolactin (●). The levels of casein mRNA in this experiment were determined by cDNA excess hybridization.

possible to calculate the number of molecules synthesized per minute per cell and therefore, to analyze the rate of synthesis of casein mRNA in the presence and absence of prolactin (Guyette et al. 1979). Converting the number of molecules of casein mRNA per cell to the rate of accumulation (molecules mRNA per min per cell, dC/dt) allows the determination of the rate of transcription (T) by plotting the rate of accumulation of casein mRNA versus the concentrations of casein mRNA at any given time (C_t). Using the expression $dC/dt = T - DC_t$, the value of T can be estimated by extrapolation back to the Y axis. The slope of each line yields the value of D or the rate of degradation, where D is related to the half-life of the mRNA by the first order decay expression, $D = \ln 2/t_{1/2}$. Using this method of analysis the transcription rates for casein mRNA were estimated to be 11 ± 0.88 and 5.25 ± 0.64 molecules per minute per cell in the presence and absence of prolactin, respectively. Thus, the addition of prolactin increased the rate of casein mRNA transcription approximately twofold above control. This increase in transcription was not, however, sufficient to account for the experimental accumulation of casein mRNA (sevenfold above control). Thus, these data suggested that in addition to increasing the rate of transcription, prolactin might also increase the half-life of casein mRNA. The values estimated for the half-life of casein mRNA in the presence and absence of prolactin were 92 (49–825 h) and 5.4 (4.2–7.8 h), respectively. (The values in parenthesis represent the range of $t_{1/2}$ values predicted for one standard deviation from the mean.) This represents a 17-fold change in the half-life of casein mRNA, and, coupled with the twofold change in its rate of synthesis, accounts for the observed accumulation of casein mRNA following prolactin addition.

Because of the wide range of values predicted for casein mRNA half-life in the presence of prolactin and the possibility that prolactin may be acting mitogenically at the later time points, these half-life estimates are best interpreted conservatively. Thus, analysis by this method indicated that the addition of prolactin to the mammary gland organ culture increased the rate of casein mRNA transcription by approximately twofold, whereas the mRNA half-life increased from 5 h to a value of at least 49 h or greater. This dual effect of prolactin on both the rates of transcription and degradation of casein mRNA could account for the observed accumulation of casein mRNA seen during organ culture. These results also suggest that prolactin regulation of casein gene expression may be a complex process requiring multiple signals which may control transcription as well as the processing and degradation of casein mRNA. Finally, using these values for the rates of casein mRNA transcription and turnover in the presence of prolactin a steady state level of 88 000 molecules of casein mRNA per cell would be predicted. This value is in excellent agreement with the maximal level of casein mRNA actually estimated per cell during midlactation (Rosen and Barker 1976).

The half-life and transcription rate estimates obtained for casein mRNA appear to be reasonable approximations. They are similar to the reported half-lifes and transcription rates of several other abundant mRNAs in specialized cells including ovalbumin mRNA (Harris et al. 1975; Palmiter 1974), globin mRNA (Aviv et al. 1976), and vitellogenin mRNA (Burns et al. 1978). Although the relative differences among these parameters in the presence and absence of prolactin are valid approximations, their absolute values should be interpreted conservatively since they are based upon a number of assumptions:

1) Notably, that transcription rates can be assessed accurately be extrapolation to the Y axis.

2) That no marked increase in epitheilial cell content occurs during the time in culture. Previous results have suggested that a slight increase in the content of epithelial cell DNA occurs in midpregnant mouse explants after 48 h in culture with prolactin, insulin, and hydrocortisone (Owens et al. 1973). These half-life and transcription rate values should therefore be confirmed by a more direct method which measures the rate of transcription and turnover of pulse-labeled casein mRNA sequences.

B. Hormonal Effects at the Transcriptional and Posttranscriptional Levels

The study of specific mRNA transcription is complicated by the fact that each mRNA represents only a small fraction of the initial RNA transcripts. Thus, hybridization in solution to a specific cDNA will result in a high background level of radioactivity owing to both nuclease-resistant secondary structure in the radioactive RNA and due to trapping during trichloroacetic acid (TCA) precipitation. This level may reach 1% of the input radioactivity, making it impossible to detect specific transcripts of unique gene sequences, which may represent only 0.1% of the rapidly labeled RNA. To overcome this problem cDNA can be covalently attached to an inert matrix and several cycles of hybridization performed, or, alternatively, RNase may be used to digest any RNA not present as a true hybrid with the immobilized cDNA.

To make an accurate judgement concerning the effects of hormones on the rate of transcription it is necessary to label for a time much shorter than the half-life of the RNA in question. The half-life of casein mRNA in the presence of prolactin has been estimated to be greater than 49 h and approximately 5 h in the absence of prolactin. Thus, a labeling pulse of 30 min was considered sufficiently short to accurately indicate alterations in the rate of transcription. Approximately 1×10^6 cpm of pulse-labeled RNA (50 μg) extracted after different times of exposure to hormone was hybridized to casein cDNA–cellulose synthesized by the method of Venetianer and Leder (1974). The percentage of casein transcripts was calculated from the specifically hybridized radioactivity as described (Guyette et al. 1979). The results from these experiments are shown in Fig. 4. In the absence of prolactin a low percentage of casein transcripts was observed, the mean = 0.09% (11 experiments). During the first 30 min of exposure to prolactin (when

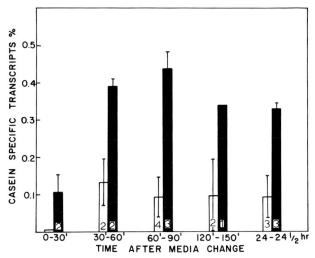

Fig. 4. Prolactin-mediated increase in casein transcription. The percentage of casein specific transcripts was determined following hybridization of organ culture RNA to casein and ovalbumin cDNA cellulose as described in the text and (Guyette et al. 1979). Cultures were performed in the presence of insulin, hydrocortisone and prolactin (IMF, solid bars), or in the presence of insulin and hydrocortisone alone (IF, open bars). The numbers within each bar denote the number of determinations performed. The numbers on the abcissa are the time of the uridine pulse, relative to the 48 hr media change. The percentage of casein specific transcripts (mean ± SEM) were IMF = 0.38 ± 0.02 and IF = 0.09 ± 0.02, respectively, indicating a 4.2-fold difference ($p < 0.01$; the determination excludes the 0–30-min data).

[^{3}H]uridine was added simultaneously with the hormone), there was no measurable effect on the transcription of casein mRNA. However, if a 30-min pulse was initiated 30 min after prolactin addition, a large increase in casein specific transcripts was detectable by the end of the hour. Thus, after the addition of prolactin, a 30-min lag occurred before a maximal increase in the rate of transcription was detected. This increased level of transcription, mean = 0.38% (9 experiments), appears to remain constant for at least 24 h. The determinations in the presence and absence of prolactin were significantly different at a $p < 0.01$. Thus, an approximate fourfold increase in casein transcription occurred after the addition of prolactin. These results are in good agreement with the twofold increase in the rate of synthesis determined indirectly by measuring the accumulation of casein mRNA at different time intervals in culture.

This two- to fourfold increase in the rate of transcription of casein mRNA is not sufficient to account for the 7- to 13-fold increase in the level of casein mRNA usually observed following a 24-h exposure to prolactin. Thus, these studies suggest that prolactin may have a dual mode of action leading not only to an increase in the rate of transcription but also altering the half-life of casein mRNA. To account for the difference in the transcription rate of two- to fourfold and the mRNA accumulation of 7- to 13-fold it is possible that prolactin may also act to increase the half-life of the mRNA either by increasing the efficiency of RNA processing in the nucleus, or perhaps by stabilizing the mature mRNA. In order to test this hypothesis the half-life of casein mRNA has been measured directly via pulse-chase experiments in organ culture. These experiments were performed using a modification of the glucosamine–uridine method as originally described by Levis and Penman (1977). The use of RNA synthesis inhibitors, such as actinomycin D, which have been reported to affect mRNA processing and turnover, was avoided. In this method cultures were pretreated for 1 h with 20 mM glucosamine to deplete the uridine triphosphate (UTP)pool. Glucosamine was then removed and a 1-h pulse, with [^{3}H]uridine employed to label newly synthesized RNA. Following the pulse, fresh culture media was added containing 20 mM glucosamine to inhibit any further [^{3}H]uridine triphosphate UTP incorporation and 5 mM unlabeled uridine and cytidine to prevent any reutilization of radioactive uridine. After 1 h the chase was continued with unlabeled uridine and cytidine alone to minimize potential glucosamine toxicity. Under these conditions analysis of the total RNA specific activity, which represented predominantly ribosomal RNA, revealed a half-life of greater than 90 h either in the presence or absence of prolactin. However, analysis of casein specific transcripts indicated that the half-life of casein mRNA increased from a value of 1.1 h in the absence of prolactin to 28 h after prolactin addition (Guyette et al. 1979). These results are qualitatively similar to those obtained by the analysis of casein mRNA accumulation and indicated that in the presence of prolactin casein mRNA stability was greatly increased. The difference in the absolute half-life obtained using these two methods may reflect a limited toxicity of the glucosamine and unlabeled nucleosides used in the pulse-chase procedure (Levis and Penman 1977). Thus, by two independent methods it appears that prolactin, after 30–60 min, increases the rate of casein gene transcription two- to four-fold, and that this effect coupled with a 17- to 25-fold increase in casein mRNA half-life results in the observed accumulation of casein mRNA.

The question still remained, however, whether hydrocortisone was able to increase the rate of casein gene transcription. In agreement with the previous in vivo data, hydrocortisone in combination with prolactin was demonstrated to be necessary for the maximal accumulation of casein mRNA (Table 1; Matusik and Rosen 1978). However, when hydrocortisone was omitted prolactin was still capable of increasing the levels of casein mRNA (Devinoy et al. 1978; Matusik and Rosen 1978). In fact, as shown in Table 1 the relative increase in casein mRNA, compared with the insulin baseline for insulin and prolactin was at least as great as the relative increase over the insulin and hydrocortisone baseline seen with all three hormones. These data suggested that in the absence of hydrocortisone prolactin maximally increased the rate of casein transcription.

In order to directly test this hypothesis the following experiment was performed. Following the initial 48 h in the presence of insulin and hydrocortisone, the tissue was divided into four experimental groups: insulin alone, insulin with prolactin, insulin with hydrocortisone, and all three hormones.

Table 1. Permissive Effect of Hydrocortisone in Mammary Gland Organ Culture[a]

Hormonal milieu	Casein-specific transcripts (%)	Casein mRNA (molecules/cell)
IF	0.1 ± 0.006 ($N = 3$)	50
IMF	0.33 ± 0.02 ($N = 3$)	2040
I	0.03 ± 0.01 ($N = 2$)	20
IM	0.3 ± 0.07 ($N = 2$)	780

[a] Mammary gland explants were incubated in the presence of insulin and hydrocortisone for 48 hr. At that time the media was changed to one containing insulin alone (I), insulin with prolactin (IM), insulin with hydrocortisone (IF), or insulin with hydrocortisone and prolactin (IMF). The incubation proceeded for an additional 24 h, at which time the tissue was labeled for 30 min with [³H]uridine and the RNA extracted and analyzed for the percentage of casein specific transcripts as described (Guyette et al. 1979). The values are expressed as mean ± S.E.M. where N equals the number of determinations at each time. The levels of casein mRNA were determined by RNA excess hybridization 48 h after the media change (Matusik and Rosen 1978).

Tissue was incubated for an additional 24 h and, following a 30-min pulse with [³H]uridine, RNA was extracted and assayed for the percentage casein specific transcripts (Table 1). Prolactin resulted in approximately 0.3% casein-specific transcripts both in the presence and absence of hydrocortisone. Although hydrocortisone did not potentiate the prolactin-mediated increase in casein mRNA transcription, its presence resulted in a 3.3-fold increase in casein-specific transcripts between insulin–hydrocortisone treated explants and those exposed to insulin along (Table 1). This may indicate that hydrocortisone can increase the rate of casein mRNA transcription compared with the insulin control but that it cannot further increase casein transcription once it is maximally induced by prolactin. However, these results must be viewed with caution for the following reasons:

1) Cell death in mammary gland organ culture occurs more rapidly in the presence of insulin alone. Thus, hydrocortisone may have a stabilizing effect on the tissue which results in an apparent increase in casein mRNA transcription.

2) The value of 0.03% casein specific transcripts approaches the level of sensitivity of this assay method, and thus the significance of the difference between 0.03 and 0.1% is questionable.

Thus, hydrocortisone may increase casein mRNA levels through a posttranscriptional mechanism, perhaps by increasing the efficiency of processing or by stabilizing the mature mRNA. Multihormonal responsive systems in general may exhibit such regulation, with peptide and steroid hormones interacting at several levels within the cell to control gene expression.

Comparable experiments have been performed with progesterone, which antagonizes the stimulatory effect of prolactin both in vivo and in mammary gland organ culture (Matusik and Rosen 1978). At a dose of 1 μg/ml of progesterone, added simultaneously with prolactin, approximately a 50% inhibition of casein mRNA accumulation was observed (Matusik and Rosen

Table 2. Antagonistic Effect of Progesterone on Casein Gene Transcription in Mammary Gland Organ Culture[a]

Hormonal mileiu	Total RNA-specific activity (cpm/μg)	Casein-specific transcripts (%)	Inhibition (%)
IF	14 400	0.061	—
IMF	27 700	0.340	—
IF + P (1 μg/ml)	26 650	0.048	—
IFM + P (1 μg/ml)	25 360	0.096	72

[a] The conditions of the experiment were identical to those described in Table 1. The same abbreviations are used with progesterone (P) added at the concentration shown.

1978). At this dose there was no change in the casein mRNA levels in the insulin–hyrocortisone and insulin–hydrocortisone–progesterone controls, indicating the lack of a nonspecific toxic effect of progesterone. When a comparable experiment was performed to measure the rate of casein mRNA synthesis, progesterone at 1 μg/ml was able to inhibit by approximately 70% the prolactin-mediated increase in casein transcription (Table 2). Furthermore there was no decrease in total RNA synthesis as determined by measuring the total RNA-specific activity in the control (IF + P) and treated (IFM + P) groups. Thus, both mRNA accumulation and pulse-labeling experiments produced comparable results. These data suggest that progesterone is able to antagonize selectively prolactin-mediated casein transcription and is consistent with the pleiotropic effects of this steroid on the mammary gland. These results, however, should not be interpreted to mean that progesterone is competing directly at the transcriptional level to block prolactin action. Instead, progesterone may be acting at a very early step in the prolactin pathway to prevent the increase in peptide hormone-mediated gene expression. Some recent experiments from our laboratory have demonstrated that if progresterone addition is delayed for 4 h after prolactin is added, no inhibitory effect of the steroid is observed. Under these conditions 4 h of exposure to prolactin followed by 20 h with insulin and hydrocortisone alone resulted in the same increase in casein mRNA levels observed with prolactin present for the entire 24-h period. By performing experiments of this type it may be possible to identify the site of progesterone action. Until more is known concerning the possible mediators of prolactin action and its precise mechanism of action, however, it will be impossible to delineate the exact mechanism of progesterone antagonism.

V. Future Directions: Characterization of the Cloned Milk Protein Structural Genes

An understanding of the organization of the milk protein structural genes and their primary transcription products is a prerequisite for further studies on the mechanism of hormonal regulation of gene expression in the mammary gland. This knowledge then may allow the elucidation of the mechanism of action of peptide and steroid hormones on gene transcription, and potentially on mRNA processing. Since mRNA processing may be related to mRNA turnover, these studies may also help identify those factors regulating mRNA half-life. As the first step in this process, the three major rat casein structural genes and the α-lactalbumin gene have been cloned and total DNA and mRNA precursor mapping studies have been initiated (Richards et al. 1979).

Total poly(A)-containing RNA, which is composed of approximately 60% casein and 20% α-lactalbumin mRNAs, was isolated from 10-day lactating rat mammary tissue. Complementary DNA synthesized from this fraction was used to synthesize double-stranded DNA using AMV reverse transcriptase and the self-priming ability of the cDNA. A single stranded nuclease (S_1) was then used to nick open the double-stranded DNA, which was sedimented on a neutral sucrose gradient. The DNA larger than 600 base pairs in length, that is, sufficient to code for α-lactalbumin mRNA and the larger casein mRNAs was selected. This double-stranded DNA fraction was inserted into the PstI site of cloning vector pBR322 by dC–dG tailing with terminal transferase. Forty nanograms of recombinant plasmid were then used to transform E. coli strain X1776, which resulted in an efficiency of 3×10^4 colony forming units per microgram of plasmid. Two hundred transformants were finally screened by colony hybridization using cDNA probes prepared against partially purified fractions of the 15 and 12S casein mRNAs and the 10S α-lactalbumin mRNA.

Following preliminary identification of the individual clones the plasmid DNAs were isolated and cut with the restriction enzyme EcoRI (Fig. 5). Following agarose gel electrophoresis (Fig. 5B) the DNA was transferred to a nitrocellulose filter and the inserts identified by hybridization with [32P]cDNAs prepared against the four milk protein mRNAs (Fig. 5A). Autoradiography (Fig. 5C) revealed the presence of DNA inserts ranging in size from 400 to 1200 nucleotide pairs. Additional restriction mapping experiments were performed using the same method as those just described in order to construct preliminary restriction maps of each cloned DNA. Further characterization of these clones was accomplished by hybrid-arrested cell-free translation (Paterson et al. 1977) and by Northern transfer analysis (Alwine et al. 1977) in which each of the 32P-labeled, nick-translated, cloned DNAs were hybridized to a filter containing the covalently bound milk protein mRNAs.

These studies have resulted in the identification of cloned DNAs coding for each of the α, β, and γ casein mRNAs and α-lactalbumin RNAs. The availibility of these probes now permits more detailed DNA mapping experiments and identification of putative casein mRNA precursors. In addition the individual cloned DNAs can be used to study the possible coordinate induction of these mRNAs and the factors regulating their individual processing and turnover. Preliminary data suggest that the three caseins may exist as a unique gene cluster. The possibility remains, therefore, that they may be transcribed and processed as a single unit. Studies involving the cloning and characterization of genomic DNA fragments and isolation of primary transcription products should also be forthcoming now that these indi-

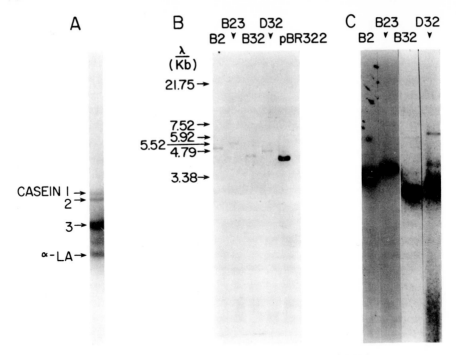

Fig. 5. Characterization of rat casein and α-lactalbumin cloned DNAs.

A Electrophoresis of the four milk protein mRNAs (3 μg) on a 1.5% agarose gel containing 6 *M* urea and 0.025 *M* critic acid, pH 4.2.

B Characterization of *Eco*RI digested plasmid DNAs on a 1% agarose gel using a BRL horizontal slab gel apparatus. Electrophoresis was performed as described by Stein *et al.* (1978). Approximately 0.2 μg of milk protein clones B2, B23, B32, and D32 and 1 μg of the parent plasmid pBR322 were digested and electrophoresed as shown. The positions of *Eco*RI digested λDNA standards run in a parallel slot are shown by the arrows. The gels in both **A** and **B** were stained with ethidium bromide and photographed by transillumination with short wave ultraviolet. **C** Southern analysis of cloned milk protein structural genes. The DNA was transferred to nitrocellulose filters as described by Southern (1975) and hybridized to [³²P]cDNA synthesized from the mRNAs shown in **A**.

vidual cloned structural DNAs are available. These experiments may ultimately yield valuable insight into the mechanisms of peptide hormone action and the complex hormonal interactions that control milk protein gene expression in the mammary gland.

Acknowledgments. The authors wish to thank Mr. John Rodgers for the analysis of the rates of casein mRNA transcription and turnover. This work was supported by grant NIH-CA-16303. J.M.R. is a recipient of a research career development award NIH-CA-00154. All bacterial transformations were carried out according to the NIH Guidelines in an approved P-3 containment facility.

References

Alwine JC, Kemp DJ, Stark GR (1977) Proc Natl Acad Sci USA 74: 5350

Amenomori Y, Chen CL, and Meites J (1970) Endocrinology 86: 506

Anderson RR Turner CW (1968) Proc Soc Exp Biol Med 128: 210

Aviv H, Voloch Z, Bastos R, Levy S (1976) Cell 8: 495

Banerjee DN, Banerjee MR (1973) J Endocrinol 56: 145

Burns ATH, Deeley RG, Gordon JI, Udell DS, Mullinix KP, Goldberger RF (1978) Proc Natl Acad Sci USA 75: 1815

Chatterton RT Jr., King WJ, Ward DA, Chien JL (1975) Endocrinology 96: 861

Davis JW, Wikman-Coffelt J, Eddington CL (1972) Endocrinology 91: 1011

Devinoy E, Houdebine LM (1977) Eur J Biochem 75: 411

Devinoy E, Houdebine LM, Delouis C (1978) Biochem Biophys Acta 517: 360

Djiane J, Durand P. (1977) Nature (London) 266: 641

Djiane J, Durand P, Kelly PA (1977) Endocrinology 100: 1348

Elska A, Matsuka G, Matiash U, Nasarenko I, Jemenova N (1971) Biochem Biophys Acta 247: 430

Gaye P, Houdebine LM, Petrissant G, Denamur R (1973) In: Gene transcription in reproductive tissue, Diczfalusy E (ed). 5, Karolinska Institute, Stockholm. Vol. 5

Guyette WA, Matusik RJ, Rosen JM (1979) Cell 17: 1013

Harris SE, Rosen JM, Means AR, O'Malley BW (1975) Biochemistry 14: 2072

Horwitz KB, McGuire WL, Pearson OH, Segaloff A (1975) Science 189: 726

Houdebine LM (1976) Eur J Biochem 68: 219

Jenness R (1974) J Invest Dermatol 63: 109

Kelly PA, Tsushima T, Shiu RPC, Friesen HG (1976) Endocrinology 99: 765

Kurtz DT, Chan KM, Feigelson P. (1978) Cell 15: 743

Levis R, Penman S (1977) Cell 11: 105

Liu TMY, Davis JW (1967) Endocrinology 80: 1043

Lockwood DH, Turkington RW, Topper YJ (1966) Biochem. Biophys. Acta 130: 493.

Martial JA, Baxter JD, Goodman HM, Seeburg PH (1977a) Proc Natl Acad Sci USA 74: 1816

Martial JA, Seeburg PH, Guenzi D, Goodman HM, Baxter JD (1977) Proc Natl Acad Sci USA 74: 4293

Matusik RJ, Rosen JM (1978) J Biol Chem 253: 2343

Morishige WK, Pepe GJ, Rothchild I (1973) Endocrinology 92: 1527

Munford RE (1963) J Endocrinol 28: 1

Oka T, Topper YJ (1971) J Biol Chem 246: 7701

Owens IS, Vonderhaar BK, Topper YJ (1973) J Biol Chem 248: 472

Palmiter RD (1974) Biochemistry 13: 3606

Palmiter RD, Haines ME (1973) J Biol Chem 248: 2107

Palmiter RD, Smith LT (1973) Nature New Biol. 246: 74

Paterson BN Roberts BE, Kuff EL (1977) Proc Natl Acad Sci USA 74: 4370

Richards DA, Supowit S, Blackburn D, Rosen JM (1979) Fed. Proc., 38, 2057a

Rosen JM, Barker SW (1976) Biochemistry 15: 5272

Rosen JM, Woo SLC, Comstock JP (1975) Biochemistry 14: 2895

Rosen JM, O'Neal DL, McHugh JE, Comstock JP (1978) Biochemistry 17: 290

Roy, AK, Chatterjee, B and Deshpande, AK (1979), This volume

Shiu RPC, Kelly PA, Friesen HG (1973) Science 180: 968

Shyamala G (1973) Biochemistry 12: 3085

Simpson AA, Simpson MHW, Sinha YN, Schmidt GH (1973) J Endocrinol 58: 675

Sippel AE, Feigelson P, Roy AK (1975) Biochemistry 14: 825

Southern E (1975) J Mol Biol 98: 503

Stein JP, Catterall JF, Woo SLC, Means AR, O'Malley BW (1978) Biochemistry 17: 5763

Terry PM, Lin FK, Banerjee MR (1977) Mol Cell Endocrinol 9: 169.

Topper YJ (1970) Rec Prog Horm Res 26: 287
Turkington RW (1968) Endocrinology 82: 575
Turkington RW, Riddle M (1970) J Biol Chem 245: 5145
Venetianer P, Leder P (1974) Proc Natl Acad Sci USA 71: 3892
Vonderhaar BK (1975) Biochem Biophys Red Commun 67: 1219
Vonderhaar BK (1977) Endocrinology 100: 1423
Wynn RM, Harris JA, Chatterton RT (1976) Am J Obstet Gynecol. 126: 920

Discussion of the Paper Presented by J.M. Rosen

THOMPSON: In considering the mechanism by which the progesterone may diminish the production of casein, and in trying to decide whether it acts through its own receptor or through the glucocorticoid receptor, what about the concentration of progesterone required? It is sufficient to saturate the glucocorticoid receptor?

ROSEN: We haven't done the latter experiment; however, in the serum during mid-pregnancy there is an equal amount of both, about 200 ng/ml. In culture, we see about a 50% inhibition. We are dealing with equal amounts of 1 μg/ml in both, but I'm not sure if that means anything in terms of mechanism. We haven't done the triamcinolone experiments, which should be interesting since you cannot compete it off. It will be interesting to see whether it will have an effect. We can go to much lower doses of prolactin in these cultures. That is something that we haven't examined extensively, but we can use 10 ng of prolactin to get the same response. It will be interesting to see, as we modulate the prolactin dose, if we can use lower doses of progesterone.

WILLIAMS-ASHMAN: I would like to ask you a question about your studies hinting at cortisol having a bigger effect on the stabilization of the casein mRNA and perhaps its synthesis. As you probably know in the case of mouse mammary gland, cortisol has an important rate-limiting control in the synthesis and accumulation of spermine. Perhaps many of the effects of cortisone indeed result from polyamine accumulation. Do you have any data concerning the role of polyamines and glucocorticoid action?

ROSEN: We have done some of those experiments. It turns out that although hydrocortisone clearly increases the polyamine concentration, you cant't replace hydrocortisone with the polyamines if mRNA levels are used as an end point. The problem with all these experiments using the polyamines is that when you use the inhibitors of polyamine biosynthesis to block synthesis, they are toxic; when you add polyamine, it is also toxic. To conclude, I'm not sure than anything definitive can be said, but it appears that the hydrocortisone effect cannot be replaced by any external addition of polyamines. It is clear the casein synthesis is increased by polyamines. They could be acting on translation somehow, or tRNA structure in this system. But as far as the replacement of hydrocortisone, which is the simple model that Oka initially presented, is concerned, I think that even he now feels that he cannot say that all hydrocortisone effects can be replaced by polyamines.

MULLER: I realize that in your experiments that you are not quite ready yet to analyze the character of the transcripts, but do you get any preliminary information that would indicate that there is a range in size of the transcripts, depending upon what your corticosteroid or progesterone background is?

ROSEN: There are two ways to analyze this. One is by looking at its steady state population with the methylmercury technique; the other way is to try to look at RNA after pulse labeling. And the RNA pulse labeling is much more difficult. You have to reduce your background so that when you are looking at a gel, you are sure that the small number of counts that you see are really some sort of high molecular weight precursor. We have run experiments with short pulse times and extracted the

RNA, but the problem is to reduce the background to the point where we can really pick up something cleanly in this region of the gel. We are trying to adjust the conditions to do just that. We have also been examining chase conditions, and then maybe we can do Dr. Edelman's suggested pulse-chase experiments and look at precursor-product relationships. That's the problem; these are very difficult studies to do, to conform with theory of a precursor–product relationship, and nobody has been able to do them in any of the RNA-processing fields, because you just can't get enough of the precursor incorporated to do the kinds of pulse-chase studies necessary to follow turnover.

MOUDGIL: I am wondering why I don't see any data on the effect of estradiol in your organ culture system?

ROSEN: There is no evidence in organ culture that you need estrogen to do these kinds of experiments. These are all done in animals at 14 days of pregnancy that have already seen a fair amount of estrogen so that the priming effect of estrogen on progesterone in vivo has already occurred. That's altogether a different problem of differentiation in the glands, which I don't want to discuss at this point.

Discussants: V.K. MOUDGIL, G.C. MUELLER, J.M. ROSEN, E.B. THOMPSON, and H.G. WILLIAMS-ASHMAN.

Chapter 5

Steroid Receptor Subunit Structure

William T. Schrader, Yuri Seleznev,
Wayne V. Vedeckis, and Bert W. O'Malley

Receptors for steroid hormones have been studied in many tissues and organisms. Perhaps the best studied from a structural and functional standpoint has been the chicken oviduct progesterone receptor (for reviews, see Schrader and O'Malley 1978; Vedeckis *et al.* 1978). This receptor has been purified to apparent homogeneity and found to consist of two nonidentical progesterone-binding components. One, progestophilin A, has a molecular weight of 79 000 (Coty et al. 1979) and binds avidly to DNA (Schrader et al. 1972). The other, progestophilin B, has a molecular weight of 115 000 (Schrader et al. 1977; Kuhn et al. 1977) and binds weakly to DNA but strongly to chromatin from target cells (Schrader et al. 1972). Isolation of higher molecular weight aggregates of these proteins (Schrader et al. 1975), crosslinking studies (Birnbaumer et al. 1979), and functional tests of chromatin transcription (Buller et al. 1976) have led to the belief that A and B are subunits of a larger AB complex.

Although A and B are not identical nor interconvertible, they have considerable similarity with respect to their hormone-binding site (Hansen et al. 1976) and with respect to liberation of fragments using specific proteases (Vedeckis et al. 1979). Owing to the heterogeneity of other receptors under various experimental conditions, it was of interest to ask whether these others might also have subunit structure.

To test this idea, receptor–hormone complexes from several sources were chromatographed as described for progesterone receptor analysis (Schrader 1975). The results show heterogeneity of each receptor tested, and point to the similarities of structure among these interesting molecules.

I. Chromatographic Analysis of Chick Progesterone Receptors

Our earlier studies of chick receptors for progesterone had shown that the preparation could be resolved into two equal amounts of hormone-binding activity following ammonium sulfate precipitation (Schrader 1975). Figure 1 shows typical elution profiles of [^{3}H]progesterone–receptor complexes from diethylaminoethyl (DEAE)–cellulose, phosphocellulose, and hydroxy-

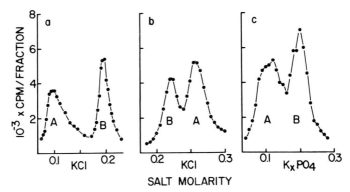

Fig. 1. Column chromatography of chick progesterone–receptor complexes. Labeled progesterone receptors, precipitated by ammonium sulfate, were chromatographed on DEAE–cellulose (**a**), phosphocellulose (**b**), or hydroxylapatite (**c**). The columns were eluted by linear KCl gradients (DEAE and phosphocellulose), as shown on the abscissa. Hydroxylapatite was eluted by a linear potassium phosphate gradient, as shown on the abscissa. Aliquots of each fraction were counted for ^{3}H using Amersham–Searle ACS counting fluid (●).

lapatite. In these tests ammonium sulfate-precipitated labeled receptor complexes were separated into three fractions and chromatographed individually on these three columns as described previously (Schrader 1975). Generally, about 70–90% of the ^{3}H adsorbed to the columns. The elution profiles show two peaks of ^{3}H in each case; these have been analyzed more rigorously elsewhere and shown to be progestophilins A and B, as indicated in Fig. 1. Since B is more acidic than A, their order of elution is reversed on phosphocellulose.

In all three cases, equal peaks of ^{3}H are obtained from A and B; this pattern is not altered by inclusion of numerous protease inhibitors, nor does it vary with time of labeling. Thus, it does not appear that the smaller A subunit is a product of B proteolysis. Since both are single polypeptide chains (Schrader and O'Malley 1978), they are apparently closely related proteins.

II. DEAE Cellulose Analysis of Other Steroid Receptors

In view of the results of Fig. 1, it was of interest to compare the progesterone receptor's behavior with that of other receptors. For this purpose, DEAE–cellulose chromatography was chosen. The results of analysis of four different receptors by this technique is shown in Fig. 2. Labeled receptors for the four hormones were precipitated by 40% saturation of ammonium sulfate and were then applied and washed through 3 ml DEAE–cellulose columns equilibrated in Buffer A. Two results can be seen in Fig. 2. First, all four receptors adsorb to and elute from DEAE–cellulose as two equal peaks of radioactivity. Second, their elution molarities with KCl are not identical from one hormone to the next, indicative of differences among the proteins themselves. However, the overall patterns are remarkably similar. It

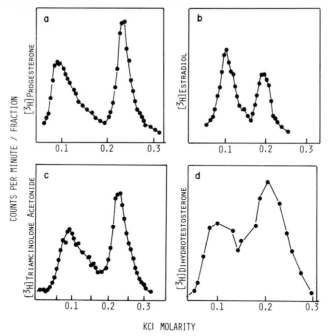

Fig. 2. DEAE cellulose column chromatography of labeled receptors for progesterone, estradiol, androgen (dihydrotestosterone) and glucocorticoid (triamcinolone acetonide). Tissue sources for these receptors were as follows: **a** progesterone, chick oviduct; **b** estradiol, rat uterus; **c** glucocorticoid, DDT-1 cell line; **d** androgen, DDT-1 cell line. Labeled receptors prepared as in Fig. 1 were chromatographed by KCl gradient elution as shown on the abscissas. Aliquots were counted for ^{3}H ($\bullet$).

should be noted here that by ammonium sulfate precipitation, the receptors would be converted to their "activated" state, that is, their DNA-binding capacity would be maximal. It is clear that heterogeneity of some sort must exist for each receptor; by analogy to the progesterone case it is an attractive hypothesis to invoke subunits to explain these findings.

Finally, the lower two panels of Fig. 2 show labeling of a cell culture extract with two different hormones. Both androgen and glucocorticoid receptors are present in these cells (Norris and Kohler, 1976), but they do not behave identically on DEAE–cellulose. Thus, even within the same cell the receptors may be structurally homologous but are not identical.

III. Receptor Analysis by DNA—Cellulose Chromatography

The most intriguing aspect of our earlier findings with progesterone receptors concerned the different affinities of subunits A and B for nuclear constituents (Schrader et al. 1972). This difference has been central to our hypothesis that the intact complex requires both B and A as a functional dimer for biologic activity (Schrader and O'Malley 1978).

The simplest initial test for such functional heterogeneity of receptor

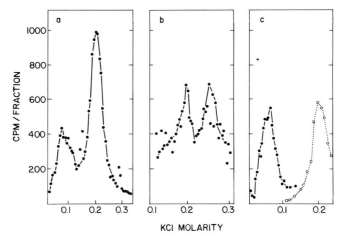

Fig. 3. DNA–cellulose chromatography. **a** and **b**, triamcinolone acetonide for rat liver and estradiol for rat uterus, respectively, were used to label receptors, which were treated as in Fig. 1. Aliquots were counted for ^{3}H (●). **c** Progesterone receptor B (●) and A (○) subunits from chick oviduct were isolated by DEAE–cellulose chromatograpy as in Fig. 2. The DNA–cellulose columns were eluted by KCl gradients, as indicated on the abscissas.

forms would be to perform DNA–cellulose chromatography. Results of these assays done for progesterone, estrogen and glucocorticoid receptors is shown in Fig. 3. Androgen receptors were not used. Figure 3C shows resolution of progesterone receptor subunits B and A on this column. In this particular experiment, B and A were separated first from each other by DEAE–cellulose (see Fig. 1) in order to confirm which one eluted at higher KCl molarity. Receptor B elutes at about 0.04 *M* KCl, whereas receptor A elutes at 0.18 *M* KCl. If an ammonium sulfate precipitate is chromatographed directly (data not shown), exactly the same profile is obtained.

Glucocorticoid and estrogen receptors were chromatographed in the other two panels. Again, following ammonium sulfate precipitation, two peaks of radioactive ligand–protein complexes were seen. The yield of glucocorticoid receptor adsorbed to the DNA was low; whether this is an experimental artifact or an unforseen aspect of its activity remains to be established. As in the progesterone case, the two peaks eluted at considerably different KCl molarities. No functional tests have yet been available to equate this behavior to that seen for progesterone receptors.

IV. Purification of Glucocorticoid Receptors of Rat Liver

In view of the similarities seen among receptors in Figs. 1–3, we speculated that a purification protocol suitable for progesterone receptor A subunit might be applicable to others as well. We used a method developed by Coty et al. (1979) that yields receptor A protein, which is apparently homogeneous as assayed by a variety of gel electrophoresis systems.

Rat liver glucocorticoid receptor was chosen, and the initial stages of the purification were carried out exactly as described for progesterone receptor. After the ammonium sulfate precipitation, labeled [³H]triamcinolone–receptor complexes were subjected to a series of column chromatograms as shown in Fig. 4.

The DEAE–cellulose elution profile is shown in Fig. 4a. As had been seen for hamster DDT-1 cell glucocorticoid receptor (Fig. 1), two peaks of ³H eluted. In analogy to the progesterone work, the earlier peak eluting at 0.1 M KCl was pooled and chromatographed on a 20-ml DNA–cellulose column. This profile is shown in Fig. 4b. The large, second peak eluting at 0.3 M KCl was pooled and subjected to the final phosphocellulose column. A sharp peak eluted, as shown in Fig. 4c.

When this entire pool was concentrated and subjected to gel electrophoresis under denaturing conditions in 0.1% sodiumdodecyl sulfate by the method of Laemmli (1970), a single faint band of protein was observed. The molecular weight of this band was 70 000, a value consistent with molecular weight estimates for the 4S glucocorticoid receptor as determined by the method of Siegal and Monty (1966).

Although only one protein band was observed, it is not possible at this time to be certain that it is the receptor and not a contaminant. The specific

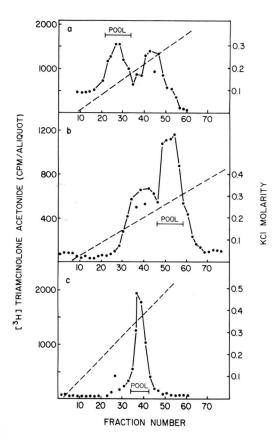

Fig. 4. Purification protocol for glucocorticoid receptors of rat liver: **a** DEAE; **b** DNA II; **c** PC II. Liver receptors were treated as described in Reference 4 for initial steps of purification. The labeled triamcinolone acetonide–receptor complexes were chromatographed successively by KCl gradient elution (– – –) on DEAE–cellulose (100 ml packed volume), DNA–cellulose (20 ml packed volume), and phosphocellulose (10 ml packed volume). Aliquots were counted for ³H (●) as in Fig. 1. At each step the fractions indicated by the bar were pooled for the subsequent steps.

radioactivity of the final pool of sample was very high (as evidenced by a lack of many protein bands on the gel). However, the theoretical specific radioactivity (cpm bound/mg protein) was not obtained. Additional studies on the protein will be needed to confirm this result.

The present information does point to the overall similarity among the various steroid receptors, as shown by electrostatic, adsorption, and purification tests. It is thus tempting to speculate that the mechanism by which these proteins act may also be common, with the differences being primarily a reflection of which genes are affected.

V. A Proposed Model for the Molecular Action of Steroid Receptors

Our current working hypothesis regarding the action of these interesting molecules is shown in Fig. 5. The concept is developed entirely from the studies of steroid action in chick oviduct. However, as the foregoing figures document, there is reason to believe the hypothesis may be a useful one for studies of other steroid receptors as well.

The functional form of the receptor is envisioned to be a dimer of two dissimilar steroid-binding subunits. The sedimentation coefficient of the dimer is about 6S, whereas the subunits are about 4S each. The dimer has nuclei-

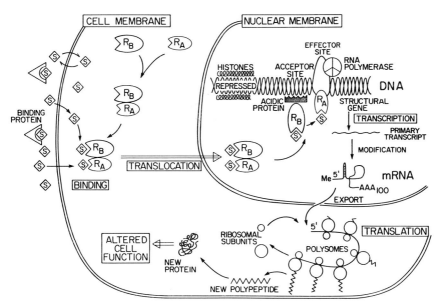

Fig. 5. A proposed model for steroid receptor function. A two-subunit receptor is envisioned to function by sequential interaction at acceptor sites via the chromatin-binding capacity of the B subunit, followed by DNA helix destabilization caused by binding of the liberated A subunit. This DNA unwinding may then promote RNA chain initiation. See text for explanatory comments.

binding activity following association with its ligand, thus causing a net accumulation of the complexes in the nucleus (Buller et al. 1975). We have found that the dimer does not bind avidly to DNA, but rather exhibits the weak DNA attraction of the B subunit (Schrader and O'Malley 1978).

Its initial interactions with the chromatin thus may be limited to the chromatin-binding activity of the B subunit. This activity has been shown in earlier studies (Schrader et al. 1972) to be target-tissue chromatin specific, and to involve adsorption to certain "acceptor sites," regions of the chromatin defined by the presence of a certain subclass of nonhistone chromosomal proteins (Spelsberg et al. 1977).

We speculate, but have no supporting evidence, that these acceptor sites are localized proximal to genes destined to be regulated by the hormone in question. There follows a dissociation of the A–B receptor complex, to yield the highly DNA-reactive A subunit. This dissociation has been found to be facilitated by the binding of the hormone (Schrader and O'Malley 1978).

The highly reactive A protein, now localized near a gene it is to regulate, has a functional DNA-binding site. In other studies, we have found that the protein has no detectable nucleotide-sequence specificity, but rather has high affinity for single-stranded DNA. Such an activity might then be able to stabilize the DNA locally in an "unwound" configuration, and thereby present a more favorable RNA chain initiation site for RNA polymerase. Thus, the presence of the receptor in the chromatin would tend to stimulate initiation of RNA synthesis at the 5' end of a gene. This concept is consistent with our earlier studies of RNA synthesis using purified receptors in vitro (Buller et al. 1976) and also with the more recent studies of ovalbumin gene induction at the initiation step (Roop et al. 1978).

The consequences of such a receptor dimeric structure are twofold. First, the model shown here would propose that hormone action would obey two-hit kinetics, since the dimer binds two moles of ligand per mole of receptor. Second, the fact that the dimer itself has occluded the DNA-binding site of the A subunit would tend to limit or prevent association to the receptor with the wrong DNA. The active subunit might only be able to express DNA regulatory activity after localization at the genes of interest as accomplished via the B "specifier" subunit (O'Malley et al. 1972). This mechanism would account for the fact that specific gene regulation can exist in the target cell for a given hormone in the face of many thousands of active genes not affected by the signal.

The chromatographic analyses presented here are indicative of considerable structural homology among steroid receptors. They thereby suggest the existence of a common functional pathway as well. However, none of these findings is of itself conclusive of this idea. It remains for receptor purification, coupled with in vitro functional tests, to establish and test these concepts more fully.

Not considered here are the many studies from other laboratories offering alternatives to this concept. Receptor "activation" and "transformation," for example, are descriptive terms at the present time for in vitro changes in receptor characteristics. To date there is no molecular explanation for these terms in any system. All of the agents so far tested that promote "activation" also promote receptor dissociation in vitro. These include ammonium sulfate, dialysis, warming, KCl, dilution, and pyridoxal phosphate. Conversely, those agents that in our hands promote "inactivation" or "deactivation" also cause dimer stabilization. These include low ionic strength, lack of hormone, fluoride ion, and molybdate ion. Thus, we also speculate that receptor "activation" and "transformation" are in fact in vitro manifestations of a simpler problem, namely, subunit–subunit assembly.

VI. Summary

Considerable similarity exists among receptors for steroid hormones from many different organisms. In the present study receptors for progesterone (chick), estrogen (rat), androgen (hamster), and glucocorticoid (rat) were compared with respect to their chromatographic behavior. Receptor–hormone complexes in crude extracts were analyzed by DEAE–cellulose, DNA–cellulose, and phosphocellulose chromatography. The study concludes that all steroid receptors tested can be resolved into multiple components.

Acknowledgments. This work was supported by NIH grants HD-07857 and HD-07495 to the Baylor Center for Population Research and Studies in Reproductive Biology. W.V.V. is recipient of a USPHS postdoctoral fellowship. Y.S. was supported in part by the Joint U.S.–U.S.S.R. Program in Myocardial Metabolism.

References

Birnbaumer ME, Schrader WT, O'Malley BW (1979) Biochem J 181: 201–213
Buller RE, Toft DO, Schrader WT, O'Malley BW (1975) J Biol Chem 250: 801–808
Buller RE, Schwartz RJ, Schrader WT, O'Malley BW (1976) J Biol Chem 251: 5178–5186
Coty WA, Schrader WT, O'Malley BW (1979) J Steroid Biochem 10: 1–12
Hansen PE, Schrader WT, O'Malley BW (1976) J Steroid Biochem 7: 723–732
Kuhn RW, Schrader WT, Coty WA, Conn PM, O'Malley BW (1977) J Biol Chem 252: 308–317
Laemmli UK (1970) Nature 227: 680–685
Norris JS, Kohler PO (1976) Science 192: 898–900
O'Malley BW, Spelsberg TC, Schrader WT, Chytil F, Steggles AW (1972) Nature 235: 141–144
Roop DR, Nordstrom JL, Tsai SY, Tsai MJ, O'Malley BW (1978) Cell 15: 671–685
Schrader WT (1975) Meth Enzymol 36: 187–211
Schrader WT, O'Malley BW (1978) In: Vol. II (O'Malley BW, Birnbaumer L. (eds) Receptors and Hormone Action. Academic Press, New York, pp 189–224
Schrader WT, Toft DO, O'Malley BW (1972) J Biol Chem 247: 2401–2407
Schrader WT, Heuer SS, O'Malley BW (1975) Biol Reprod 12: 134–142
Schrader WT, Kuhn RW, O'Malley BW (1977) J Biol Chem 252: 299–307

Siegal LM, and Monty KJ (1966) Biochim Biophys Acta 112: 346–362
Spelsberg TC, Webster R, Pikler G, Thrall C, Wells D (1977) Ann NY Acad Sci 286: 43–63
Vedeckis WV, Schrader WT, O'Malley BW (1978) In: Litwack G. (ed) Biochemical actions of hormones. Academic Press, New York, Vol. VI, pp 321–372
Vedeckis WV, Schrader WT, O'Malley BW (1979) In: Leavitt WW and Clark JH (eds) Steroid hormone receptor systems. Plenum, New York, pp 309–327

Discussion of the Paper Presented by W.T. Schrader

BAXTER: Why haven't you done tryptic digestion of the purified receptors?

SCHRADER: I knew that the endogenous protease would clip the protein to a form that still had hormone bound to it. Until we have the antibody, we do not have an assay for non-hormone-binding fragments.

BAXTER: Did you have enough material to do the experiments?

SCHRADER: Yes, but we haven't done it yet. It is a good experiment, but we just haven't gotten around to it. I think that, in our own defense, for the past couple of years it takes nanomolar amounts of receptor to study DNA binding, chromatin binding, nuclei binding, and gene transcription. It takes micromolar amounts if not millimolar amounts to do detailed protein structure work. We are trying to gear up to do that now, but it has been a matter mostly of priority.

MUELLER: Does your receptor have a difference in binding capability to supercoiled DNA versus more relaxed regular DNA?

SCHRADER: It binds to ϕX174 DNA, but there is no evidence of specificity at this stage of the game. That is, in fact, exactly the experiment that is needed to be done in order to quantitate the effect of the receptors as unwinding protein, and we are trying to gear up to do that now. We are trying to take a look at whether or not the receptor is an unwinding protein or whether or not there is a topological preference for DNA binding. I simply don't have any data on that.

EDELMAN: You probably have heard that Mary Sherman has found that in ordinary high-speed supernatant extract she can show the conversion of, I guess, all of the receptors into small forms. She has shown this with estradiol receptor, glucocorticoid receptor, and aldosterone receptor. In addition, in collaboration with us, we did a study on the relationship between the generation of the meroreceptor form of the glucocoiticoid receptor, where one cannot demonstrate an A and B form and transfer to the nucleus. Only the intact receptor transfers; this is where you allow the process to occur in a crude mixture endogenously. And to the extent that you block the conversion to the small form, you promote nuclear binding. It turns out that in her hands, none of the ordinary protease inhibitors worked in blocking this effect. But the two arginine specific proteases, leupeptin and antipain, do block the transfer; they block the conversion of the small form. Isn't it possible to block the appearance of the A form with antipain or leupeptin?

SCHRADER: We have tried those experiments, and that doesn't happen. You can use leupeptin or antipain and they don't work. Better inhibitors than leupeptin are fluoride and iodoacetamide, neither of which were shown on the slide. This is a sulfhydryl protease and it is inhibited very nicely by iodoacetamide.

EDELMAN: A question about another control on the DNA binding experiment. Have you checked the effect of what should be an inactive steroid binding protein like PBG or CBG, on the assays as another control.

SCHRADER: We do the assays in the presence of 50 μg/ml of BSA, which is a steroid binding protein of sorts. I am not sure what you are driving at.

EDELMAN: There may be an effect that simply has to do with the presence of protein steroid complex independent of whether that protein steroid complex is capable of forming a specific high affinity union with the DNA. Since you want to do it with a purified protein as a control, you don't want to do it with a crude mixture, because another protein may cross-react or interfere. It seems to me that either purified PBG or purified CBG, which now can be prepared in abundance, could be the protein of choice.

SCHRADER: They don't bind to DNA cellulose.

EDELMAN: But this is a different assay.

SCHRADER: I agree.

GELEHRTER: What is the relative affinity of the A and B protein for DNA. What is the role of the steroid in determining the DNA-binding properties of these proteins?

SCHRADER: First, the A protein experiments that I showed were done with chick progesterone receptor, which binds tritiated steroid stoichiometrically. It was not done with the hen receptor because when the A protein from the hen is purified by that protocol, you get the pure protein all right, but again it only has very low specific radioactivity and you don't know exactly what the specific radioactivity is. This results from doing titration with receptors.

Second, we have not been able to purify the apoprotein for either subunit to homogeneity. So we are not in a position at the present time to answer the questions: What is the relative affinity of the apoprotein for general DNA or what is the function of the hormone in that assay? W.A. Coty, who was a postdoctoral fellow in our laboratory, asked the question: Does the hormone have any effect on the binding of receptor to DNA in general? So he took the partially purified apoprotein consisting just of the A protein; it binds just fine to DNA cellulose. There is no detectable effect of hormone on the binding of receptor to the DNA cellulose. So there is no evidence as far as DNA binding is concerned; there is no data that I am aware of in any system to show that the hormone affects the ability of the receptor to bind to DNA directly. In chromatin studies, a number of laboratories have produced experiments to show that the hormone is required for the binding of the receptor to nucleus. What the consequences are of the ligand binding to the protein is unknown. I have looked at circular dicroism spectra after removing the ligand, etc., and then tried to reconstitute the apoprotein. We could get back the α-helix content of the protein from guanidine hydrochloride but could not show any effect on the general α-helix content. So, in the progesterone receptor case, there is no change in the charge in these proteins that we can detect. The function of the hormone, as far as I'm concerned, remains completely unknown.

STEVENS: Is there any evidence for two dissimilar subunits for the estrogen receptor?

SCHRADER: This coming Thursday, the day after tomorrow, I have to pay off a bet with Debbie Metzer, a graduate student working with Jim Clark, by taking her out for a steak dinner, regarding the subunit structure of the uterine estrogen receptor. When she does any of the ion exchange chromatographic procedures or gel filtration procedures, her findings are that the uterine estrogen receptor can also be fractionated into two nonidentical estradiol binding molecules, one with a molecular weight of approximately 80 000 and one with the molecular weight of about approximately 50 000. These two proteins seem to have the proper molecular weight to be the subunits of estradiol receptor, which Notides and Yamamoto had predicted from the conversion of the rat uterine estradiol receptor from 4S to 5S forms. The conversion of 4S to 5S form has been shown by both Yamamoto and Notides to be a second-order reaction. Notides also demonstrated a molecular weight change that was equivalent to the coupling of a 80 000 to a 50 000 molecular protein. When Debbie warms rat uterine estrogen receptor under conditions in which this 4–5S conversion takes place, the two DEAE monomeric peaks, which are 4S molecules, go down concomitantly, and there appears a dimeric structure, with larger Stokes radius and larger sedimentation coefficient, that appears concomitantly. That molecule subsequently can be dissociated under proper conditions again to yield two hormone-binding molecules from itself that were dissimilar.

STEVENS: Is this true for the so-called "activated receptor?"

SCHRADER: The activated estradiol receptor in Debbie's hands appears to be a dimer of two nonidentical hormone binding subunits. Finally we have done the same sort of ion exchange chromatography for androgen receptor in a hamster ductus deferens tumor cell line, for androgen receptor in the rat ventral prostate, and we have also done it for glucocorticoid receptor in the chick oviduct. In all of those cases we see the same behavior.

MUELLER: The possibility is that the conversion of similar subunits may not necessarily be part of the activation process.

SCHRADER: That's right. There is also, I think a dicotomy here. Allan Munck and Edwin Milgrom independently have been doing some experiments on activation of glucocorticoid receptor. If I can parapharase Allan's experiments, they are done in isolated rat thymocytes, kept at 37°C in the absence of glucocorticoid. Allan can do a very rapid translocation experiment, because the cells can instantly be lysed with magnesium, and he can therefore collect the cytoplasm separately from the nuclear compartment. When he does these experiments, he looks at the rates of receptor activation at 37°C in the cell, which I think is the most relevant activation experiment to do. What he sees is that the activation process appears to be a first-order reaction, which is not consistent with the two subunit associated reaction. This raises the possibility that there is a very early and a very fast reaction in which the receptor is changed conformationally, or phosphorylated, or you name it.

Subsequent to that, there may or may not be in vivo a dimerization reaction. One way you can put these puzzles together will be if the dimers exist in vivo in the cell; then there is some activation step and we don't know what it is.

Discussants: J.D. BAXTER, I.S. EDELMAN, T.D. GELEHRTER, G.C. MUELLER, W.T. SCHRADER, and J. STEVENS

Chapter 6

Estrogen Receptor Heterogeneity and Uterotropic Response

JAMES H. CLARK, SUSAN UPCHURCH, BARRY MARKAVERICH, HAKAN ERIKSSON, AND JAMES W. HARDIN

I. Introduction

For several years investigators have suggested that at least two forms of estrogen binding sites exist in the uterus (Best-Belpomme et al. 1970; Ellis and Ringold 1971; Erdos et al. 1969; Michel et al. 1974; Puca et al. 1971; Sanborn et al. 1971; Steggles and King 1970). One of these sites, the estrogen receptor, has been intensively investigated (Baulieu et al. 1975; Clark et al. 1978a; Gorski and Gannon 1976; Jensen et al. 1974; O'Malley and Means 1974). This receptor is a protein macromolecule that binds estrogens in a stereospecific manner and is found in the cytosol of estrogen-sensitive cells. It has a very high affinity for estradiol ($K_d \sim 10^{-9} M$) and is generally considered to exist in uterine cells at a concentration of about 20,000 sites per cell or ~ 0.5 pmol/100 μg DNA (Anderson et al. 1972; Clark and Gorski 1969; Katzenellenbogen et al. 1973). The other binding site(s) has received little attention and is often ignored or considered to result from serum albumin or α-fetoprotein (Katzenellenbogen et al. 1973). Early work indicated that dissociation of estrogen from a single binding site could not easily account for this multiplicity, and so two or more sites were proposed (Sanborn et al. 1971; Best-Belpomme et al. 1970; Erdos et al. 1969; Ellis and Ringold 1971; Puca et al. 1971). Rochefort and Baulieu (1969) had noted previously the presence of a secondary site in the uterus that bound estradiol with low affinity but very high capacity.

In addition to the above observations work from our laboratory indicated that only 50% of the [^{3}H]estradiol present in the uterus after in vivo injection or in vitro exposure is bound to the estrogen receptor (Peck et al. 1973). We have suggested that the remaining 50% is bound to secondary sites that could include lipids, serum albumin, and other proteins. Such secondary sites are usually ignored and/or considered of no physiological significance. However, the proper evaluation of such secondary binding sites is necessary for the valid measurement of estrogen receptors. In addition, an understanding of the characteristics of these secondary sites should provide insight into their possible function.

II. Estrogen Binding Sites in Uterine Cytosol

Saturation analysis of uterine cytosol over a wide range of [³H]estradiol concentrations produces the data shown in Fig. 1A (Clark et al. 1978b; Eriksson et al. 1978). Although this curve appears to consist of only one binding component, it is actually composed of two components, which can be resolved by Scatchard analysis, using the Rosenthal–Feldman method for resolution of curved Scatchard plots (Scatchard 1949; Rosenthal 1967; Feldman 1972). One of these sites, type I (Fig. 1B), conforms to the characteristics expected of the classic estrogen receptor, having a K_d of 0.8 nM with 0.6 pmol of sites per uterus (30–35 mg wet weight). In addition, this site is depleted from the cytoplasm after an injection of estradiol (Fig. 1A). The other site, type II (Fig. 1B), has a lower affinity for estradiol ($K_d \sim 30$ nM) but a higher binding capacity (>2.0 pmol/uterus). Note that type II sites do not disappear from the cytosol after an estradiol injection (Fig. 1A).

To further characterize these sites, we have used sucrose density gradient analysis of uterine cytosol (Clark et al. 1978b). This method has been used extensively for qualitative and quantitative studies of estrogen receptors in both normal and abnormal tissues. In this method cytosol is usually prelabeled with [³H]estradiol and subsequently centrifuged at high speed for 12–16 h. We anticipated that [³H]estradiol would dissociate rapidly from type II sites during this 12–16-h period and hence that very little bound hormone could be observed after centrifugation. On the other hand, dissociation

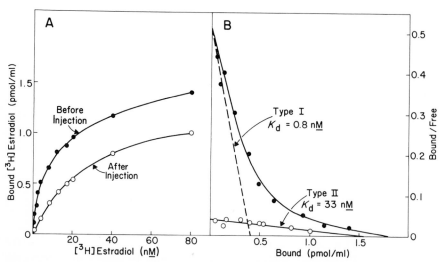

Fig. 1, A,B. Saturation analysis of estrogen binding in rat uterine cytosol

A The quantity of specifically bound [³H]estradiol was determined in uterine cytosol from noninjected rats (●) and rats injected with 2.5 μg of estradiol 60 min prior to sacrifice (○).

B Scatchard analysis of the data in A. The amount of [³H]estradiol bound to uterine cytosol from estrogen treated animals (○, type II) was subtracted from the total binding in the system (●) to yield the dashed line labeled type I.

from type I sites is quite slow ($t_{1/2} \sim 20$ h), and hence at least 50% of the bound hormone should be observed after gradient analysis. To avoid this problem and to allow the demonstration of type II, we have added [³H]estradiol to gradient fractions after centrifugation and employed the hydroxylapatite (HAP) assay to separate free from bound steroid. In Fig. 2, the prelabeling and postlabeling of receptors for sucrose gradient analysis is compared. As shown in Fig. 2A the quantity of [³H]estradiol bound in a specific manner, that is, the amount that is displaced by an excess of diethylsilbestrol (DES), is significant in both the 4S and 8S regions of prelabeled gradients. However, neither peak accounts for the predicted quantity of type II binding. Postlabeling of gradient fractions followed by a HAP assay of each fraction reveals the presence of large quantities of an estrogen binding molecule in the 4S region of the gradient. The relative quantities of

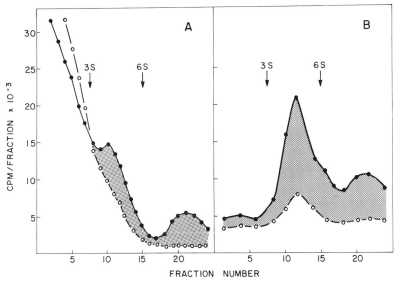

Fig. 2, A,B. Sucrose density gradient analysis of type I and II estradiol binding sites by prelabeled and postlabeled methods.

A Prelabeled gradients: Linear 5–20% sucrose gradients (4.9 ml) were prepared with a Beckman density gradient former. Sucrose solutions were prepared in TE buffer plus 1 mM dithiothreitol (DTT). Uterine cytosol (250 μl) which had been adjusted to 1 mM DTT was incubated at 4°C for 60 min with 20 nM [³H]estradiol (●) or the same concentration of [³H]estradiol plus a 100-fold molar excess of diethylstilbestrol (DES, ○). The cytosol was layered on the gradients and centrifuged at 189,000g for 16 h.

B Postlabeled gradients: Sucrose density gradients were prepared as described above. The cytosol, either labeled or unlabeled with [³H]estradiol, was centrifuged for 16 h and either 0.2- or 0.4-ml fractions were collected in tubes that contained [³H]estradiol (20 nM, final concentration). An identical gradient was fractionated into tubes that contained 20 nM [³H]estradiol plus a 100-fold molar excess of DES. The tubes were incubated for 60 min at 20°C and the measurement of specific estrogen binding was performed by the HAP assay. (Rosenthal HE (1967) Anal Biochem 20: 525) bound [³H]estradiol in the 4S and 8S region are compatible with the assumption that the 8S region is made up of type I sites while type II sites are in the 4S region. Type II sites also exist in the uterine cytosol from castrate mature rats and human and mouse mammary cancer (Panko, Watson and Clark unpublished observations).

Additional evidence that the 4S and 8S regions of gradients contain type I and II sites, respectively, was obtained by postlabeling gradients of cytosol from rats that either had or had not been injected with estradiol. Injection of 2.5 µg of estradiol into immature rats will deplete cytosol estrogen receptors (type I) but type II sites should remain in the cytosol. It is clear from Fig. 3 that the 8S region is depleted of specific binding sites, whereas the quantity of bound [³H]estradiol in the 4S region does not change. This agrees with the observation that cytosol from estrogen injected rats contains only type II sites when analyzed by the HAP assay (see Fig. 1).

Hormone specificity of type I and II binding sites was examined by postlabeling gradients as well as by direct HAP assay of high-speed cytosols. The HAP assay of high-speed cytosol showed that estradiol and DES inhibited the binding of [³H]estradiol to both types of sites, whereas progesterone, testosterone, and cortisol did not. The specificity of both sites was also shown by the postlabeled gradient technique (Fig. 4).

The presence of multiple estrogen-binding sites in other tissues was examined by the postlabeled gradient method as well as by direct HAP assay of high-speed cytosols. The vagina contains large quantities of the type II

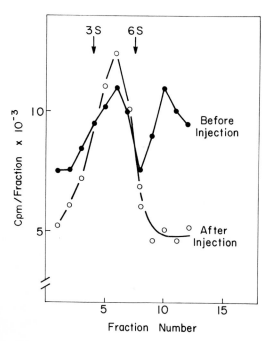

Fig. 3. Effect of estradiol injection on type I and II binding sites in uterine cytosol. Uterine cytosol was examined by the postlabeled sucrose density gradient technique, as described in Fig. 2, in noninjected rats (●) and in rats that had been injected 60 min prior to sacrifice (○).

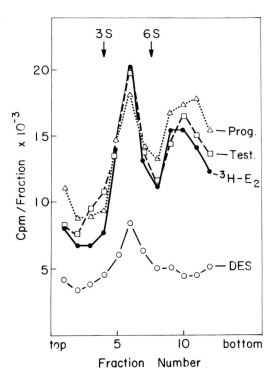

Fig. 4. Hormone specificity of type I and II estradiol binding sites. Uterine cytosol from immature rats was examined by the postlabeled sucrose density gradient technique as described in Fig. 2. Hydroxylapatite assays were performed with [³H]estradiol (20 n*M*) or [³H]estradiol plus progesterone (△), testosterone (□), and DES (○). The concentration of each added steroid was 2 *μM*.

binding site, whereas the kidney contains significant but much lower amounts (Fig. 5). This is also true for the spleen (data not shown). In these experiments the animals were injected 60 min before sacrifice with 2.5 μg of estradiol; hence, very little type I receptor is present in the cytosol fraction of the vagina. Saturation analyses of [³H]estradiol binding to serum and cytosols from spleen and kidney by direct HAP assay do not detect significant levels of type II binding.

Thus, there exists at least two general types of macromolecules that bind estrogen in a stereospecific fashion: type I, which has the properties of the classical cytosol estrogen receptor, and type II, which has a lower affinity and a higher capacity for estradiol than type I. Type II sites also differ from type I in that they do not undergo translocation from the cytoplasm to the nucleus after an estrogen injection. The implications of the existence of type II sites are far reaching. Their presence interferes with the measurement of type I sites, producing an overestimate of type I and/or an incorrect identification of type II as type I.

That type II is a macromolecule with a sedimentation coefficient of 4S also has important implications. Sucrose gradient analyses usually employ the labeling of cytosol before centrifugation. During the 12–16-h centrifugation period, hormone dissociation from type I and type II may take place and both types of binding sites will be underestimated. Depending on their rates of dissociation, a multitude of estimates of their number can be obtained.

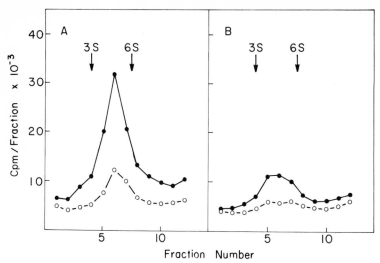

Fig. 5, A,B. Tissue specificity of type I and II estradiol binding sites. Cytosols were prepared from vagina (**A**) and kidney (**B**) and postlabeled sucrose density gradient analysis was performed as described in Fig. 2.

The HAP assay combined with postlabeling of gradient fractions eliminates this problem and results in a reproducible estimate of both type I and II binding sites.

Bound [³H]estradiol in the 4S region of gradients is usually attributed to the presence of α-fetoprotein in the immature rat uterus. Certainly this serum protein makes some contribution (Michel et al. 1974). However, in these experiments DES was used as a competitive inhibitor, and since α-fetoprotein has a very low affinity for DES, the type II binding observed cannot result from α-fetoprotein (Soloff et al. 1971). In addition, type II sites are readily observed in adult rat uteri that are devoid of α-fetoprotein (Clark et al. 1978b).

The presence of estrogen receptor in human breast tissue is used by several laboratories to determine whether endocrine ablation should be employed in the treatment of breast cancer (McGuire et al. 1975; Jensen et al. 1976). Since breast tissue cytosols are assayed by the prelabeled sucrose gradient technique, it is obvious from the results shown in Fig. 2 that much of the estrogen binding capacity of the tissue is underestimated. The method of postlabeled sucrose density gradients coupled with the HAP assay should provide a tool for a more thorough and valid estimate of estrogen-binding sites in breast tissue and thus increase the accuracy of predictability of estrogen receptor assays in breast cancer therapy.

It is possible that type II sites are extracellular binding proteins that help to accumulate estrogens. If concentrated in the extracellular spaces of the uterus, they could act as effective estrogen concentrating agents, maintaining local or organ levels of free estrogen at high concentrations relative to the blood. As discussed above, approximately half of the [³H]estradiol in the uterus after in vivo injection or in vitro incubation is bound to the estrogen receptor (Peck et al. 1973). The remaining estradiol may be distributed between type II sites and other sites such as serum albumin. Such a mixture of binding sites with variable affinities and capacities may create an extracellular estrogen buffer. Alternatively, type II sites may represent precursors to type I sites. Thus, one could envision a cytoplasmic reserve of low affinity macromolecules that serve to replace type I sites that have been translocated to the nucleus.

In summary, uterine cytosol contains two types of specific estrogen binding macromolecules. Type I conforms to the characteristics of the high-affinity–low-capacity estrogen receptor, which undergoes translocation to the nucleus. Type II sites have a lower affinity and greater capacity for estrogen than type I sites and do not undergo translocation. Whether type II sites exist in various forms is unknown. The possible functions and significance of these sites will be discussed in the summary of this paper.

III. Two Types of Estrogen Binding Sites in Uterine Nuclei

As discussed previously, an injection of estradiol will cause the depletion of type I sites from uterine cytosol (see Fig. 1). This depletion is accompanied by the accumulation of these sites in the nucleus and represents the well-known cytoplasmic to nuclear translocation phenomenon. As shown in Fig. 6, analysis of nuclear fractions for estrogen binding sites by the [³H]estradiol exchange assay reveals a complex picture that also involves at least two sites. One conforms to the type I site, which was depleted from the cytosol and is undoubtedly identical to the classically described estrogen receptor. When the quantity of the second site, which we will call nuclear type II, is subtracted from the total quantity of nuclear bound hormone as measured by exchange, one obtains the amount bound to type I. Scatchard analysis of the nuclear type I sites reveals a K_d of 0.60 nM and a maximal number of sites of 0.36 pmol/ml (Fig. 7). These values do not differ significantly from those obtained for the cytosol receptor, type I, which is depleted by estrogen treatment and conforms to the usually accepted properties of the estrogen receptor.

The nuclear type II sites do not appear identical to the cytosol type II sites and display cooperative binding behavior with a Hill coefficient of approximately 2 (Fig. 8). These measurements were made with crude nuclear pellets; therefore, nuclear type II sites might result from cytoplasmic contamination. However, this does not appear to be the case since purified nuclear preparations also contain these sites (Fig. 8). In this experiment immature rats were injected with 2.5 μg of estradiol 1 h before assay, and nuclei were isolated as described in the caption to Fig. 8. Specific estrogen binding was measured by [³H]estradiol exchange. Both types of sites are present in nuclei after estradiol treatment; however, neither of these sites appeared in noninjected animals (Fig. 8). Since the cytosol type II sites do not undergo

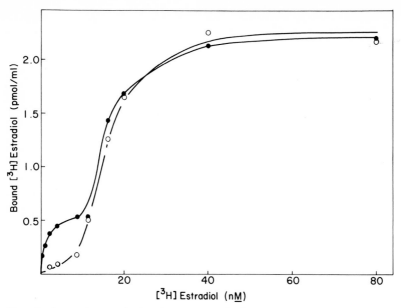

Fig. 6. Saturation analysis of estrogen binding in the nuclear fraction of the rat uterus. The quantity of specifically bound [³H]estradiol was determined by incubating uterine nuclear fractions at 4°C (○) and 37°C (●) for 30 min. Immature rats were injected with 2.5 μg of estradiol 60 min before sacrifice.

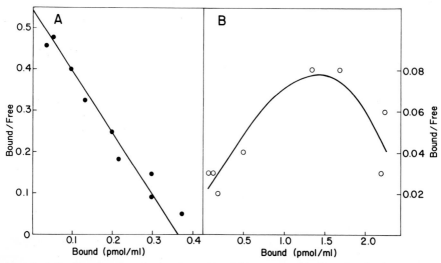

Fig. 7, A,B. Scatchard analysis of type I and II estrogen binding sites in the rat uterine nuclear fraction.

A Scatchard analysis for type I nuclear sites ($K_d = 0.6$ nM). The amount of [³H]estradiol bound to type I sites was obtained by subtracting the amount of [³H]estradiol bound at 4°C from that observed at 37°C in Fig. 6.

B Scatchard analysis of type II nuclear sites. The quantity of bound [³H]estradiol that was observed at 4°C in Fig. 6 was used here.

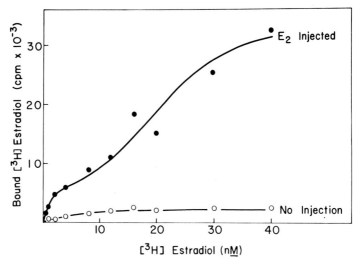

Fig. 8. Saturation analysis of type I and II binding sites in purified uterine nuclei. Purified uterine nuclei were prepared from noninjected (○) and estradiol-treated (●) immature rats (2.5 μg estradiol, 60 min prior to sacrifice). The quantity of specifically bound [³H]estradiol was determined by incubating the nuclei with [³H]estradiol or [³H]estradiol plus a 100-fold molar excess of DES.

depletion, it appears unlikely that the nuclear type II sites are derived from them. At this time, the relationship between these two classes of type II sites is unknown.

Nuclear type II sites are hormone and tissue specific. Organs such as spleen and kidney, which are not generally considered estrogen targets, do not have any measurable quantities of this binding site. The data shown in Fig. 9 were obtained with purified nuclear preparations from the uterus and spleen of estradiol treated rats. From these data, it is clear that spleen nuclei will bind larger quantities of [³H]estradiol; however, all of this binding is of a nonspecific type. Hormone specificity was examined in purified nuclear preparations by exposing them to [³H]estradiol alone or [³H]estradiol plus competitor under exchange conditions. Diethylstilbestrol inhibits binding to both uterine sites, whereas nonestrogenic hormones have no inhibitory effect (data not shown). Thus, the nuclear type II sites also display a binding specificity for estrogens similar to type I estrogen receptors.

These results indicate that at least two types of specific estrogen binding sites can be found in estrogen target cell nuclei. One of these, type I, corresponds to the classical estrogen receptor and is probably derived from the cytosol type I site. A second site, nuclear type II, does not appear to be derived from the cytosol type II site, but it is found in the nucleus after estrogen injection. The physiological significance of the second nuclear site is unknown; however, as with studies of cytosol, the presence of these sites in the nuclear fraction has important implications with respect to the validity of receptor measurement. Valid estimates of type I binding must take into account the contribution made by the presence of nuclear type II. Assays

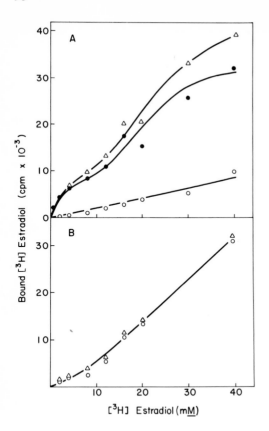

Fig. 9,A,B. Estradiol binding in purified nuclei from the uterus and spleen. **A** Uterine nuclei were prepared and non-specific binding (○) was subtracted from total binding (△) to yield the quantity of specifically bound [3H]estradiol (●). **B** Nuclei from spleen were prepared and analyzed as described in **A**.

to differentiate the two sites are readily accomplished by saturation analyses of nuclear fractions via exchange at both 37 and 4°C. At 37°C [3H]estradiol exchanges with steroid bound to type I sites and binds to type II sites; hence, at 37°C both sites are measured. Since exchange of occupied type I sites occurs very slowly at 4°C, only type II sites are measured at this temperature. Subtraction of type II sites as measured at 4°C from those sites measured via exchange at elevated temperature (types I plus II) yields the contribution made by type I alone. This evaluation is a necessity if accurate assessments of individual receptive sites are required. Previous reports from our laboratory and others have probably overestimated the quantities of type I sites in nuclear fractions. As pointed out above, this overestimate is also true for assays of cytosol receptor employing hydroxylapatite or other protein adsorbant procedures. Fortunately, the error introduced by the presence of type II sites makes an insignificant contribution to an estimate of type I sites when the concentration of [3H]estradiol is below 10 nM. The vast majority of laboratories have used this or lower concentrations to assess type I sites, and thus the quantity of type I sites has not been overestimated to a significant degree.

IV. Relationship of Type I and II Sites to Uterotropic Response

The apparent stimulation of type II sites in uterine nuclei by estradiol indicates that these sites may be involved in the mechanism of action of estrogen. If these sites are involved, one would expect that they should correlate with estrogen-induced uterotropic responses. In the past we have used the differential uterotropic response pattern to estradiol and estriol to test the assumption that type I sites are involved in producing growth of the uterus (Anderson et al. 1975; Clark et al. 1977). The results from our laboratory and others are summarized in Table 1. From these data it is apparent that estriol is a weak estrogen when injected, and this weak estrogenicity is correlated with failure of the receptor–estriol complex to be retained in the nucleus for extended periods. Conversely, estradiol is a potent estrogen that causes true growth of the uterus, and these effects are correlated with the long-term retention of the estrogen receptor in the nucleus. Therefore, it should be possible to use these two hormones to examine the biological relevance of type II sites by correlating their presence in nuclei with estrogen stimulation of uterine growth.

In these experiments mature ovariectomized rats were injected with either estradiol or estriol and sacrificed at various times after the injection. The nuclear levels of type I and type II estrogen binding sites were determined by saturation analysis. The data presented in Fig. 10A illustrated that the nuclear retention patterns of type I sites and elevations in nuclear type II estradiol binding sites were very similar. Treatment with estradiol resulted

Table 1. Effects of Estradiol and Estriol on Early and Late Uterotropic Responses

Response	Comparison of estradiol (E_2) and estriol (E_3)
Initial nuclear accumulation of receptor hormone complex	$E_2 = E_3$
Long-term retention of receptor hormone complex by the nucleus after an injection	E_2, longer than 6 h E_3, shorter than 6 h
Early uterotropic events: RNA polymerase I and II activity, template activity, histamine mobilization, water imbibition	$E_2 = E_3$
Late uterotropic events: sustained and elevated RNA polymerase I and II activity, sustained RNA polymerase initiation sites, RNA + DNA synthesis, cellular growth	$E_2 \gg E_3$
True uterine growth after paraffin implant of hormone	$E_2 = E_3$
Receptor occupancy in the nucleus after paraffin implant of hormone	$E_2 = E_3$

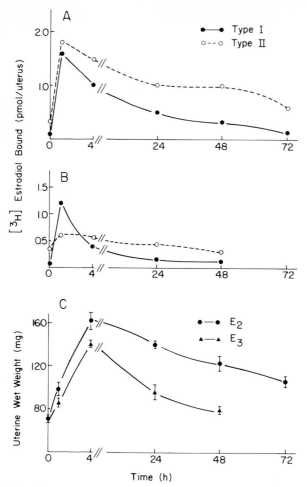

Fig. 10,A–C. Temporal effects of estradiol (**A**) and estriol (**B**) injection on uterine wet weight (**C**) and nuclear type I (●) and type II (○) estrogen binding sites. Mature ovariectomized rats were treated with 10 μg of estradiol or estriol and sacrificed at various times following injection. The quantity of specifically bound [³H]estradiol was determined by saturation analysis of uterine nuclear fractions at 37°C for 30 min.

in maximum levels of both type I and type II sites at 1 h, which gradually declined to near control levels by 72 h postinjection. Both estradiol and estriol treatment resulted in equivalent levels of type I sites at 1 h after injection, (compare Fig. 10A and B) and corresponding increases in uterine wet weight at 4 h (see Fig. 2C). However, estradiol caused type I sites to be retained for a longer period of time than estriol and stimulated elevated levels of type II sites for 48 h. Estradiol also stimulated significant increases in the wet weight of the uterus during the 24–72-h period after the injection, whereas estriol failed to cause a significant increase in uterine weight at these times. This failure of estriol to stimulate true growth of the uterus is correlated with the inability of this estrogen to induce long-term nuclear retention of type I sites or to increase the levels of nuclear type II estrogen binding sites above control levels (Fig. 10C).

If the long-term retention of type I sites is a requirement for the stimulation of nuclear type II sites, then estriol should also elevate these sites when it is maintained at a constant level. This can be accomplished by implanting paraffin pellets that contain the hormone. As shown in Table 1 and in Fig. 11, estradiol or estriol produced equivalent growth of the uterus in the immature rat when administered as a paraffin implant (Clark et al. 1977; Martucci and Fishman 1977). Therefore, if the elevation in nuclear type II sites is related to the estrogenic stimulation of true uterine growth, we would predict that estriol implants would cause a stimulation of these sites. The data presented in Fig. 12 support this hypothesis.

These experiments were performed in mature ovariectomized rats that received paraffin implants containing estradiol or estriol. The animals were sacrificed 48 h later, and the concentration of nuclear estradiol binding sites was determined. The high dose of estriol stimulates elevated levels of type I and II sites in a fashion similar to that seen in the animals treated with a low dose of estradiol. The response is somewhat reduced but qualitatively similar. The low dose of estriol has little effect on either site. The differences observed between the potency of estradiol and estriol in this study probably result from the sum of several interacting factors, which include the following two: estriol is more rapidly cleared from the blood than is estradiol; estriol has a lower affinity for the estrogen receptor and the release rate of hormone from the pellets may differ. The important point, however, is that sustained occupancy of the type I sites by estriol causes an elevation in type II sites.

These elevations in nuclear type II sites were also correlated with estradiol or estriol stimulation of true uterine growth (Table 2). In rats treated with estradiol implants or implants containing the high dose of estriol, type

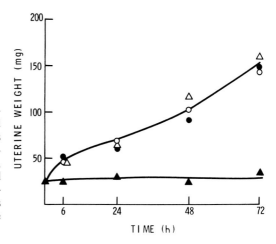

Fig. 11. Failure of estriol to antagonize uterine growth induced by estradiol. Immature rats were implanted with paraffin pellets that contained estradiol (●), estriol (○), estradiol + estriol (△), or no hormone (▲). The animals were sacrificed at various time intervals and the uterine weights were determined.

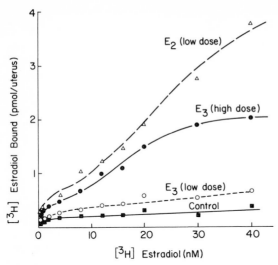

Fig. 12. Saturation analysis of estrogen binding sites in the nuclear fraction of the rat uterus. Mature ovariectomized rats were implanted with a high (4 pellets) or low (1 pellet) dose of estradiol or estriol and were sacrificed 48 h following hormone administration. The pellets weighed approximately 10 mg and contained 10% (w/w) estradiol or estriol. The quantity of specifically bound [³H]estradiol was determined by incubating uterine nuclear fractions at 37°C for 30 min.

II sites and uterine weight were stimulated. In contrast, the low-dose estriol implants did not alter either of these end points significantly. The response obtained with the high-dose estriol implants is in sharp contrast to the results obtained following a single injection of estriol, which demonstrates that estriol fails to increase either the numbers of nuclear type II estrogen binding sites or to stimulate true uterine growth.

We conclude that the growth response of the rat uterus elicited by estrogen may require two estrogen binding entities: type I sites, which undergo translocation to the nucleus, and type II sites, which are stimulated by the presence of type I sites. The role of secondary sites in the nucleus or cytosol is not known; however, it is possible to make the suggestions contained in the following summary.

Table 2. Effect of Estradiol and Estriol Containing Paraffin Implants on Uterine Wet Weight and Nuclear Type II Sites[a]

Treatment	Uterine wet weight (mg)	Type II site (pmol/uterus)
Control (4)[b]	87.3 ± 3.4[c]	0.4
Estradiol (1)	212.0 ± 11.7	8.0
Estriol (1)	82.3 ± 3.9	0.5
Estriol (4)	180.0 ± 8.8	4.2

[a] Mature ovariectomized rats were implanted with paraffin pellets containing 10% by weight of estradiol or estriol and sacrificed 48 h following treatment. The pellets weighed approximately 10 mg.
[b] Number in parenthesis is the number of pellets implanted in each animal.
[c] Values represent the MEAN ± SEM for uteri obtained for 6–8 animals.

V. Summary

Cytosol. Type II sites do not undergo translocation to the nucleus but may be involved in the retention of estrogens within the uterus. Relatively large quantities of such secondary sites could create an estrogen-rich environment for binding to type I sites, which in turn translocate estrogen as a receptor–hormone complex to the nucleus. Type I sites are identical to the classically described cytoplasmic estrogen receptor. Since we do not know whether type II sites are intra- or extracellular, they may act in this buffer capacity in either or both cellular compartments. Another possibility is that type II sites are precursors of type I sites. The ability of estradiol to elevate the level of nuclear type II sites may represent a phase in the replenishment cycle for cytosol type I sites.

Nuclear. Cytosol type I sites accumulate in the nucleus as receptor steroid complexes. These complexes bind to specific acceptor sites, which in turn activate or stimulate the formation of type II sites. These secondary sites do not appear to be translocated, but instead may be chromosomal proteins which are present in the nucleus of uterine cells at all times. The function of these sites is not known; however, they may be: (1) components of the nuclear processing machinery which interact with the receptor estradiol complex and bring about recycling or replenishment of the receptor; (2) integral components of the mechanisms which control RNA and/or DNA synthesis; (3) components of the "off reaction" involved either in turning off receptor stimulated events or in removing hormone and/or receptor from the nucleus; (4) an amplification mechanism to amplify the nuclear events initiated by binding of the receptor to acceptor sites.

References

Anderson J, Clark JH, Peck EJ Jr. (1972) Biochem Biophys Res Commun 48: 1460–1467

Anderson J, Peck EJ Jr, Clark JH (1975) Endocrinology 96:160

Baulieu EE, Atger M, Best-Belpomme M, Corvol P, Courvalin JC, Mester J, Milgrom E, Robel P, Rochefort H, DeCatalogne D (1975) Vitam Horm 33: 649

Best-Belpomme M, Fries J, Erdos T (1970) Eur J Biochem 17: 425

Clark JH, Gorski J (1969) Biochim Biophys Acta 192: 508

Clark JH, Peck EJ Jr (1976) Nature 260: 635

Clark JH, Peck EJ Jr (1977) In: O'Malley BW, Birnbaumer L. (eds) Hormone action: Steroid hormone receptors. Academic Press New York, pp 383–410

Clark JH, Paszko Z, Peck EJ Jr (1977) Endocrinology 100: 91

Clark JH, Peck EJ Jr, Hardin HW, Eriksson H (1978a) In: O'Malley BW, Birnbaumer L (eds) Hormone Action: Steroid hormone receptors. Academic Press New York, pp 1–31

Clark JH, Hardin JW, Upchurch S, Eriksson H (1978b) J Biol Chem 253: 7630

Ellis DJ, Ringold HJ (1971) In: McKerns KW (ed) The sex steroids. Appleton-Century-Crofts, New York, p 73

Erdos T, Bessade R, Fries J (1969) FEBS Lett 5: 161

Eriksson H, Upchurch S, Hardin JW, Peck EJ Jr., Clark JH (1978) Biochem Biophys Res Commun 81: 1

Feldman HA (1972) Anal Biochem 48: 317
Gorski J, Gannon F (1976) Ann Rev Physiol 38: 425
Jensen EV, Mohla S, Gorell TA, DeSombre ER (1974) Vitam Horm 32: 89
Jensen EV, Smith S, DeSombre ER (1976) J Steroid Biochem 7: 911
Katzenellenbogen HA, Johnson HJ Jr, Carlson KE (1973) Biochemistry 12: 4091
Martucci C, Fishman J (1977) Endocrinology 101: 1709
McGuire WL, Carbone PP, Vollmer EP (eds) (1975) Estrogen receptors in human
    breast cancer. Raven Press, New York
Michel G, Jung I, Baulieu EE, Aussel C, Uriel J (1974) Steroids 24: 437
O'Malley BW, Means AR (1974) Science 183: 610
Peck EH Jr, Burgner J, Clark JH (1973) Biochemistry 12: 4596
Puca GA, Nola E, Sica V, Bresciani F (1971) Biochemistry 10: 3769
Rochefort H, Baulieu EE (1969) Endocrinology 84: 108
Rosenthal HE (1967) Anal Biochem 20: 525
Sanborn BM, Rao BR, Korenman SG (1971) Biochemistry 10: 4955
Scatchard G (1949) Ann NY Acad Sci 51: 660
Soloff MS, Creange JE, Potts GO (1971) Endocrinology 88: 427
Steggles AW, King RJ (1970) Biochem J 118: 695
Wittliff JL, Beatty BW, Savlor ED, Patterson WB, Cooper RA Jr (1976) In: St Ar-
    neault G, Band P, Israel L (eds) Recent results in cancer research. Springer-Ver-
    lag, Berlin, p 59

Discussion of the Paper Presented by J.H. Clark

O'MALLEY: Why can't the cytoplasmic and nuclear type II be the same receptor? It could be made in cytoplasm. Second, is it specific to reproductive tissues? Have you tried this in a male tissue, like sperm?

CLARK: I can't get any cooperation on the sperm deal. Regarding the first question, I don't know. It seems unlikely that they are identical because of their differences in saturation binding parameters; however, these differences could be due to in vitro artifacts. There are circumstances when the level of nuclear type II is great, and little can be found in the cytosol—the reverse is also true. At the present time the exact relationships are not understood.

MUELLER: Have you looked into the Scatchard plot of nuclear type II and cytoplasm type II. Are they the same?

CLARK: No, the cytoplasmic is linear and the nuclear is curvilinear.

MUELLER: My real question is: What explanation do you have for the kind of bell-shaped Scatchard plot? It seems to me that there is some kind of cooperativity.

CLARK: A sigmoid saturation curve shows up as a bell-shaped function on a Scatchard plot. Cooperative reactions will have these characteristics; however, several artifacts can also cause them. One of them is receptor degradation, or failing to come to equilibrium at low concentration of ligand will also give you an artifactual sigmoid curve. When the ligand is very low, you might simply be degrading the sites in the absence of ligand, and that will give you a sigmoid saturation curve.

STEVENS: Could you comment on the possible relationships that you think might exist among the type II binding sites that you described, the salt-resistant nuclear binding sites, and the empty sites that McGuire has described.

CLARK: I had some data on salt resistance. However, Dr. O'Malley told me to cut the talk short and since he decides on my salary, I threw those slides away. The salt resistant sites still live. At first, we thought the answer was simple: what we had been measuring and calling salt insoluble sites were simply type II sites. That is what we thought. I wish that had been true, because it would simplify a lot of things. The type I sites display salt resistance and so do type II sites. I have only two comments to make about McGuire's work. Once upon a time Dan Medina and I tried exactly the same experiments in tumor lines that are nonestrogen dependent. We checked for unoccupied sites in the nuclei of these animals and found none. So it is not a general phenomenon, but I am not opposed to the idea that they could exist. There is one problem in saying that they are unoccupied. What McGuire did was run an exchange assay at 4° and 37°C and consider binding observed at 4°C to be due to unoccupied sites. But the fact of the matter is that it is very difficult to know for sure that these represent unoccupied sites. They may be occupied by ligand that has a low affinity and a fast off rate; if the ligand has a fast off rate, it will look very much like one is measuring unoccupied sites.

LANDAU: In terms of the generality of type II cytosol receptor, I want to mention some rather striking similarities in what I found while working with Dr. Plapinger. We found virtually identical Scatchard plots with both types of receptors in cytosol of hypothalamus, but only the type I receptor was found in the pituitary and only type II receptor was found in the cerebral cortex. We also found that type II receptor didn't go into the nucleus. Only type II receptor was found in the cortex, and there is no nuclear binding in the cortex. We also found that RU2858, which does not bind to α-fetoprotein, did not bind to cytoplasmic type II receptor.

CLARK: With respect to the generality, I mentioned the estrogen-sensitive human breast cancer; it is also present in rat lactating mammary glands, the rat vagina, and the mammary tumors that are sensitive to estrogen in the mouse.

LANDAU: We looked at rats and guinea pigs, and the type II receptor was unique to guinea pigs.

MUELLER: Is the type II receptor occupied by estradiol under physiological conditions?

CLARK: It is not likely that it is occupied by estradiol in vivo at physiological concentrations. I personally don't think that this is its function. I think its function is to recognize the complex, and the concentration of the complex would certainly be high enough in the nucleus to be recognized. The levels of estradiol required to bind to type II sites would not generally be considered physiological. However, the question of what is physiological is a difficult one. We don't really know about microenvironments in the cells—much less the concentrations of hormones present in these compartments.

MUELLER: There seems to be a general concept of stability of the receptors which has permeated the literature on hormone action. We seem to think about these receptors as if they were a piece of granite. I wonder if we couldn't begin to look at them dynamically for what they do.

CLARK: That is an excellent point, and it is one that all of us have banged around so much that we've gotten a whole host of people believing it. The point is that you never quite know what is going on in the tissue. This is easily demonstrated by injecting titriated estradiol into an animal and determining the counts bound to the receptor. If there were 10,000 counts, only 5000 of them will be bound in the receptor. The remaining 5000 counts were bound to something and have undergone dissociation during the preparation of cytosol and nuclei. If you do the experiment, which we have done, where there is no homogenization and no chance for dissociation, then you do see type I and type II present after a physiological injection of estradiol.

Discussants: J.H. CLARK, I.T. LANDAU, G.C. MUELLER, B.W. O'MALLEY, and J. STEVENS

Chapter 7

Analysis of the Avian Progesterone Receptor with Inhibitors

Virinder K. Moudgil, Hideo Nishigori,
Thomas E. Eessalu, and David O. Toft

I. Introduction

Some insight into the function of steroid hormone receptors might be gained by studying their ability to interact with various cellular constituents. This approach has been used with many systems in attempts to identify the sites of receptor interaction in the target cell nucleus, and binding of progesterone receptors to both DNA (Schrader et al. 1972) and to chromatin proteins (O'Malley et al. 1973; Steggles et al. 1971; Spelsberg et al. 1972) has been demonstrated. Since the receptor is a relatively large protein and is believed to be regulatory in function, it is possible this protein interacts with one or more chemical signals or effectors within the cell. Studies to identify such interactions are most readily approached using cell-free systems. However, it is difficult to establish the biological significance of cell-free interactions, and care must be taken to eliminate artifacts and to correlate results with in vivo observations. Previous studies from our laboratories have demonstrated an interaction between the avain progesterone receptor and ATP (Moudgil and Toft 1975, 1977; Toft et al. 1977). This interaction was observed using ATP–Sepharose chromatography and was shown to be reversible and dependent upon ionic strength. While the functional significance of this ATP binding remains obscure, it may represent an important step in the mechanism of steroid hormone action. In an effort to provide additional definition of the receptor–ATP interaction, inhibitors of this process have been sought.

Over the past few years, some compounds have been identified that appear to be useful inhibitors of steroid receptor proteins (Lohmar and Toft 1975; Toft et al. 1976; Nishigori et al. 1978). Four such compounds are o-phenanthroline, rifamycin AF/013, aurintricarboxylic acid, and pyridoxal 5'-phosphate. These agents do not interfere with the steroid binding process, but block the cell-free interaction of receptor with isolated nuclei or with ATP immobilized on Sepharose. Efforts are being made to describe the mechanism of action of these inhibitors since this information should provide additional clues as to the normal interactions of receptors with cellular components.

In this chapter, we have reviewed our recent studies on receptor inhibitors and have concentrated on two of the above inhibitors, pyridoxal 5′-phosphate and aurintricarboxylic acid, plus another more recently identified inhibitor, sodium molybdate. These compounds were selected because each appears to have a different inhibitory action on the receptor molecule.

II.   Binding of ATP to the Progesterone Receptor

For the past few years we have been investigating the interaction of various nucleotides with the avian progesterone receptor (Moudgil and Toft 1975). Our interest in nucleotides arose from investigations on the progesterone receptor from cow uterus, where we observed that addition of ATP to crude cytosol slightly enhanced progesterone binding by the receptor. To determine whether this effect involved the direct binding of ATP to the progesterone receptor, we applied the method of affinity chromatography. The ATP was covalently attached to Sepharose-4B through its ribose via a 6-carbon spacer. When receptor preparations containing [³H]progesterone were passed through columns of ATP–Sepharose, a majority of the complex was adsorbed to the affinity resin and could be eluted with 1.0 M KCl (Moudgil and Toft 1975). This interaction was shown to be reversible, ionic in nature, and to have a preference for ATP over other nucleoside triphosphates, AMP and cAMP. In these studies, the binding of progesterone receptor to ATP–Sepharose was demonstrated using receptor preparations that were fractionated by ammonium sulfate precipitation.

In contrast, freshly prepared cytosol has little or no affinity for ATP–Sepharose. Since ammonium sulfate fractionated receptor is in an "activated state" in that it can bind oviduct nuclei in a cell-free system at 4°C (Buller et al. 1975a,b), the possibility that the progesterone receptor may first require an activation step before binding to ATP can occur, has been tested (Miller and Toft 1978).

III.   Receptor Activation

It is now well known that steroid receptors that are initially extracted in the tissue cytosol do not have the capacity to bind to nuclear sites. However, the nuclear binding capacity can be acquired in vitro by incubating the receptor (in the presence of hormone) at elevated temperatures (e.g., 23°C) or in the presence of various salts (e.g., 0.3 M KCl). After this activation process, the receptor–hormone complex is able to bind not only to isolated nuclei or chromatin (Lohmar and Toft 1975; Buller et al. 1975a; Spelsberg et al. 1971), but also to DNA–cellulose (Schrader et al. 1972), phosphocellulose (Schrader et al. 1975), and ATP–Sepharose (Miller and Toft 1978). As mentioned earlier, previous studies have demonstrated the application of ATP–Sepharose chromatography for the measurement of progesterone receptor that has been activated by treatment either at elevated temperature or with

high salt (Miller and Toft 1978). Figure 1 illustrates ATP–Sepharose binding following temperature activation. Cytosol that had been incubated with
[³H]progesterone at 0°C and had not been incubated at room temperature
showed little binding to ATP. However, subsequent incubation of receptor
at 23°C greatly enhanced receptor binding to ATP–Sepharose. The activated portion of receptor is quantitatively adsorbed to ATP–Sepharose and
can be subsequently eluted with 1 *M* KCl. Using this method of receptor

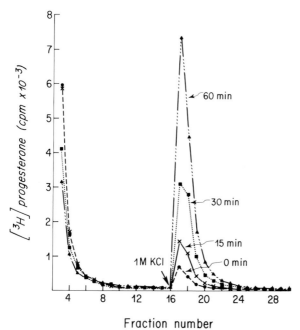

Fraction number

Fig. 1. The effect of room temperature activation on the binding of progesterone receptor to ATP–Sepharose. Oviducts were obtained from 3–4-week-old White
Leghorn chicks that received daily injections of 5 mg diethylstilbestrol in sesame oil
for a period of 2–4 weeks. The oviducts were homogenized in 4 volumes of buffer
that contained 40 m*M* Tris–HCl, 12 m*M* thioglycerol, and 10% glycerol (pH 8.0), and
the cytosol fraction was obtained as previously described (Lohmar and Toft 1975;
Toft et al. 1976). Aliquots (0.5 ml) of cytosol were incubated at 4°C for 2 h with
20 n*M* [³H]progesterone (50 Ci/mmol; from New England Nuclear) and were then incubated at room temperature for periods indicated. Samples of progesterone receptor thus treated were fractionated on columns of ATP–Sepharose. ATP–Sepharose
containing 4–5 μmol ATP/ml packed Sepharose was prepared as described previously (Moudgil and Toft 1975, 1977). After the column was washed with 15 ml
buffer that contained 0.01*M* Tris–HCl, 1 m*M* EDTA, 12 m*M* monothioglycerol,
10 m*M* KCl, and 10% (v/v) glycerol, pH 8.0 (at 25°C) (TETG), the adsorbed receptor
was eluted with TETG buffer that contained 1 *M* KCl. One-milliliter fractions were
thus collected. Aliquots (0.05 ml) from each fraction were transferred to scintillation
vials and 5 ml of scintillation fluid consisting of toluene (Fisher), and Scintiprep I
(Fisher), 160 ml/liter were added to determine radioactivity by a liquid scintillation
(48% efficiency). The amount of ATP binding by the 60-min sample represents 42%
of the total progesterone receptor applied to the column. (Miller JB, Toft DO (1978)
Biochemistry 17:173–177, courtesy of the American Chemical Society.)

activation, the maximum extent of receptor binding to ATP–Sepharose was found to be between 40 and 60% of the total cytosol receptor.

The binding of receptor to ATP–Sepharose was also increased by high salt treatment The binding was very low using cytosol incubated in 0.01 M KCl but rose to a maximum in cytosol incubated with 0.5 M KCl. Cytosol incubated with 1 M KCl produced ATP binding, which was comparable to cytosol incubated with 0.5 M KCl (data not shown).   Under these conditions, 80–100% of the progesterone–receptor complex is able to bind to ATP–Sepharose.   The increased ATP binding observed after temperature or salt activation could not be accounted for by a change in hormone binding activity as determined through charcoal adsorption assay.   The ATP–Sepharose method is much more convenient and quantitative than is the measurement of receptor activation by use of nuclear binding methods. Therefore, we have used this procedure for most of our studies on the effects of inhibitors.

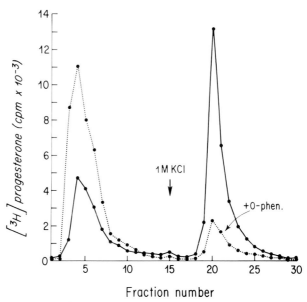

Fraction number

Fig. 2. Effect of o-phenanthroline on the binding of progesterone receptor to ATP–Sepharose.   The cytosol was prepared and fractionated by ammonium sulfate precipitation (45% of saturation) as previously described (Moudgil and Toft 1975).   Ammonium sulfate-precipitated receptor from hen oviduct was dissolved in buffer [10 mM Tris–HCl, 12 mM thioglycerol, and 20% glycerol (pH 8.0)], and preparations that contained 200 μl of receptor, 12 nM [^{3}H]progesterone, and 5 mM o-phenanthroline were incubated at room temperature for 30 min.   The final concentration of glycerol in the sample was 15%, and volume was made up to 0.5 ml with the above buffer. Following this incubation, the samples were brought to 4°C for 2 h before layering on 1 ml ATP–Sepharose columns.   The columns were washed with buffer that contained 0.01 M KCl and then with the same buffer containing 1 M KCl.   Fifteen fractions (0.6 ml/fraction) were collected in each case at a flow rate of 2 ml/min.   A sample without inhibitor served as the control.   (Toft D, Lohmar P, Miller J, Moudgil V (1976) J Steroid Biochem 7:1053–1059, courtesy of Pergamon Press.)

IV. Treatment with Inhibitors

In our initial studies, two agents, *o*-phenanthroline and rifamycin AF/013, were identified as inhibitors of the progesterone receptor. These compounds were shown to block the binding of activated progesterone receptor to nuclei in a cell-free system (Lohmar and Toft 1975). Figures 2 and 3 illustrate the effect of these compounds in blocking the binding of the progesterone–receptor complex to ATP–Sepharose. In these experiments, the progesterone receptor from hen oviducts was activated by precipitation with ammonium sulfate. The redissolved receptor was then fractionated on columns of ATP–Sepharose in the presence or absence of inhibitor. *o*-Phenanthroline is a metal chelator, and its inhibitory action suggests that the progesterone receptor may be a metalloprotein. On the other hand, rifamycin AF/013 is an antibiotic derivative that has been shown to inhibit the activity of DNA and RNA polymerases (Meilhac et al. 1972; Tsai and Saunders 1973). At the concentrations used here, neither of these inhibitory agents alters the binding of progesterone to the receptor, but they apparently interfere with another site on the receptor that functions in the binding to nuclei and to ATP–Sepharose.

V. Treatment with Aurintricarboxylic Acid

Another compound that has been shown more recently to inhibit the progesterone receptor is the triphenylmethane dye, aurintricarboxylic acid ATA), which blocks the binding of progesterone receptor to nuclei, DNA–cellulose, and to ATP–Sepharose (Moudgil and Eessalu 1978). Its effect on ATP–Sepharose binding is illustrated in Fig. 4. The concentration depen-

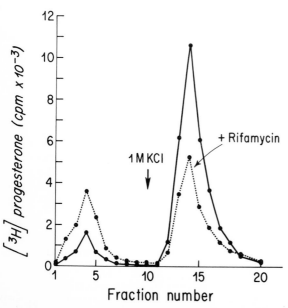

Fig. 3. The effect of rifamycin AF/013 on the binding of progesterone–receptor complex to ATP–Sepharose. Ammonium sulfate-fractionated receptor was dissolved and labeled with [³H]progesterone as described in Fig. 2. The receptor complex was treated with rifamycin AF-013 (100 μg/ml) for 2 h at 4°C. A sample with no inhibitor served as the control. The chromatography procedure was similar to that described in Fig. 2. (Toft D, Lohmar P, Miller J, Moudgil V (1976) J Steroid Biochem 7:1053–1059, courtesy of Pergamon Press.)

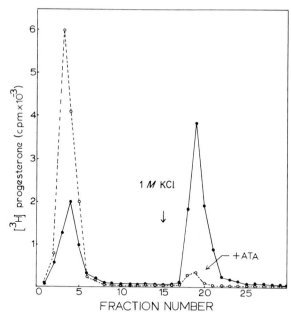

Fig. 4. The effect of aurintricarboxylic acid (ATA) on the binding of progesterone–receptor complex to ATP–Sepharose. Ammonium sulfate-fractionated receptor was dissolved and labeled with [³H]progesterone as described in Fig. 2. The receptor complex was treated with 0.1 m*M* ATA (Sigma Chemical Co.) for 2 h at 4°C. A sample with no inhibitor served as a control. The chromatography procedure was as described in Fig. 2.

dency of this inhibition is similar to the concentrations required to block nuclear binding. Figure 5 illustrates the inhibition of nuclear binding over a range of ATA concentrations. As with the other agents we have tested, ATA does not disrupt progesterone binding at the concentrations used. While the mode of action of ATA in this system remains unknown, it is possible that the inhibitor binds to a site on the activated receptor that is critical to ATP–Sepharose binding. If this were the case, one might expect that the activated receptor would show a greater sensitivity toward ATA as compared with nonactivated receptor.

When preparations containing cytosol progesterone receptor are first activated by elevation of temperature in the presence of [³H]progesterone, treated with ATA, and then allowed to incubate with isolated nuclei, the ATA causes a significant inhibition in the uptake of the complex. However, if ATA is added to the cytosol receptor preparations, and removed prior to the activation, no such inhibition occurs (Fig. 6). These results indicate an apparently selective action of ATA on the activated form of progesterone receptor.

Aurintricarboxylic acid could bring about its effects by complexing to the receptor protein, ATP–Sepharose, or certain sites in the nuclei. Such possibilities were tested by first incubating the nuclei or ATP–Sepharose with 0.2 m*M* ATA and then removing the free ATA. Subsequently, activated receptor complexes were allowed to incubate with inhibitor-treated nuclei or

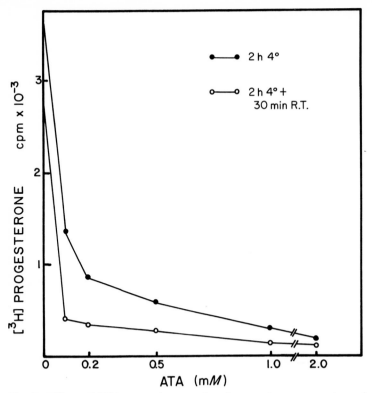

Fig. 5. Effect of ATA on nuclear uptake of progesterone receptor. Ammonium sulfate-fractionated receptor was complexed with 10 nM [³H]progesterone for 2 h at 4°C. The preparation was divided into two groups and both treated with different concentrations of ATA for 2 h at 4°C, but one received an additional 30 min incubation at room temperature. The extent of nuclear binding was then assessed by incubating samples with isolated oviduct nuclei for 60 min at 4°C. The nuclei were isolated from hen oviduct tissue by the method of Spelsberg et al. (1974). The nuclear assay consisted of incubation of progesterone receptor complexes pretreated with or without ATA and aliquots containing hen oviduct nuclei. The details of the procedures for determining nuclear binding of progesterone receptor have been reported elsewhere (Lohmar et al. 1975).

chromatographed over ATA-treated ATP–Sepharose columns. Binding of progesterone receptor to nuclei or the affinity resin was observed with values comparable to those seen in control experiments without ATA treatment (data not shown). These results suggest that ATA is interacting with the receptor preparation rather than the affinity resin or nuclear sites.

VI. Treatment with Pyridoxal 5′-Phosphate

It has been shown that the binding of activated receptor to ATP–Sepharose can be blocked by the addition of pyridoxal 5′-phosphate to the activated receptor preparation (Nishigori et al. 1978). This is illustrated in Fig. 7. The ATP–Sepharose binding is completely inhibited by the addition of 5 mM pyridoxal 5′-phosphate. The concentration dependency of this effect

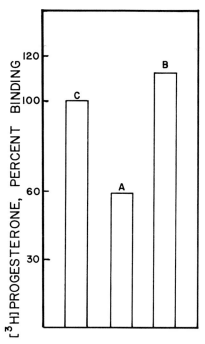

Fig. 6, A–C. ATA treatment before or after receptor activation. Hen oviduct cytosol samples containing progesterone receptor were complexed with 10 nM [³H]pro-
gesterone and divided into three groups. Groups I and II were activated by warming
at 23°C for 1 h. Following this, group II was incubated with 0.2 mM ATA at 4°C for
1 h (**A**) while group I was incubated with 10 mM Tris–HCl buffer, pH 8.0 and served
as a control (**C**). Group III was exposed to 0.2 mM ATA at 4°C for 1 h. All three
groups were passed over columns of Sephadex G-75 to remove the free inhibitor.
Group III was then brought to 23°C for 1 h (**B**). Isolated nuclei were subsequently
added and nuclear binding assays were performed as described in methods. **C** control; **A** ATA after activation; **B** ATA before activation.

is illustrated in Fig. 8. Previous studies have shown this effect to be rather
specific for pyridoxal 5'-phosphate since little or no inhibition was observed
using 5 mM pyridoxal, pyridoxine, pyridoxamine, or pyridoxamine 5'-phos-
phate (Nishigori et al. 1978). The inhibition by pyridoxal 5'-phosphate is
likely to result from the formation of a Schiff base between the inhibitor and
a critical lysine of the receptor molecule, although an indirect effect of the
inhibitor has not been ruled out. The inhibition can be reversed by adding
an amine (e.g., 0.1 M Tris–HCl), but an irreversible inhibition results from
the addition of sodium borohydride, which would reduce the Schiff base to
form a more stable covalent bond (Nishigori et al. 1978). These principles
have been applied to compare the sensitivity of nonactivated and activated
receptor to pyridoxal 5'-phosphate. Nonactivated and activated receptor
samples were treated for 1 h with 1 or 3 mM pyridoxal 5'-phosphate. Any
inhibitor complexes thus formed were then made irreversible by the addition
of 5 mM sodium borohydride. Finally, excess pyridoxal 5'-phosphate was
complexed by the addition of 0.1M Tris–HCl. The nonactivated receptor
sample could then be heat activated without the possibility of further inhibi-

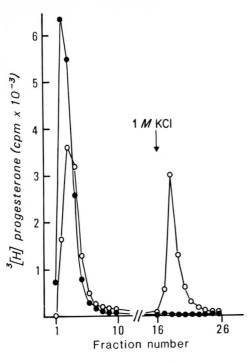

Fig. 7. The effect of pyridoxal 5′-phosphate on the binding of progesterone receptor to ATP–Sepharose. Cytosol from hen oviducts was fractionated with ammonium sulfate. The precipitate was dissolved and dialyzed in barbital buffer and then labeled with 10^{-8} M [³H]progesterone for 3 h at 4°C. A portion of this was treated with 5 mM pyridoxal 5′-phosphate for 18 h at 4°C and the treated (●) and untreated (○) samples were fractionated on ATP–Sepharose. In this untreated sample, the first column peak represents primarily unbound hormone, whereas the bound receptor complex is eluted with 1 M KCl. The amount of receptor complex was equal in both groups as determined by charcoal adsorption assay. (Nishigori H, Moudgil VK, Toft D (1978) Biochem Biophys Res Commun 80:112–118 courtesy of Academic Press, Inc.)

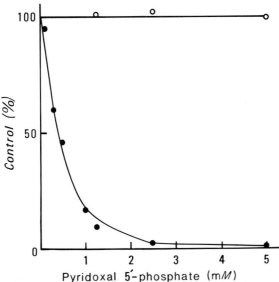

Fig. 8. The effect of different concentrations of pyridoxal 5′-phosphate on the progesterone–receptor complex and its binding to ATP–Sepharose. Samples of [³H]progesterone–receptor complex were treated with pyridoxal 5′-phosphate and the extent of ATP-Sepharose binding (●) was measured as illustrated in Fig. 7. The total progesterone binding activity (○) was also measured in a portion of each sample by the charcoal adsorption method. A sample without pyridoxal 5′-phosphate treatment served as the control. (Nishigori H, Moudgil VK, Toft D (1978) Biochem Biophys Res Commun 80:112–118 courtesy of Academic Press, Inc.)

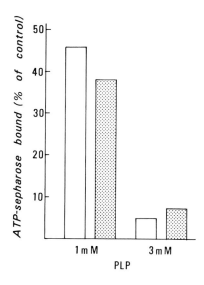

Fig. 9. Pyridoxal 5'-phosphate inhibition of nonactivated and activated progesterone receptor. Chick oviduct cytosol was labeled with $10^{-8}M$ [^{3}H]progesterone, and a portion of this was activated by incubation for 1 h at 23°C. Pyridoxal 5'-phosphate (PLP) (1 or 3 mM) was added to nonactivated (not stippled) and activated (stippled) samples for 1 h at 4°C. The inhibition was "fixed" by addition of NaBH$_4$ for 1 h at 4°C and then excess PLP was inactivated by the addition of 0.1 M Tris–HCl. The nonactivated receptor sample was then incubated for 1 h at 23°C. The degree of PLP inhibition was determined by measuring the extent of receptor binding to ATP–Sepharose. A control sample was included that was activated at 23°C for 1 h and treated with NaBH$_4$ and Tris–HCl, but not PLP. The binding of this to ATP–Sepharose was determined and designated as 100%.

tion by pyridoxal 5'-phosphate. Figure 9 illustrates the results of this study. It is evident that pyridoxal 5'-phosphate does inhibit the nonactivated receptor and the sensitivity of this receptor form toward the inhibitor appears to be identical to that of activated receptor. Therefore, the interaction that presumably occurs between pyridoxal 5'-phosphate and the receptor does not require the activated conformation of the receptor molecule. This suggests that ATA and pyridoxal 5'-phosphate affect the receptor at different sites or by different mechanisms.

VII. Treatment with Sodium Molybdate

Nielson et al. (1977a,b) have shown that the addition of molybdate stabilizes the glucocorticoid receptor from rat thymocytes. We have observed similar results with the avian progesterone receptor (Nishigori and Toft 1980). Furthermore, we have found that the addition of molybdate blocks the heat induced activation of the receptor molecule. This is illustrated in Fig. 10. In this experiment, the cytosol–[^{3}H]progesterone complex was incubated at 23°C for 1 h, and the degree of activation was assessed by binding the receptor to ATP–Sepharose. When 50 mM sodium molybdate was added to cytosol before receptor activation for 30 min, the extent of ATP–Sepharose binding was low and comparable to a nonactivated control preparation. However, molybdate did not directly block receptor binding to ATP–Sepharose. When it was added after activation at 23°C, no inhibitory effect was observed. While molybdate had no effect on the receptor preparation that was already activated, it was a potent inhibitor of ATP–Sepharose binding when added to the preparation before activation at 23°C.

Therefore, the mode of action of molybdate is quite different from that of pyridoxal 5'-phosphate and it seems to interfere in some way with the process of receptor activation. Whether this is related to the stabilizing effect

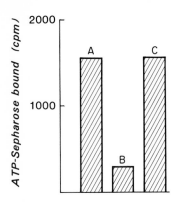

Fig. 10. Inhibitory effect of sodium molybdate before or after receptor activation. Chick oviduct cytosol was labeled with 10^{-8} M [^{3}H]progesterone and divided into three portions: (**A**) incubated at 23°C for 1 h; (**B**) incubated with 50 mM MoO$_4$ for 30 min at 4°C and then 1 h at 23°C; (**C**) incubated for 1 h at 23°C and then 30 min at 4°C with 50 mM MoO$_4$. Following incubation, the activated receptor complex in each sample was quantitated by the extent of receptor bound to ATP–Sepharose.

of molybdate is unknown. The concentration dependency of this molybdate effect is shown in Fig. 11. As with the other inhibitors, sodium molybdate has no effect on the extent of steroid binding. However, the action of this inhibitor in blocking receptor activation is quite unique. An understanding of the mechanism of this inhibition may clarify the receptor activation process.

VIII. Discussion

The above results illustrate five inhibitors that can block the binding of progesterone receptor to ATP–Sepharose. Table 1 summarizes our results with these compounds. The effects of these inhibitors have been measured by receptor binding to either nuclei, ATP–Sepharose, DNA–cellulose, or phosphocellulose. Our studies are not yet complete and the absence of a number (binding method) in Table 1 only indicates that this method has not yet been tested. However, it appears that these agents generally block interactions that are characteristic of the activated receptor form.

The modes of action of these inhibitory agents are not yet clear, however, it is evident that ATA, pyridoxal 5′-phosphate, and sodium molybdate act by

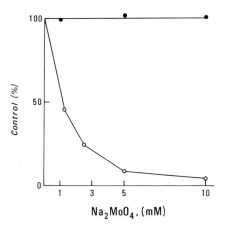

Fig. 11. The effect of different concentrations of sodium molybdate on the progesterone–receptor complex and its binding to ATP–Sepharose. Chick oviduct cytosol was labeled with $10^{-8}M$ [^{3}H]progesterone for 2 h at 4°C. Samples then were incubated for 1 h at 4°C with or without sodium molybdate. Incubation was then continued for 30 min at 23°C. The samples were tested for progesterone binding (○) by the charcoal method and also for the extent of ATP–Sepharose binding (●). The results are expressed as percentage of control (no molybdate).

Table 1. Chemical Inhibitors of Progesterone Receptor Binding Processes

Inhibitor	50% Effective concentration (mM)	1. Nuclei 2. ATP-S 3. DNA-C 4. Phos-C	Reversible
Rifamycin AF/013	0.1	1, 2	?
o-Phenanthroline	1.0	1, 2	?
Aurintricarboxylic acid	0.1	1, 2, 3	?
Pyridoxal phosphate	0.5	2	Yes
Pyridoxal phosphate, NaBH$_4$	0.5	1, 2, 3, 4	No
Sodium molybdate	1.0	1, 2, 3, 4	Yes

somewhat different mechanisms. Aurinticarboxylic acid is a triphenylmethane dye that has been shown to bind to and inhibit some nucleotide binding proteins such as RNA polymerase (Blumenthal and Landers 1973), QB replicase (Blumenthal and Landers 1973), and certain ribosomal proteins (Grollman and Stewart 1968). Since the receptor appears to show a greater sensitivity toward ATA after activation, it is very possible that the inhibitor binds the receptor at regions that become exposed with activation. Therefore, actual characterization of ATA binding to the receptor might provide additional information on the molecular changes that occur during receptor activation and the regions of receptor that are involved in nuclear interactions.

Unlike ATA, pyridoxal 5'-phosphate does not appear to interact selectively with activated or nonactivated receptor. Pyridoxal 5'-phosphate is active on highly purified receptor preparations (Nishigori et al. 1978), and it is very likely that this inhibitor directly binds to the receptor molecule. However, its sites of interaction are exposed prior to receptor activation, and it may only secondarily affect binding regions that are exposed with activation.

Sodium molybdate is a third type of inhibitor, which only acts prior to receptor activation. Its action on receptor may be direct or indirect, but it is of considerable interest since it appears to block the activation process. Although molybdate is a known inhibitor of phosphatases (Nielsen et al. 1977a,b), other possibilities exist and its action on receptors remains unknown. We have found that this inhibition can be easily reversed when molybdate is removed by dialysis or by salt precipitation of the receptor. Further studies on this and other receptor inhibitors should provide valuable information on the various sites of receptor interaction.

IX. Summary

When the avian progesterone receptor is activated by incubation at elevated temperatures, it acquires the ability to bind to nuclei and to ATP–Sepharose. A variety of chemical agents have been identified which can block these interactions of the activated receptor. Five such agents are o-phenanthroline, rifamycin AF/013, aurintricarboxylic acid, pyridoxal 5'-phosphate, and sodium molybdate. The three latter compounds have been studied in detail. Whereas aurintricarboxylic acid is a potent inhibitor of the activated receptor, it appears to have little or no affect on the nonactivated receptor form. This inhibitor therefore may interact with a region of the receptor that becomes exposed during the activation process. On the other hand, pyridoxal 5'-phosphate does not distinguish between the activated and nonactivated receptor and inhibits both receptor forms. The third agent, sodium molybdate, is inhibitory only when it is added to the receptor preparation before activation at elevated temperature. This agent does not block the binding of receptor to ATP–Sepharose directly, but it appears to interfere with the receptor activation process. Therefore, these three inhibitory agents seem to act through different mechanisms, and they should represent useful probes for receptor characterization.

Acknowledgments. The technical assistance of Nancy McMahon, Vernon Summerlin, and Bridget Stensgard is greatly appreciated. This work was supported by the National Institutes of Health grants AM-20214, HD-9140-J and AM-20893-01A1.

References

Blumenthal T, Landers TA (1973) Biochem Biophys Res Commun 55:680–688
Buller RE, Toft DO, Schrader WT, O'Malley BW (1975a) J Biol Chem 250:801–808a
Buller RE, Schrader WT, O'Malley BW (1975b) J Biol Chem 250:809–818B
Grollman AP, Stewart ML (1968) Proc Natl Acad Sci USA 61:719–725
Lohmar PH, Toft DO (1975) Biochem Biophys Res Commun 67:8–15
Meilhac M, Tysper Z, Chambon P (1972) Eur J Biochem 28:291–300
Miller JB, Toft DO (1978) Biochemistry 17:173–177
Moudgil VK, Eessalu TE (1978) Fed Proc 37:3420–3426
Moudgil VK, Toft DO (1975) Proc Natl Acad Sci USA 72:901–905
Moudgil VK, Toft DO (1977) Biochim Biophys Acta 490:447–488
Nielsen CJ, Sando JJ, Vogel WM, Pratt WB (1977a) J Mol Chem 252:7568–7578a
Nielsen CJ, Vogel WM, Pratt WB (1977b) Cancer Res 37:3420–3426B
Nishigori H, Moudgil VK, Toft D (1978) Biochem Biophys Res Commun 80:112–118
Nishigori H, Toft D (1980) Biochemistry 19:77–83
O'Malley BW, Schrader WT, Spelsberg TC (1973) Adv Exp Med Biol. 36:174–196.
Schrader WT, Toft DO, O'Malley BW (1972) J Biol Chem 247:2401–2407
Schrader WT, Heuer SS, O'Malley BW (1975) Biol Reprod 12:134–142
Spelsberg TC, Steggles AW, O'Malley BW (1971) J Biol Chem 246:4188–4197
Spelsberg TC, Steggles AW, Chytil F, O'Malley BW (1972) J Biol Chem 247:1368–1374
Spelsberg TC, Knowler JT, Moses HL (1974) Meth Enzymol 31:263–279
Steggles AW, Spelsberg TC, O'Malley BW (1971) Biochem Biophys Res Commun 43:20–27
Toft D, Lohmar P, Miller J, Moudgil V (1976) J Steroid Biochem 7:1053–1059
Toft DO, Moudgil VK, Lohmar PH, Miller J. (1977) Ann NY Acad Sci 386:29–42
Tsai MJ, Saunders G (1973) Proc Natl Acad Sci USA 70:2072–2076

Discussion of the Paper Presented by V.K. Moudgil

JAFFE: Can you estimate how much pyridoxal phosphate you have bound per molecule.

MOUDGIL: No, we haven't done any kinetics at this moment. The results I presented are only preliminary in nature, and we do not have information on the stoichiometry of the reaction.

JAFFE: Does the material in the control sample that is activated and does not bind to ATP–sepharose bind to nuclei?

MOUDGIL: No, it won't. The point that I wanted to make was that the binding of ATP to the receptor could be a measure of receptor activation. It has some advantages over the conventional methods. That is you don't need hormones. If you get any activation, you'll find ATP binding; if there is no activation, there is no binding to ATP, whereas in the case of nuclear binding you get a lot of nonspecific interaction. These are some advantages of using ATP–Sepharose binding as a measure of receptor activation.

STEVENS: You showed that the molybdate inhibited the activation of the receptor, but what you showed was the activation at 23°C. You didn't say anything about whether it would inhibit the salt activation of the receptor.

MOUDGIL: It would.

Discussants: R. JAFFE, V.K. MOUDGIL, and J. STEVENS

Glucocorticoids

Chapter 8

Molecular Mechanisms of Glucocorticoid Action: An Historical Perspective

Thomas D. Gelehrter

In this chapter I will try to provide a selective, personal account of one aspect of the development of the field of glucocorticoid hormone action. I will concentrate on the contributions of one individual, the late Gordon M. Tomkins (Fig. 1), to the development of our understanding of molecular mechanisms of glucocorticoid action.

Gordon Tomkins was the first to exploit tissue culture model systems for the study of glucocorticoid hormone action. In the mid-1960s, he and Brad Thompson at the National Institutes of Health established the HTC cell line, a line of rat hepatoma cells in tissue culture (Thompson et al. 1966). The use of cell culture provided a number of important advantages, including the ability to study a pure population of cells responsive to a given hormone, to manipulate the environment in a fashion not possible in intact animals, and to study the direct effects of individual hormones either in isolation or combination. Although the HTC cell line falls far short of becoming the *E. coli* of animal cells, it nonetheless has become (along with other similar hepatoma lines) a standard model for the study of glucocorticoid action (Gelehrter 1976; Thompson and Lippman 1974).

A second major contribution was Tomkins' emphasis on genetic approaches to understanding hormone action. Despite the success of genetic approaches in unraveling biological regulatory mechanisms in prokaryotes, surprisingly genetic approaches have been little utilized in the studies of hormone action (Gelehrter 1976; Swank et al. 1973). In recent years we have witnessed a significant increase in the number of genetic analyses of glucocorticoid action, not only in hepatoma cells but also in lymphoma cells (Seifert and Gelehrter 1979; Thompson et al. 1977; Bernhard 1976; Yamamoto et al. 1976).

A third major contribution was the introduction of innovative models to explain glucocorticoid hormone action (Garren et al. 1964; Tomkins et al.

Fig. 1. Gordon M. Tomkins (1926–1975). (Photo courtesy of Millicent Tomkins.)

1965; Tomkins 1975). While most of the scientific community readily and uncritically embraced models derived from prokaryotic systems to explain hormone action, Tomkins introduced a novel posttranscriptional repressor model to explain glucocorticoid action, and to explain the anomalous results obtained with the inhibitor actinomycin D (Garren et al. 1964; Tomkins et al. 1969). The value of the model has been its stimulation of a large number of experiments, which in turn have yielded interesting new insights into hormone action. Not only did such a model keep investigators honest, but it forced them to consider their data from new perspectives. Tomkins had that rare gift of being able to deal creatively with anomalous results. Rather than dismissing such anomalies in favor of data that supported generally accepted hypotheses, he was always ready to innovate and challenge current dogma with new hypotheses and models.

Gordon Tomkins' contribution to science, however, transcends his work on experimental systems, approaches, and models. Far more important was his personal magnetism, his ability to inspire his colleagues and particularly young scientists with his zest for science and for life. Several of us at this conference had the good fortune to work with Tomkins, and few others were untouched by his influence. As shown in the fine line drawing (Fig. 2) by Gordon's talented wife Millicent, music played a very important role in Tomkins' life. Among scientists there are a very few whose creativity, whose insight, and whose enthusiasm truly make music; Gordon Tomkins was such a man.

The senior authors of the next two chapters have made important contributions to the field of corticosteroid action. John Baxter is a product of the University of Kentucky and Yale University Medical School. Following postdoctoral training with Gordon Tomkins at the National Institutes of

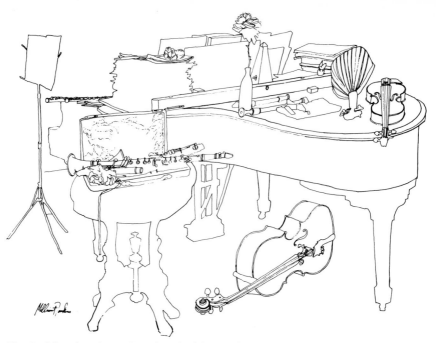

Fig. 2. Line drawing. (Reprinted with permission of the artist, Millicent Tomkins.)

Health he accompanied him to the University of California, San Francisco, where he is currently Professor of Internal Medicine. John has made major contributions to our understanding of the role of glucocorticoid receptors in hormone action. Recently he has turned his attention to the regulation of growth hormone synthesis by glucocorticoids and thyroxin, and, employing sophisticated techniques of recombinant DNA technology, has made major advances in our understanding of hormone action at the molecular level.

Brad Thompson, unlike some of the earlier speakers at this meeting, is a real Texan. A native of Houston and graduate of Rice University, he is a devotee of the stories of J. Frank Dobie and the paintings of Frank Remington. Following his medical training at Harvard, he too worked with Gordon Tomkins at the National Institutes of Health, where he was responsible for beginning the "famed" HTC cell line. Brad also was among the first to exploit genetic approaches for understanding hormone action, utilizing both HTC cells and recently human leukemic cells.

References

Bernhard HP (1976) Int Rev Cytol 46: 289–325
Garren LD, Howell RR, Tomkins GM (1964) Proc Natl Acad Sci USA 52: 1121–1129
Gelehrter TD (1976) N Engl J Med 294: 522–526, 589–595, 646–651
Seifert SC, Gelehrter TD (1979) J Cell Physiol 99: 333–342

Swank RT, Paigen K, Ganschow RE (1973) J Mol Biol 81: 225–243
Thompson EB, Tomkins GM, Curran JF (1966) Proc Natl Acad Sci USA 56: 296–303
Thompson EB, Lippman ME (1974) Metabolism 23: 159–202
Thompson EB, Aviv D, Lippman ME (1977) Endocrinology 100: 406–419
Tomkins GM, Gelehrter TD, Granner D, et al. (1969) Science 189: 1474–1480
Tomkins GM (1975) Science 189: 760–763
Yamamoto KR, Gehring U, Stamfer MR, Sibley CH (1976) Rec Prog Horm Act 32: 3–32

Chapter 9

Multiple Actions of Glucocorticoids Studied in Cell Culture Systems

E. Brad Thompson, Aniko Venetianer, Thomas D. Gelehrter, Gordon Hager, Darryl K. Granner, Michael R. Norman*, Thomas J. Schmidt, and Jeffrey M. Harmon

I. Introduction

Glucocorticoids produce differing effects in various cells both in vivo and in vitro. The wide spectrum of effects of these steroids baffled endocrinologists for some time until the unifying concept of steroid- and tissue-specific receptors provided a means of recognizing those cells likely to respond (Thompson and Lippman 1974; King and Mainwaring 1974; Yamamoto and Alberts 1976). This clearing in the mystery, however, has proved to be more valuable for the sex steroids and for cells selected for resistance to growth inhibition by glucocorticoids than it has for glucocorticoid-sensitive cells and tissues in general. In fact, the majority of cells and tissues tested have been found to contain receptors for glucocorticoids. Nevertheless, specific cellular responses to these steroids remain varied, and the problem of explaining the differences in response in different cell and tissue types remains. These responses are as widely disparate as induction of a limited number of peptides or suppression of a limited number of functions (Ivarie and O'Farrell 1978) and cell lysis and death (Claman 1972; Baxter and Harris 1975). At present it is not clear whether there is a single unifying mechanism that can account for such widely differing effects. It does seem that in the vast majority of the systems studied cytoplasmic receptors seem to be required. Beyond that, however, the mechanisms remain obscure (Makman et al. 1971; Munck 1971; Turnell and Burton 1975; Borthwick and Bell 1975; Munck and Leung 1977; Nicholson and Young 1978). In the case of several well-studied induced functions, application of glucocorticoids in concentrations that saturate receptors can account for the increase of accumulation of functional messenger RNA for the peptide in question (Ramanarayanan-Murthy et al. 1976; Martial et al. 1977; Hofer et al. 1978; Kurtz et al. 1978). The number of such examples is limited, however, and the molecular mechanisms by which the accumulation takes place have still not been

*Permanent address: Department of Chemical Pathology, King's College Hospital Medical School, London, England.

worked out. In the case of cells that are inhibited or even lysed by steroids the mechanism of action is even more obscure, although in these cases also, receptor occupancy by glucocorticoid correlates well with cell inhibition.

Some years ago we concluded that it would be of advantage to develop cultured cell systems in which to explore the actions of steroids, thus avoiding the many complications of experiments in whole animals. In this report we will describe our recent experiments involving three cell lines, L cells, HTC cells, and CEM cells. These three lines were chosen for investigation because no single cell line affords a model for all of the steroid-mediated effects, and because each of them offers a special advantage to the investigator. L cells are a line of transformed fibroblasts (Earle, 1943) known to contain receptors for several classes of steroids (Jung-Testas et al. 1976). These cells have long been employed as a model system both for the inhibitory effects of steroids on cell growth and macromolecular synthesis and for studies of the biochemistry of steroid receptors. The HTC cells comprise a line of rat hepatoma cells (Thompson et al. 1966), which has proved to be a useful model system for studies concerning hepatic enzyme induction by glucocorticoids (Thompson 1979). Finally we have utilized CEM cells as a model system for studies of steroidal effects in human lymphoblastic leukemia (Norman and Thompson 1977). The CEM cell line was originally derived from a female patient with acute lymphoblastic leukemia (Foley et al. 1965).

As will be seen, our approach with each of these cells has been to try to combine biochemical and cell biological methods to derive information about the mechanism of cellular responses to steroid hormones.

II. Steroid Effects in L-Cell Variants

L Cells were originally derived from the inguinal perimammary area of mouse skin by Earle (1943). They have been shown to be growth inhibited in the presence of glucocorticoids, and it has been shown that when L cells resistant to steroids were isolated, they often lacked the normal amount of active receptors for this class of steroids (Hackney et al. 1970; Lippman and Thompson 1974). In probing steroid action, however, one of the more difficult problems has been to identify the exact steps by which steroids act after their binding to the receptor and after the steroid receptor complex has translocated to the nucleus of the sensitive cell. There have been sporadic reports of receptor–positive cells that were, nevertheless, insensitive to steroids (Levisohn and Thompson 1972; Lippman et al. 1974a; Breslow et al. 1978), and indeed in studies of various animal and human malignancies many examples of receptor–positive, hormonally insensitive cells have been found (McPartland et al. 1977; Thompson et al. 1977a; Konior-Yarbro et al. 1977). We therefore were interested in trying to develop or identify steroid-resistant cells that contain normal glucocorticoid receptors in hopes thereby of obtaining cells that could be used as cell genetic and biochemical reagents

Table 1. Characteristics of Binding of [³H]Dexamethasone in Sensitive and Resistant L Cells[a]

Clone	Whole cell uptake — Fraction of [³H]dexamethasone localized in crude nuclear pellet:[b] $\dfrac{\text{Nuclear [³H]dexamethasone}}{\text{Nuclear + cytoplasmic [³H]dexamethasone}}$	Number of experiments	Cell-free binding — Dissociation constant[c] ($M \times 10^{-9}$)	Concentration of receptor sites[c] (pmol/mg)
A₉HT (parent)	0.39	5	2.0	0.49
11	0.41	4	1.8	0.53
15	0.52	6	N.D.[d]	0.45
21	0.37	4	N.D.	0.54
25	0.49	7	1.8	0.54
30	0.40	4	0.6	0.36
SLB82 − 17R⁺	0.46	4	1.7	0.6

[a] Source: Venetianer A, Bajnoczky K, Gal A, Thompson EB (1978) Som Cell Genet 4: 513–530.

[b] Cells were grown to high density, removed by trypsin–EDTA, and washed three times with phosphate buffered saline. The distribution of specifically bound [³H]dexamethasone in the crude nuclear pellet and crude cytosol was determined according to Sibley and Tomkins (1974).

[c] The specific binding of [³H]dexamethasone to receptors in particle free cell extracts was determined using a dextran-coated charcoal competitive binding assay. Values given are the average of 3–5 experiments. K_d values were calculated from the Scatchard plots of the binding data (one or two experiments) using increasing concentrations of dexamethasone from 0.4 nM to 100 nM.

[d] N.D.: Experiment was not done.

to explore the steps beyond the formation of the steroid receptor complex. One of the systems in which encouraging results have been obtained is L cells (Venetianer et al., 1978).

A large number of steroid resistant L cells were isolated in the following way. Cells were exposed to the mutagens N-methyl-N'-nitro-N-nitroso-guanidine (MNNG) or ethylmethane sulfonate (EMS). They were then washed free of mutagen, placed in growth medium for 96 h, and then plated in 10^{-6} M dexamethasone phosphate in growth medium. Cell concentrations were employed that allowed evaluation of individual colonies arising in the tissue culture dishes. After about 10–12 days, rapidly growing colonies were picked from the dishes, and the clones were expanded and screened for the presence of glucocorticoid receptors. Examples of receptor-containing clones obtained from both EMS- and MMNG-treated cells were taken for analysis. The resulting colonies of steroid-resistant cells then were compared with steroid sensitive parents for several properties: plating efficiency, inhibition of mitosis, inhibition of thymidine uptake, and colony size. In each case the resistant colonies were found to show reduced sensitivity to the inhibitory effects of glucocorticoid. Although their ability to clone was completely insensitive to the steroid, they showed some variability in their sensitivity with respect to the other parameters. The stability of the clones was demonstrated by dividing each after it had been grown continuously for 6 months in medium containing micromolar dexamethasone phosphate and then continuing growth in medium that either contained the steroid or lacked it. After six months in the absence of the steroid, the colonies were retested and still showed only slight alteration from their original steroid resistance.

The initial screening for steroid receptors had shown that the clones described here contained binding sites for the hormone. They were therefore chosen for more detailed analysis, since they potentially represented cells with the phenotype of interest. Table 1 describes the characteristics of binding of tritiated dexamethasone in the resistant L cells and in their sensitive parent. Two methods of analysis were used: (1) the whole cell uptake technique, which allows one to determine the fraction of steroid–receptor complex found in the nuclear and cytoplasmic cell fractions; (2) cell-free assay of the glucocorticoid receptor activity in the soluble cytoplasmic fraction of cell extracts. As the data show, none of the six clones chosen for analysis differs substantially from the parental line with respect to either the fraction of dexamethasone capable of nuclear transfer in the whole cell assay or the concentration of receptor sites found in the cytoplasmic assay. In addition, from Scatchard analysis of the cell-free binding data, the dissociation constant of the steroid bound shows that in the four clones analyzed, the affinity of steroid for its receptor sites in the cytoplasm is the same as that of the wild type (Fig. 1). Thus, these glucocorticoid-resistant fibroblasts appear to contain receptors that by standard criteria are normal. Analysis of their biochemistry and physiology has just begun. However, one essential experiment in distinguishing among various possibilities to ex-

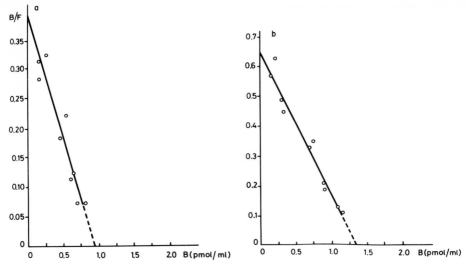

Fig. 1 a,b. Scatchard plot of binding data for [³H]dexamethasone in cytoplasmic extracts of (**a**) A₉HT cells and (**b**) clone 25. Binding was determined using dextran-coated charcoal competitive binding assay. B/F, bound/free; F, free, in pmol/ml. (Venetianer A, Bajnoczky K, Gal A, Thompson EB (1978) Som Cell Genet 4: 515–530)

plain their properties has been carried out. This employs the methods of somatic cell hybridization. In this technique different cells are chosen that contain mutually complimentary functions so that hybrids between them can be chosen in selective medium. By hybridizing cells of a steroid-responsive phenotype with cells of a nonresponsive phenotype, one can carry out a simple but basic test for the dominant or recessive properties of the phenotype in question. Accordingly, hybrids were made between sensitive parents and the receptor positive insensitive clones. The results of this experiment are shown in Table 2. As can be seen, the hybrids between the sensitive parent A₉HT and the resistant clone SLB82-17R⁺ or between parent LB82 and resistant clone 25 were sensitive to the effects of dexamethasone on mitosis and thymidine uptake. Thus, it seems clear that the mechanism by which these cells have achieved dexamethasone resistance is not one involving a transdominant inhibitor of the action of the steroid.

A quite unexpected result that was obtained in further analysis of these hybrids was that of overlapping resistance of the cells to the action of non-glucocorticoidal steroids. Table 3 shows the effect of various steroids on tritiated thymidine uptake, expressed as percentage of inhibition. The parent clonal A₉HT, as can be seen, is sensitive to a variety of steroids. It is most sensitive to steroids that are potent glucocorticoids: dexamethasone, prednisolone, corticosterone, aldosterone (which at these concentrations has glucocorticoid activity), hydrocortisone, and cortisone. It is also somewhat sensitive to inhibition by progesterone and 17 β-estradiol. Examining the dexamethasone resistant clones 15, 11, and 25, in particular, one can see

Table 2. Glucocorticoid Responses in Sensitive (Receptor⁺) × Resistant (Receptor⁺) L Cell Hybrids[a]

| | Chromosomes | | | Inhibition by 10^{-6} M dexamethasone | | Receptors[d] |
	Mean	Range	Age[b]	Mitosis[c] inhibition (%)	[³H]Thymidine uptake[c] inhibition (%)	
PARENTS						
Sensitive						
LB82	55			41	N.D.[e]	+
A₉HT	51	46–55		34	48	+
Resistant						
25 (selected from A₉HT)	51	47–54		No inhibition	14	+
SLB82 – 17R⁺	52	50–54		11	N.D.	+
HYBRIDS						
A₉HT × SLB82 – 17R⁺						
Clone 5	91	79–106	75	20	45.5	N.D.
LB82 × 25						
Clone 2	91	77–107	67	28	29	+
Clone 3	90.5	78–101	67	32	36	+
Clone 4	93	83–98	68	30.5	37	N.D.

a Source: Venetianer A, Bajnoczky K, Gal A, Thompson EB (1978) Som Cell Genet 4: 513–530.
b Age indicates days after fusion before chromosomes were counted.
c Experiments were done 2–3 months after cell hybridization. Results represent the average of two independent experiments.
d The presence of cytoplasmic glucocorticoid receptors was determined by cell-free competitive binding assay.
e N.D.: experiment not done.

Table 3. The Effect of Various Steroids on [^{3}H]Thymidine Incorporation (% inhibition)[a,b]

Clone	Dexa-methasone	Prednis-olone	Corticos-terone	Aldos-terone	Hydro-cortisone	Cortisone	Proges-terone	17β-Estradiol	17-α-Methyl-testosterone
A$_9$HT	48	53	52	50	42	37	23	19	8
11	29	27	25	25	16	12	11	8	0
15	36	27	28	8	24	20	3	0	13
21	37	2	9	0	3	0	0	0	0
25	14	0	19	10	6	6	3	2	0
30	26	0	15	13	7	10	22	12	28

[a] Each value represents the average of six independent experiments. For details see Venetianer et al. (1978).
[b] All the steroids were used at a concentration of 1×10^{-6} M.

that whereas there is some loss of sensitivity to dexamethasone by this biochemical parameter, there is also marked loss of sensitivity to both the progesterone and the estradiol effects. Clone 21 is remarkable in that it is still relatively sensitive to thymidine inhibition by dexamethasone but is quite insensitive to inhibition by all the remaining steroids tested. The overlapping effect of selection for glucocorticoid resistance with progesterone resistance is somewhat ambiguous since progesterone can bind glucocorticoid receptors. However, the loss of sensitivity to the estrogen is remarkable, suggesting that the cells have developed some method of resistance involving a general steroid-sensitive step, one that recognizes both estrogen receptor complexes and glucocorticoid receptor complexes. The extremely high concentration of estrogen used in these experiments, $10^{-6} M$, far exceeds that necessary to saturate estrogen receptors, and the possibility that at such high pharmacologic levels there is some binding of estrogen to the glucocorticoid receptor needs to be tested.

These receptor-positive clones of steroid-resistant L cells therefore represent the first examples in a fibroblast-derived cell line of steroid-resistant cells containing apparently normal receptors. They therefore appear to offer opportunities for exploration of the later steps in the inhibitory actions of glucocorticoids. The inhibitory effect of estradiol in these cells and the unexpected estradiol cross resistance between cells selected for dexamethasone resistance needs to be explored further. If the estradiol inhibitory effect is to be attributed to partial occupancy of the glucocorticoid receptor, a paradox is immediately apparent, namely, that the quantity of glucocorticoid receptor and affinity of the receptor for dexamethasone is unaltered in these cells. Thus, it seems more likely that some as yet undefined step accounts for the double resistance conferred upon cells selected in high concentration of glucocorticoid.

III. HTC Cell Variants

Rat hepatomas and cultured lines derived from them have been widely used for studies of glucocorticoid effects both in vivo and in vitro (Potter et al. 1967; Pariza et al. 1976). It is known that different hepatomas and different lines of hepatoma cells in culture have different phenotypes with respect to their responses to glucocorticoid hormones. Because it has been difficult to devise selective pressures, however, direct isolation of variants for steroid responsiveness from sensitive lines, as with L cells or lymphoid cells, has proved difficult. Thus, classic somatic cell genetic experiments have not been applicable to these inducible systems. The HTC cells are one particular line of rat hepatoma derived from Morris hepatoma 7288C, a hepatoma induced chemically in the inbred Buffalo rat strain (Odashima and Morris 1966). We isolated the line known as HTC cells in 1964 (Thompson et al. 1966), and it has proved to be of considerable value in the investigation of the mechanism of steroid hormone induction of enzymes in general and of tyrosine aminotransferase in particular (Thompson 1979). These cells are known to contain glucocorticoid receptors that follow the classic pattern, and it has been shown that occupancy of these receptors by steroid is coordinate with the induction of the enzyme (Baxter and Tomkins 1970; Baxter et al. 1972). The ability of various glucocorticoid analogs to induce the enzyme corresponds to their ability to bind to the receptor (Samuels and Tomkins 1970). As in other systems it appears that nuclear translocation is an essential element for the steroid receptor complex to induce the enzyme (Baxter et al. 1972). Induction of the enzyme results from increased enzyme synthesis (Granner et al. 1970) and an increase in its translatable mRNA (Hofer et al. 1978; P. Olsen and D.K. Granner unpublished observations 1979). Furthermore, various studies over the years have demonstrated a variety of other glucocorticoid specific responses in these cells. An estimate of the limit of the changes in HTC cell peptides following these steroids was provided recently by studies in which two-dimensional gel electrophoresis was employed to display, as nearly as possible with current technology, all of the alterations in peptides following these hormones (Ivarie and O'Farrell 1978). This method showed that following steroid treatment perhaps there were a dozen or so changes in the cytoplasmic peptides.

Specific glucocorticoid-sensitive events in these cells include the following: induction of tyrosine aminotransferase by direct of application of steroids as given above, additional induction of this enzyme by N^6, $0^{2'}$-dibutyryl adenosine $3':5'$-monophosphate (dibutyryl cyclic AMP) following pretreatment of the cells with glucocorticoid (Granner 1976; Granner et al. 1977), induction of phenylalanyl tRNA (Yang et al. 1974; Lippman et al. 1974b), induction of glutamine synthetase (Kulka et al. 1972; Crook et al. 1978), induction of mouse mammary tumor virus after the cells have been infected by the virus (Ringold et al. 1975, 1978; Yamamoto et al. 1978), suppression of plasminogen activator synthesis and release by induction of an inhibitor (Carlson and Gelehrter 1977: Seifert and Gelehrter 1978), suppression of cyclic AMP phosphodiesterase activity (Manganiello and Vaughan, 1972), inhibition of α-aminoisobutyric acid uptake (AIB) (McDonald and Gelehrter 1977; Kelly et al., 1978), and induction of a cell surface factor that increases cell adhesion (Ballard and Tomkins 1969). Thus, HTC cells represent a well studied example of hepatoma in which in theory it would be of great interest to examine the coordinate control of a number of these steroid mediated responses. This has recently been made possible by our isolation through direct cloning, in the absence of known selective pressure, of HTC cells resistant to glucocorticoids with respect to tyrosine aminotransferase induction.

Such clones, we found, arose rarely in the wild-type population of HTC cells and in fact had to be derived by serial subcloning and screening of the original wild-type cells under conditions which had been described previously (Aviv and Thompson 1972; Thompson et al. 1977a). The tyrosine aminotransferase noninducible clones were found to be quite stable with respect to their altered phenotype. They contain basal enzyme that is enzymatically and immunologically active, and they show no evidence for carrying a tyrosine aminotransferase peptide altered in its enzymic activity but still antigenically active. Thus, as with the wild-type cells, the amount of enzymic activity appeared to be a clear reflection of the actual quantity of enzyme protein in the cells. It was therefore concluded that they possessed the gene for the enzyme in an unaltered form and were producing structurally normal tyrosine aminotransferase. The rate of decay of the enzyme appeared to be normal, since if cycloheximide was added to the cells, the loss of enzymic activity from the variant clones appeared to proceed at the same rate as did that of wild-type clones. A variety of steroids in high concentration were applied to the cells, and no evidence for steroid responsiveness was found with any glucocorticoid, including those even more potent than dexamethasone. Kinetic studies of enzyme induction showed that there was no fleeting induction of the enzyme followed by rapid deinduction nor was there delayed induction. The cells could be grown in a variety of media, stored frozen, and thawed without alteration of this phenotype. These facts have been presented in detail (Thompson et al. 1977a).

Of immediate interest of course was the question as to the steroid receptor content of these cells. They were duly analyzed for the quantity of glucocorticoid receptor that they possessed as well as for its ability to undergo translocation to the nucleus after application of steroid at 37°C in whole cells. Data from a representative selection of several of the noninducible clones are shown in Fig. 2; all show receptor quantity in the cytoplasm as high as or greater than that of the inducible clone. Nuclear transfer experi-

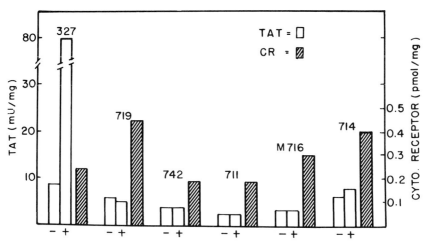

Fig. 2. Comparison of tyrosine aminotransferase (TAT) levels, basal (−) and induced by an 18-h exposure to 10^{-6} M dexamethasone (+), with cytosol steroid–receptor content (CR, hatched bars). The numbers over the sets of bars refer to the clone. 327 is the inducible clone HTC-327; the others are low- to noninducing clones. Enzyme activity is expressed as mu ($\equiv$nmol product formed per min) per mg protein. (From Thompson, E.B., Aviv, D, and Lippman, M.E. (1977) Endocrinol 100: 406–419

ments in whole cells showed that under standard conditions the quantity of receptor steroid complex in the nucleus of the cells was equivalent to that of control cells with the exception of one clone. That clone, which showed partial induction of tyrosine aminotransferase, appeared to show reduced nuclear transport but has not been explored in further detail. Thus, we concluded that these cells, for some reason other than a lesion in their overall receptor for steroid, had acquired the phenotype of noninducibility. They therefore represented a unique opportunity to study, in subclones from a single cell line, questions of the nature of control of multiple functions under glucocorticoid control.

The questions to be asked were the following. Was the loss of tyrosine aminotransferase induction an independent event, or was it associated with loss of other steroid responses as well? If so, was there loss of all steroid-sensitive functions in these cells; that is, was a single postreceptor control function altered, resulting in total loss of responsiveness? If only some responses were lost, was the pattern consistent or variable?

We chose for examination six steroid-sensitive functions inherent to the HTC cell plus a seventh, which could be introduced into the HTC cell. The three inherent inducible functions were induction of tyrosine aminotransferase by glucocorticoids, the reinduction of tyrosine aminotransferase by cyclic nucleotides following pretreatment with glucocorticoids, and the induction of glutamine synthetase. The three inherent inhibited functions were inhibition of uptake of AIB, inhibition of plasminogen activator synthesis and release, and inhibition of cyclic AMP phosphodiesterase. The "added" function was that of the induction of mouse mammary tumor virus in cells deliberately infected by the virus. When several clones were tested for these seven responses, a specific and constant pattern was quickly found. Some responses were lost while others remained intact (Thompson et al. 1979).

A. Inducible Responses in HTC Variants

Among the inductive responses, tyrosine aminotransferase induction by cyclic nucleotides either in the absence or presence of glucocorticoids was uniformly missing. A summary of a typical experiment showing this result is given in Table 4. In all of the four clones tested, a variety of other experiments produced the same result, that this tyrosine aminotransferase was unaffected by treating the cells with dibutyryl cAMP, with or without prior exposure to glucocorticoids. Expression of this enzyme was suppressed to low basal levels in these variants, despite hormonal treatments. Next tested were other inducible functions.

Glutamine synthetase is induced by glucocorticoids in wild-type HTC cells. Data have been elicited suggesting that glutamine synthetase and tyrosine aminotransferase induction are closely linked. These data result from experiments infecting wild-type HTC cells with mouse mammary tumor virus; after such infection it has been noted that often the inducibility

Table 4. Nonresponse of Tyrosine Aminotransferase to Cyclic Nucleotide in HTC Variants[a]

	Control	DBC	DEX	DEX + DBC
Variant clone				
M714I	1.0	1.2	1.2	1.4
M714H	1.75	1.5	1.2	1.55
719C	1.6	1.8	2.1	2.1
268E	1.05	1.0	2.1	2.3
Wild-type clone				
921	12.9	17.3	40.8	66.2

[a] Example of an experiment testing "reinduction" of tyrosine aminotransferase. Cells in monolayer culture were treated with 5×10^{-7} M dexamethasone (DEX) for 2 days, with 10^{-3} M dibutyryl cyclic AMP for 4 h (DBC), or with 5×10^{-7} M dexamethasone for 2 days plus 10^{-3} M dibutyryl cyclic AMP for 4 h (DEX plus DBC). The monolayers then were harvested and assayed for tyrosine aminotransferase. Results shown are enzyme specific activity (nmol product/min/mg protein), averages of duplicate samples.

of this pair of enzymes was affected in the infected cells. In every case described, in those cells in which the induction of one enzyme was altered, both enzymes were affected in the same way, either increased or decreased (Yamamoto et al. 1978). In the spontaneous variants studied here, however, glutamine synthetase was always inducible, showing clearly that the control of the two enzymes need not be linked. In addition, we infected wild-type HTC cells and several clones of the variants with mammary tumor virus, and then compared the inducibility in each clone and in the average of all the clones with and without virus infection. We found that in the wild-type cells infection with mouse mammary tumor virus had no effect on the inducibilty of either glutamine synthetase or tyrosine aminotransferase, although infection with the virus did seem to lower the average basal activity of the enzymes.

The induction of glutamine synthetase in the variants was unaffected by the presence of the virus, just as in the wild type. In most cases the lack of induction of tyrosine aminotransferase in the variants was also unaffected by virus. The failure of viral infection to alter the synthetase and the aminotransferase in most wild-type and variant cells appears to argue against tight linkage of control for the two enzymes and against the view that mouse mammary tumor virus either carries or associates with a DNA region especially involved in glucocorticoid control of other functions.

In one of 32 virus-infected variant subclones, however, inducibility of the latter enzyme returned, although the absolute levels of the enzyme remained depressed (Table 5). Whether this clone represents an occasional spontaneous "revertant" or a direct effect of viral infection is not certain as yet. Favoring the latter possibility, however, is the evidence that the noninducible clones have remained stable in culture over several years under nonselective conditions. If there were a high rate of spontaneous return of inducibility, one might expect the inducible phenotype to return in the mass culture; yet it has not done so during several years' growth. Therefore, this

Table 5. Tyrosine Aminotransferase in MTV-Infected Subclones of HTC Cell Variants[a,b]

Subclones of	Basal	+DEX	Fold
M714(6)	1.8 ± 1.6	1.8 ± 1.7	0.9 ± 0.1
719C(10)	2.2 ± 2	1.8 ± 1.6	0.9 ± 0.3
268E(6)	0.6 ± 0.6	0.75 ± 0.4	1.6 ± 0.6
Subclone 268E$_{38}$[c]	1.2	8.5	7

[a] Preliminary screening data on 23 MTV-infected subclones from three different noninducible HTC variant clones. In most cases, the subclones were noninducible, and these data are presented as averages ±SD. The average fold induction was calculated from the fold induction for each clone and averaged separately. Number of subclones tests in parentheses. Tyrosine aminotransferase specific activity in control cells = "Basal," in cells treated with 10^{-6} M dexamethasone for 18 h = "+DEX." The large standard deviations are due to variation in enzyme levels from subclone to subclone.
[b] Further data may be found in Thompson et al. (1979).
[c] The single inducible MTV-infected subclone 268E$_{38}$ is shown separately. It has been tested repeatedly and appears to be stably inducible at about the enzyme levels shown.

clone is a candidate for an effect of viral infection on tyrosine aminotransferase induction.

The reverse effect, that of the noninducible tyrosine aminotransferase phenotype on virus induction, was also tested in the infected variants. In the limited number of examples tested so far, the virus itself was inducible in the variant clones. This result, however, must be interpreted with caution. Since the method of infection probably results in multiple copies of the virus being integrated in the cell, it could be possible that some viral genes were rendered uninducible by the same lesion that caused the cell to be noninducible for tyrosine aminotransferase, while other copies of the virus in other loci were fully inducible, thus masking the silent copies.

B. Repressible Responses in HTC Variants

The endogenous inhibited responses were studied next. Two of the three were unaffected in the tyrosine aminotransferase variants. Inhibition of AIB uptake and inhibition of plasminogen activator proceeded as in the normal wild-type HTC cells. Phosphodiesterase, however, was found to be unresponsive to glucocorticoids. The suppression of phosphodiesterase in wild-type HTC cells had only been briefly described previously. Therefore, we undertook several studies to examine the time course and dose response of the effect following both the low K_m and high K_m forms of the enzyme. In addition the basic enzyme kinetics in wild-type HTC cells were examined. Over a wide variety of concentrations and time points the variant HTC cells were found to be unresponsive to glucocorticoids with respect to suppression of phosphodiesterase activity. An example of such an experiment is shown in Table 6. As is shown, a wild-type clone shows suppression of the enzyme, whereas none of the variant clones tested suppress. Concentrations of dexamethasone as high as 12 μM were employed with the same result. The variants also express a higher level of the enzyme generally. Most interestingly clone 921, which was arbitrarily chosen as "wild type"

Table 6. Failure of Dexamethasone to Suppress Cyclic Nucleotide Phospho-diesterase in Several HTC Variants, Expressed in Terms of cAMP PDE Activity[a]

Clone	−DEX	+DEX
E-11–18	145	67
268E	1051	1108
263E	451	447
M714H	840	1051
719C	1013	1228

[a] Duplicate plates of cells were treated with vehicle only (−DEX) or with 10^{-7} M dexamethasone (+DEX) for 18 h, harvested and assayed for enzyme activity as described in Thompson et al. (1979). Shown here are results from cAMP PDE specific activity (pmol/min^{-1} mg protein^{-1}) at 5×10^{-6} M cAMP concentration. Similar results were obtained using 1.5×10^{-4} M cAMP as substrate. Results are the averages from duplicate plates of cells; individual samples never exceeded the average by $>10\%$.

with respect to tyrosine aminotransferase induction, also contained steroid-insensitive phosphodiesterase. Furthermore, this clone also proved to have steroid-insensitive plasminogen activator activity. Clone 921 was derived in parallel with the aminotransferase variants and was not their direct parent. It therefore represents an independent variant and proves that there is no tight linkage between the aminotransferase and phosphodiesterase phenotypes.

These interesting cell variants have therefore supplied us with a number of important results concerning glucocorticoid action in these cells. A summary of their responsiveness is given in Table 7. The aminotransferase variant cells contain low basal tyrosine aminotransferase, which is not inducible by glucocorticoids, nor by dibutyryl cyclic AMP following glucocorticoids. They contain normally inducible glutamine synthetase, and when they have been infected by mammary tumor virus, the virus becomes an inducible function. Their high levels of plasminogen activator and their uptake of α-aminoisobutyric acid are suppressed by steroids as in wild type. However, their phosphodiesterase activity is constitutive. Clone 921, on the other hand, contains steroid-sensitive aminotransferase but steroid-insensitive phosphodiesterase and plasminogen activator. From these results, one may conclude that in all probability there are at least some functional glucocorticoid receptors in these cells. It is possible that total receptor binding of steroid has remained normal, and total nuclear transfer has remained normal, but that a specialized subset of receptors specifically responsible for tyrosine aminotransferase induction and phosphodiesterase suppression or plasminogen activator suppression are no longer performing properly. It is also possible that putative acceptors for steroid receptor complexes, molecules that have been proposed as responsible in some way for the proper association of steroid receptor complex with specific regions of the genome (Spelsberg et al. 1972; Puca et al. 1974; Webster et al. 1976; Mainwaring et al. 1976; Simons et al. 1976), exist in multiple classes and in the appropriate variants, that the class(es) responsible for the steroid responses of tyrosine aminotransferase, phosphodiesterase, and plasminogen activator have been

Table 7. Responses to Glucocorticoids by Variant HTC Cells[a]

Cell	TAT				GS		PA		PDE		AIB: Uptake +steroid
	Basal	+Steroid	+dBcAMP	+Steroid +dBcAMP	Basal	+Steroid	Basal	+Steroid	Basal	+Steroid	
Wild-type	+	↑	±↑	↑↑	+	↑	+	→	+	→	→
Variants											
268E	+	No↑	No↑	No↑	+	↑	+	→	+	No→	→
M714	+	No↑	No↑	No↑	+	↑	+	→	+	No→	→
719C	+	No↑	No↑	No↑	+	↑	+	No↓	+	No→	→
921	+	↑	↑	↑↑	+	↑	+		+	No→	
MTV-infected variant subclones											
M714H 719C 268E* }	+	No↑	Not done	Not done	+	↑	Not done		Not done		Not done
* Exception 268E₃₈	+	↑	Not done	Not done	+	↑	Not done		Not done		Not done

[a] Table taken in part from Thompson et al. (1979).

lost or rendered functionally inoperable. Or it may be that specific control of the various corticosteroid responses is exerted neither at the level of receptor nor at the level of the putative acceptor, but at the level of the genes responsive to the hormone receptor complex. This would require that alterations in subtle gene structure or in proteins masking certain areas of the genome render the key sites for specific inductions or suppressions unresponsive to steroid receptor complex. In any case, these cells support the view that response to glucocorticoid hormones is controlled in a noncoordinate fashion and that one aspect of the response in the same cell type can be lost without another. Clearly the exploration of the cell biology and biochemistry of these variants and others like them offers considerable promise in analysis of glucocorticoid effects.

IV. CEM Cells: A Tissue Culture Model for Glucocorticoid-Sensitive Human Lymphoblastic Leukemia

Much of the pioneering cell genetics concerning the mechanism of action of glucocorticoids has come from mouse lymphoid cells, and they, together with normal rodent thymocytes, have been used extensively to study the biochemistry of steroid action. In these cells it has been known for decades that steroids cause cell lysis (Dougherty and White, 1945). By selecting steroid-resistant clones of mouse lymphoid lines it has been shown that the chief lesion found in the resistant survivors is loss of functional receptors (Yamamoto et al. 1976; Pfahl et al. 1978; Bourgeois and Newby 1977). Furthermore, the occupancy of receptor and movement of steroid receptor complex to the nucleus, as with induced systems, has been shown in the thymocyte system to correlate closely, although not in all cases perfectly, with the biological response of cell lysis and a variety of biochemical parameters inhibited by the steroids as well (Makman et al. 1967, 1968; Mosher et al. 1971; Munck and Wira 1971; Munck et al. 1972). The functions inhibited include glucose uptake, thymidine uptake, thymidine incorporation into DNA, uridine uptake and (to a much smaller extent) uridine incorporation into RNA, and amino acid incorporation and uptake. The intensively studied mouse lymphoma line S49 was shown to spontaneously develop resistance to glucocorticoids at a rate of 3×10^{-5} (Sibley and Tomkins 1974; Yamamoto et al. 1976; Bourgeois et al. 1978). Later it was pointed out that this is indeed a high rate for a diploid gene locus if one assumes the "usual" spontaneous rate for a single gene of one event per million per generation. When a second cell line, W7, was explored, it was found to contain initially twice the number of receptors per cell as did S49 and to yield spontaneous resistant cells extremely rarely ($<1.6 \times 10^{-9}$) when selected in dexamethasone at 10^{-6} to $10^{-5} M$. However, when stepwise selection was carried out at an intermediate level of dexamethasone followed by a high level ($10^{-5} M$) of the steroid, variants were obtained. It was found that at the intermediate level of steroid, cell killing proceeded efficiently, resulting in the survival of resistant clones occurring at a frequency of up to 2×10^{-6}. These clones for the most part contained approximately half the receptors as did wild-type W7. When the partially resistant clones, in turn, were subjected to high-level dexamethasone, fully resistant clones were selected, occurring at least as often as had the initial semiresistant clones (frequencies of up to 1.5×10^{-5}). The fully resistant clones were found to have lost most or all of their remaining receptors. From these results it was argued that W7 represented a cell line diploid for the receptor gene and that S49 was haploid for this gene (Bourgeois and Newby 1977). Attempts to reconstruct cells containing various levels of receptor showed that it was difficult to predict the degree of receptors per cell in the hybrids from the sum of the receptors in the parents. The degree of steroid sensitivity of the hybrids was also variable, though all were sensitive to some extent (Yamamoto et al. 1976; Gehring and Thompson, 1979). In a large number of sensitive (receptor+) by resistant (receptor−) lymphoid cell crosses, the resulting hybrids have been found to be steroid sensitive (Gehring and Thompson, 1979).

Thus, in the mouse systems the conclusion has been that for the vast majority of steroid-resistant cells loss of functional receptors has been the mechanism by which resistance has been acquired. Whether this is always, often, or even at all due to mutations in the structural gene(s) for receptors has not been shown unequivocally. It is possible that spontaneous stable alterations at other loci also could result in loss of receptor function. Final proof must await better methods of identifying the actual glucocorticoid receptors peptide and their structural genes.

Several years ago, we decided to examine human leukemias for glucocorticoid receptors. Many, but not all of these diseases are exquisitely corticosteroid sensitive (Claman, 1972; Goldin et al., 1971). Among those that do respond to steroid treatment, resistance to the hormones often develops when these compounds are used as the sole therapy (Vietti et al. 1965; Leikin et al 1968). Our initial studies showed that in childhood acute lymphoblastic leukemia, receptors are to be found in the leukemic blasts, occupancy of these receptors correlates with cell inhibitory effects of the hormones, and that in all of the previously untreated patients studied the ALL blasts were receptor positive and steroid sensitive. After repeated remissions on combined chemotherapy, a few patients became clinically resistant. In vitro their blasts were steroid resistant, and their cytoplasm contained few or no cytoplasmic receptors for glucocorticoids (Lippman et al. 1973). Many such surveys of both normal and leukemic human cells have since been carried out. Their conclusions have been that while in some instances steroid resistance in the human occurs owing to simple loss of receptor function, in most cases studied that is not so in normal and leukemic human lymphoid and myeloid cells. The presence of receptor (measured by competable steroid binding) and its ability to undergo nuclear translocation does not necessarily predict that those cells will be steroid sensitive (Crabtree et al. 1979). In some leukemias also, the quantity of receptor sites per cell does not correlate well with cell sensitivity to steroids (see discussion and references in Schmidt and Thompson, 1978; Bell, 1979). On the other hand, in acute lymphoblastic leukemia the quantity of cellular glucocorticoid receptors seems to be of important prognostic value. The lymphoblasts of null cell ALL, a disease known to be more responsive to therapy and to show a more favorable prognosis than other forms of the illness, were shown to contain on the average much higher numbers of glucocorticoid receptors per cell than T cell leukemic lymphoblasts. The T cell form of ALL is known to be more toxic and to have a less favorable prognosis. Furthermore, within the null and T cell classes of leukemia, the receptor levels found in an individual's blasts seem to have predictive value for the course of the illness in the individual concerned (Konior-Yarbro et al. 1977).

In human cells, therefore, it seems that the correlation of steroid resistance with steroid-receptor content is more complicated than that seen in the mouse lymphoid model systems, where the common phenotype of resistance was R⁻. The problem in human leukemias is one of finding a test that clearly identifies functional receptors as opposed to those detected by the rather gross tests of steroid binding and overall nuclear transfer. Perhaps whole-cell-binding studies measure spurious sites. Obviously, the mechanisms of steroid resistance in human lymphoid cells remain to be found. Such questions as rates at which resistance is acquired in sensitive populations are unanswered. For these reasons it seemed important to begin to develop model systems that corresponded more directly to the state of affairs in humans, namely, cultured human leukemic cells. We have identified a line of such cells, and in the remainder of this chapter will discuss their properties.

By screening a number of human leukemic T and B cell lines provided by Dr. Dean Mann of the National Cancer Institute, we were able to identify a cell line that arose from the peripheral lymphoblasts of a female patient with ALL, a cell line that appeared to offer many of the properties one would wish for in a model system of human ALL. The CEM cells in the uncloned state were found to be growth inhibited by glucocorticoids. Cloning without selection of this wild-type line at once gave rise to fully sensitive and fully resistant clones. Several of the steroid-sensitive clones were picked for further examination. It was first necessary to demonstrate that these cells responded to steroids in a manner similar to the animal model systems, in order to apply the rationale for steroid action developed in those systems. Accordingly, we established the basic facts of steroid specificity and dose dependence in CEM cells for inhibition of macromolecular synthesis, growth, and cloning (Norman and Thompson 1977; Thompson et al. 1977b). We then examined the cells for the presence of glucocorticoid receptors, and found highly specific receptor sites that appeared to undergo nuclear transfer and exhibited the usual properties of high specificity and limited capacity for the appropriate steroid. The average number of receptor sites per cell was found to be approximately 20 000 at saturating levels of dexamethasone. Nuclear transfer estimated by the technique used in our laboratory was 40% at 37°C after 1-h incubation of whole cells with radiolabeled steroid. Occupancy of the receptors corresponded with inhibition of biochemical and growth parameters. By direct microscopic examination both at the light and electron microscope level, we found that CEM cells do not show progressive damage soon after the addition of steroid. They appear relatively normal for 18–24 h after hormone is added and then begin to be lysed rapidly in a process that appears to affect the population on a probability basis. This loss of cell viability could be quantitated most accurately by cell cloning experiments. The dose response curve for loss of cloning ability in the presence of steroid is shown in Fig. 3.

These results led us to wonder whether the lethal effects of steroids on these cells were in some way cell cycle specific. Many, though not all, treatments that stop cell growth cause cessation of growth in the G_1 phase of the cell cycle. Such treatments include withdrawal of essential amino acids, addition of cyclic nucleotides to S49 lymphoma cells, omitting serum from the medium of serum-dependent cells, etc. Also, in human leukemic patients treated with steroids, indirect measurements had suggested that the leukemic blasts were blocked in G_1 (Lampkin et al. 1969, 1971).

It seemed both of basic biologic interest and of potential therapeutic concern therefore to find whether a similar effect took place in CEM cells. Indeed, we found that cells did accumulate in the G_1 phase of the cell cycle following steroid (Harmon et al 1979). By use of flow microfluorometry we were able to examine the time course and dose response of the altered cell cycle in CEM cells following dexamethasone. An example of the results ob-

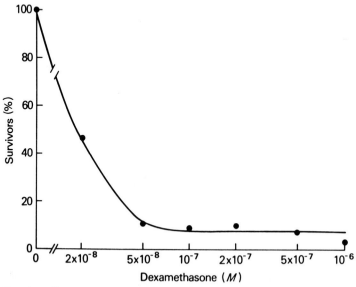

Fig. 3. Effect of increasing dexamethasone concentration on cloning efficiency (% survivors) of CEM clone 7 cells. Cells treated for 48 h with steroid were plated in agarose gel over feeder layers, allowed to grow to visible clones, and counted. Approximately one-half of the untreated cells formed colonies, and this number was set at 100% for the ordinate.

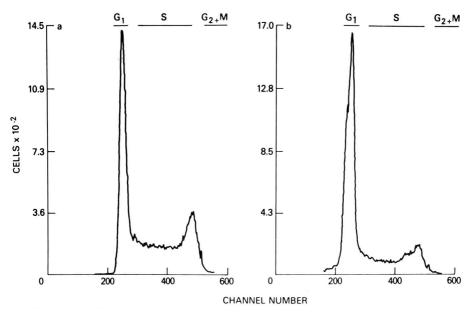

Fig. 4, a,b. Influence of dexamethasone on cell cycle distribution. Cells were grown in the presence (b) or absence (a) of dexamethasone for 48 h and then analyzed by flow microfluorometry. Abscissa: Channel No. is linear function of DNA per cell. Ordinate: A total of 10^5 cells was counted for each histogram and scale difference in the two panels reflects the fact that display of data is normalized to give a uniform maximum peak height (Harmon JM, Norman MR, Fowlkes BJ, Thompson EB (1979) J Cell Physiol 98: 267–278).

tained with flow microfluorometry is given in Fig. 4. As can be seen there is a relative diminution of cells in the S and (G_2 + M) portions of the cell cycle and a relative accumulation of cells in the G_1 phase of the cell cycle. The onset of this response was close to the time at which loss of cloning efficiency was seen, namely, at around 24 h. Furthermore, the concentrations of dexamethasone responsible for increasing G_1 accumulation and loss of cloning efficiency are quite similar and were concentrations that caused increasing occupancy of receptors. We therefore wondered whether the accumulation of cells in G_1 represented or was coincident with some lethal event imposed upon the cells by the steroid.

Several experiments were designed to explore this question. First, cells were simply treated with steroid until G_1 accumulation had reached a high proportion. Then the hormone was withdrawn from the culture, the cells washed, and the culture followed to see whether those cells trapped in G_1 were capable of reentering the cell cycle. They were not. At earlier time points after steroid, when G_1 accumulation and loss of viability are beginning to be detectable, the experiment could not be done so simply. Colcemide was therefore employed to block the entry of additional cells into G_1 after the removal of steroid. The protocol of a typical experiment was as follows. Cells were treated for 36 h with dexamethasone. To half the culture of steroid-treated cells colcemide was added between 24 and 36 h.

Control cultures treated with colcemide only showed that the cells in G_1 at the time of addition of the mitotic blocker were capable of leaving it, and that no further cells could enter G_1. However, in the steroid–colcemide-treated cells a fraction of the cells in G_1 were unable to leave in the 12-h time period following addition of the mitotic blocker. The proportion of cells remaining in G_1 during that time was similar to the fractional loss of cloning efficiency. These results indeed suggested that accumulation of cells in G_1 was reversible only slowly, if at all, and that there was a correlation between G_1 accumulation and loss of cloning ability.

To see whether cells could recover from G_1 block over a long time period, a different experimental protocol was followed. This took advantage of the fact that cells which are allowed to incorporate bromodeoxyuridine (BUDR) into their DNA during replication and which are then exposed to fluorescent light are killed. This effect can be intensified and nonspecific toxic consequences of BUDR exposure minimized by following the period of BUDR incorporation with a brief exposure to the fluorescent dye Hoechst 33258. The dye and the BUDR act synergistically to intensify the light-killing effect. If cells caught in G_1 by dexamethasone can later escape and return to the growth cycle, they should be protected from the S-phase-specific incorporation of BUDR and subsequent light killing. The result of an experiment testing this is shown in Fig. 5. Cells were treated for 48 h with the steroid. During the latter 24 h of treatment, half the culture was also exposed to BUDR, with Hoechst 33258 added for the final 3 h. Then the two cultures were plated and their cloning efficiencies determined. The data in the figure show that there was no protective effect of the steroid against the BUDR–Hoechst–light killing regimen. We conclude that glucocorticoids kill these cells by a process that results in their permanent arrest in a state in which they contain a G_1-like quantity of DNA. These results confirm the in vivo observations of G_1 arrest of leukemic blasts by glucocorticoids, show that this is a direct effect of the steroids on the cell, and suggest that the block is permanent.

All the above observations are of potential clinical relevance. The fact that prolonged (>24 h) exposure to the steroid is required to kill the cells points up the importance of maintaining high circulating levels of glucocorticoid in the patient and is consistent with the observation that high-dose parenteral steroid therapy has been found to be the most efficacious way to utilize these hormones as antileukemic drugs (Goldin et al. 1971). The cell-cycle action of the steroid also is of potential importance owing to the fact that steroids are nowadays often used in combination with a variety of cycle-active drugs. We have investigated the combined effects of prednisolone plus methotrexate, vincristine or 6-mercaptopurine (6-MP) on CEM cells. Antagonism between the steroid and 6-MP was observed, and our data suggested that this resulted from protection of the cells against the steroid by S-phase arrest as a consequence of the 6-MP administration (Norman et al., 1978). Other possibilities, however, have not been excluded. Finally,

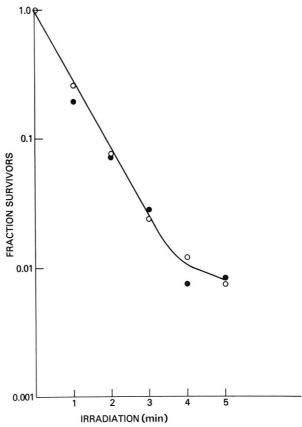

Fig. 5. Effect of dexamethasone on photosensitivity of BrdUrd substituted cells. Parallel cultures were incubated in the presence (●) or absence (○) of 10^{-6} M dexamethasone for 48 h.   BrdUrd (2×10^{-7} M) was added to each culture for the final 24 h followed by 3 h in the presence of 1.5 μg/ml 33258 Hoechst, and irradiated by exposure to a standard fluorescent light source.   Survivors are expressed on the basis of unirradiated samples, normalizing for the toxicity of dexamethasone in the dexamethasone dexamethasone-treated culture.   Survival of the dexamethasone-treated culture was 7.3%.   (Harmon JM, Norman MR, Fowlkes BJ, Thompson EB (1979) J Cell Physiol 98: 267–278)

CEM cells have been examined for their rate of acquisition of resistance to corticosteroids and for the receptor phenotype of the resistant cells.

As was discussed above, in the rodent models, lymphoid cells selected in culture for steroid resistance usually lacked receptors.   The transplantable rat lymphosarcoma P1798, on the other hand, acquired resistance in vivo, without total loss of receptors.   The original observation of a reduction in receptor concentration in resistant tumors was interpreted as indicating a contamination of the resistant, R⁻ cells with some sensitive, R⁺ cells (Kaiser et al., 1974).   New data have led to a reinterpretation of the observation, however; at least some resistant P1798 tumors are R⁺ (Nicholson and Young

1978; D. A. Young, in Bell, 1979.) And as noted before, the cells of many human leukemias have been found to be receptor positive. In addition, the usual outcome of treatment of glucocorticoid-sensitive leukemia with steroids alone is only temporary remission, followed eventually by a steroid-resistant state. Among the many possible reasons for this outcome is the possibility that the leukemic blasts have acquired resistance to the direct effects of the hormone by adaptation or by selection. Obviously, understanding of the nature of steroid resistance in human leukemias would be useful. The CEM cells may serve as a valuable paradigm for these processes.

In a series of Luria–Delbruck fluctuation tests, we have observed that cells of the steroid-sensitive clone CEM C7 acquire stable resistance to 10^{-6} M dexamethasone by independent spontaneous events occurring at a high frequency, approximately 1×10^{-5} (Harmon et al. 1979). Such a high rate could result from mutations occurring at many sites, any of which affect the steroid sensitivity of the cells, to mutations at a "normal" rate occurring in a site that is functionally haploid, or to mutations at a diploid site that is a "hot spot." This rate could also be the result of a stable, nonmutational, nongenetic event. Regardless of the mechanism, if the shift to steroid resistance occurs in vivo at a similar rate, it is easy to see how this would go far to explain the commonly observed development of steroid resistance.

Some 60 or so independent, steroid-resistant clones have been grown out and screened for their steroid-receptor phenotype. All contain measureable receptors, but fewer receptor sites per cell than wild-type cells as assayed in the whole-cell binding assay, and in every case there is less than normal fractional nuclear transfer of that receptor which remains (Harmon et al. 1979). In this respect, therefore, resistant CEM cells differ sharply in their receptor phenotype from that seen in the rodent cell lines. The presence of receptor in the resistant CEM cells suggests that they may be more akin to the commonly seen receptor-positive steroid-resistant cells obtained from patients. Experiments are underway to examine in more detail the biochemistry of the receptors in typical resistant clones.

V. Summary and Conclusions

Experiments with three different types of steroid-sensitive systems have been described. Each system provides certain advantages in investigating glucocorticoid actions. Steroid-resistant L cells have been isolated that contain apparently normal glucocorticoid receptors, qualitatively and quantitatively. Both these cells and the HTC cell variants described represent interesting systems in which to explore the postreceptor steps of steroid actions. The HTC cell variants show dissociation of various functions affected by corticosteroids in a single cell line. They point up the complexity and significance of the steps subsequent to steroid–receptor binding. The CEM cells appear to be a useful model system for studying the lethal effects of corticosteroids in human lymphoblastic leukemia. Of course, CEM cells are from a single patient, and other cultured human leukemic cells must be explored before broad generalizations can be made. However, we hope that our results with this model have some relevance and that they will encourage similar studies with other such lines.

Acknowledgments. Dr. Granner's research was supported by Veterans' Administration research funds and by United States Public Health Service Grant AM 24037; Dr. Gelehrter's research was supported by National Institutes of Health Grant CA 22729 and a Faculty Research Award from the American Cancer Society. Dr. Norman was a Fogarty International Fellow. Dr. Schmidt's research was supported by National Cancer Institute Postdoctoral Fellowship 1 F 32 CA 05447-02; Dr. Harmon's research by Damon Runyon Postdoctoral Fellowship DRG 116-FT.

References

Aviv D, Thompson EB (1972) Science 177: 1201–1203
Ballard PL, Tomkins GM (1969). Nature 224: 344–345
Baxter JD, Harris AW (1975) Transplant. Proc. 7: 55–65
Baxter JD, Tomkins GM (1970) Proc Natl Acad Sci USA 65: 709–715
Baxter JD, Rousseau GG, Benson MC, Garcea RL, Ito J, Tomkins GM (1972) Proc Natl Acad Sci USA 69: 1892–1896
In Bell PA and Borthwick NM (eds) (1979) Proceedings of the Seventh Tenovus Workship, Cardiff Wales, 217 pp.
Borthwick NM, Bell PA (1975) FEBS Lett. 60: 396–399
Bourgeois S, Newby RF (1977) Cell 11: 423–430
Bourgeois S, Newby RF, Huet M (1978) Cancer Res 38: 4279–4284
Breslow JL, Epstein J, Fontaine JH (1978) Cell 13: 663–669
Carlson SA, Gelehrter TD (1977) J Supramol Struct 6: 325–331
Claman HN (1972) N Engl J Med 287: 388–397
Crabtree GR, Smith KA, Munk A (1979) In: Thompson EB, Lippman ME (eds) Steroid Receptors and the Management of Cancer, Vol. I. CRC Press, West Palm Beach, Florida pp. 81–97
Crook RB, Louie M, Deuel TF, Tomkins GM (1978) J Biol Chem 253: 6125–6131
Dougherty T, White A (1945) Am J Anat 77: 81–116
Earle WR (1943) J Natl Cancer Inst 4: 165–212
Foley G, Lazarus H., Farber S, Uzman BG, Boone B, McCarthy R (1965) Cancer 18: 522–529
Gehring U, Thompson EB (1979) In: Baxter JD, Rousseau GG (eds) Glucocorticoid Hormone Action. Springer-Verlag, Berlin Heidelberg New York, 399–421.
Goldin A, Sandberg J, Henderson E, Newman J, Frei E, Holland J (1971) Cancer Chemother. Rep. 55: 309–507
Granner DK (1976) Nature 259: 572–573
Granner DK, Thompson EB, Tomkins GM (1970) J Biol Chem 245: 1472–1478
Granner DK, Lee A, Thompson EB (1977) J Biol Chem 252: 3891–3897
Hackney JF, Gross SR, Aronow L, Pratt WB (1970) Mol. Pharmacol 6: 500–512.
Harmon JM, Norman MR, Fowlkes BJ, Tompson EB (1979) J Cell Physiol 98: 267–278
Harmon JM, Norman MR, Thompson EB (1979) In: Thompson EB, Lippman ME (eds) Steroid Receptors and the Management of Cancer, Vol. II. CRC Press, West Palm Beach, Florida, pp 113–129.
Hofer E, Land H, Sekeris CE, Morris HP (1978) Eur J Biochem 91: 223–229
Ivarie RD, O'Farrell PH (1978). Cell 13: 41–55
Jung-Testas I, Bayard F, Baulieu EE (1976) Nature 259: 136–138
Kaiser N, Milholland RJ, Rosen F (1974) Cancer Res 34: 621–626

Kelly DS, Becker JE, Potter VR (1978) Cancer Res 38: 4591–4600
King RJB, Mainwaring WIP (1974) Steroid–Cell Interactions.   University Park Press, Baltimore, 430 pp
Konior-Yarbro GS, Lippman ME, Johnson GE, Leventhal BG (1977).   Cancer Res 37: 2688–2695
Kulka RG, Tomkins GM, Crook RB (1972) J Cell Biol 54: 175–179
Kurtz DT, Chan K-M, Feigelson P (1978) J Biol Chem 253: 7886–7890
Lampkin BC, Nagao T, Mauer AM (1969) J Clin Invest 48: 1124–1130
Lampkin BC, Nagao T, Mauer AM (1971) J Clin Invest 50: 2204–2214
Leikin SL, Brubaker, Hartmann JR, Murphy ML, Wolff JA, Perrin E (1968) Cancer 21: 346–351
Levisohn SR, Thompson EB (1972) Nature [New Biol.] 235: 102–104
Lippman ME, Thompson EB (1974) J Biol Chem 249: 2483–2488
Lippman ME, Halterman R, Leventhal B, Perry S, Thompson EB (1973) J Clin Invest 52: 1715–1725
Lippman ME, Perry S, Thompson EB (1974a) Cancer Res 34: 1572–1576
Lippman ME, Yang SS, Thompson EB (1974b) Endocrinology 94: 262–266
Mainwaring WIP, Symes EK, Higgins SJ (1976) Biochem J 156: 129–141
Makman MH, Dvorkin B, White A (1968) J Biol Chem 243: 1485–1497
Makman MH, Nakagawa S, White A (1967) Rec Prog Horm Res 23: 195–227
Makman MH, Dvorkin B, White A (1971) Proc Natl Acad Sci USA 68: 1269–1273
Manganiello V, Vaughan M (1972) J. Clin. Invest. 51: 2763–2767
Martial JA, Baxter JD, Goodman HM, Seeburg PH (1977) Proc Natl Acad Sci USA 74: 1816–1820
McDonald RA, Gelehrter TD (1977) Biochem Biophys Res Commun 78: 1304–1310
McPartland RP, Milholland RJ, Rosen F (1977) Cancer Res 37: 4256–4260
Mosher KM, Young DA, Munck A (1971) J Biol Chem 246: 654–659
Munck A (1971) Perspec Biol Med 14: 265–289
Munck A, Leung K (1977) In: Pasqualini JR (ed) Receptors and Mechanism of Action of Steroid Hormones, Part II.   Marcel Dekker, New York pp 311–397
Munck A, Wira C (1971) In: Raspé G (ed) Advances in the Biosciences, Vol 7: Schering Workshop on Steroid Hormone Receptors.   Pergamon Press, Oxford, pp 301–330
Munck A, Wira C, Young DA, Mosher KM, Hallahan C, Bell PA (1972) J Steroid Biochem 3: 567–588
Nicholson ML, Young DA (1978) Cancer Res 38: 3673–3680
Norman MR, Thompson EB (1977) Cancer Res 37: 3785–3791
Norman MR, Harmon JM, Thompson EB (1978) Cancer Res 38: 4273–4278
Odashima S, Morris HP (1965) In: Yoshida T (ed) Gann Monograph I, Proceedings of the U.S.–Japan Joint Conference on Biological and Biochemical Evaluation of Malignancy in Experimental Hepatomas.   Kyoto, Nov. 4–5, pp. 55–64 and 7 plates
Pariza, Kletzien RF, Butcher FR, Potter VR (1976) Adv Enzyme Regul 14: 103–115
Pfahl M, Sandros R, Bourgeois S (1978) Mol Cell Endocrinol 10: 175–191
Potter VR, Watanabe M, Becker JE, Pitot HC (1967) Adv Enzyme Res 5: 303–316
Puca GA, Sica V, Nola E (1974) Proc Natl Acad Sci USA 71: 979–983
Ramanarayanan-Murthy L, Colman PD, Morris HP, Feigelson P (1976) Cancer Res 36: 3594–3599
Ringold GM, Yamamoto KR, Tomkins GM, Bishop JM, Varmus HE (1975) Cell 6: 299–305
Ringold GM, Shank PR, Yamamoto KR (1978) J Virol 26: 93–101
Samuels HH, Tomkins GM (1970) J Mol Biol 52: 57–74
Schmidt TJ, Thompson EB (1978) In: Sharma RK, Criss WE (eds) Endocrine Control in Neoplasia.   Raven Press, New York, pp 263–290
Seifert SC, Gelehrter TD (1978) Proc Nat Acad Sci USA 75: 6130–6133

Sibley C, Tomkins T (1974) Cell 2: 221–227

Simons SS, Martinez HM, Garcea RL, Baxter JD, Tomkins GM (1976) J Biol Chem 251: 334–343

Spelsberg TC, Steggles AW, Chytil F, O'Malley BW (1972) J Biol Chem 247: 1368–1374

Thompson EB (1979) In: Baxter JD, Rousseau GG (eds), Mechanism of Action of Glucocorticoid Hormones. Springer-Verlag, Berlin pp 203–211

Thompson EB, and Lippman ME (1974) Metabolism 23: 159–202

Thompson EB, Tomkins GM, Curran JF (1966) Proc Natl Acad Sci USA 56: 296–303

Thompson EB, Aviv D, Lippman ME (1977a) Endocrinology 100: 406–419

Thompson EB, Norman MR, Lippman ME (1977b) Rec Prog Horm Res 33: 571–615

Thompson EB, Granner DK, Gelehrter TD, Hager GL (1979) In: Ross R, Sato G (eds), Hormones and Cell Culture, Cold Spring Harbor Conferences on Cell Proliferation, Vol. 6, pp 339–360.

Turnell RW, Burton AF (1975) Mol Cell Biochem 9: 175–189

Venetianer A, Bajnoczky K, Gal A, Thompson EB (1978) Som Cell Genet 4: 513–530

Vietti TJ, Sullivan MP, Barry DH, Haddy T, Haggard M, Blattner R (1965) J Pediatr 66: 18–26

Webster RA, Pikler GM, Spelsberg TC (1976) Biochem J 156: 409–418

Yamamoto KR, Alberts BM (1976). Annu Rev Biochem 45: 721–746

Yamamoto KR, Gehring U, Stampfer MR, Sibley CH (1976) Rec Prog Horm Res 32: 3–32

Yamamoto KR, Ivarie RD, Ring J, Ringold GM, Stallcup MR (1978) In: Litwack G (ed), Biochemical Actions of Hormones, Vol 5. Academic Press, New York.

Yang SS, Lippman ME, Thompson EB (1974) Endocrinology 94: 254–261

Discussion of the Paper Presented by E. B. Thompson

McGinnis: Is there any correlation between the inductionless cells that you have shown us today and the effects of concanavalin A on the elimination of TAT induction. Has the Con-A inhibition introduced a new step in the mechanism of glucocorticoid action?

Thompson: We have not looked specifically in those cells to see whether that basal level of TAT is affected by Con-A; however, in normal phenotype cells it is. I have no evidence however that rapid inactivation by Con-A is involved in steroid action. Unfortunately, I can't give you a hint; I have no idea as to what the mechanism is.

Stevens: Have any of you looked specifically to see whether Don Young's nuclear fragility idea holds up to any of these lymphocyte lines that you are looking at? That's one general question.

Thompson: The first experiment didn't work.

Stevens: Has it been tried on the S49 or CEN cells?

Thompson: We have not tried it.

Gelehrter: You suggested that there is a possibility that glucocorticoids may only work if the cells are in G_1, or can move into G_1. Did you try the experiment where you load up the cells in G_2 with colcemide and then give them dexamethasone and then try to assay for colonizing ability?

Thompson: We had that one result that suggested an apparent protection by the 6-mercaptopurine which appeared to block the cells in S phase.

Mueller: With respect to the apparent G_1 state that your cells go into, have you looked to see whether or not the actual DNA is fragmented at an early state?

Thompson: You mean extract DNA from the cells, see if it is in little pieces or not?

Mueller: Yes.

Thompson: No, I'm trying to think—we have not done that specific experiment.

MUELLER: It might be worthwhile, if it were that way you wouldn't expect much recovery. I forgot your time intervals. How long did it take before the cells actually appear to go through some sort of deterioration?

THOMPSON: The cells begin to accumulate at 24 h, and at that time all that you can see is that they are slightly smaller. Thereafter, there is recruitment; apparently more and more of the cells begin to look frayed out.

GELEHRTER: Daryl Fanestyl reported last year that the aldosterone–receptor or the binding of aldosterone to its receptor could be blocked not only by low molecular weight protease inhibitors but by substrates of the proteases, and has extended those observations apparently to the glucocorticoid receptor in HTC cells, estrogen binding to α-fetoprotein, androgen binding with its receptor. Do any of you have comments about that, or what that might be suggesting?

EDELMAN: The effects are unquestionable. I think that the facts are correct. We have a variety of protease inhibitors, including those that bind covalently to proteases which will block receptor binding. Whether the block occurs at the site of the location at the binding site, or whether the block occurs as a result of recognizing some sort of common sequence that is found in receptors and then unfolding the protein, is completely unknown. Because it requires very high concentration, it may turn out to be somewhat limited, because there may be many other proteins that it may have similar effects on. At the moment it is an interesting observation that could lead to new approaches to receptor purification. Because in principle, if it turned out that this sequence was common to a variety of receptors, for example, involved in the capability of receptors to produce protein–protein aggregates, which is what proteases recognize, then this might be a way of radio labeling receptors covalently.

GELEHRTER: Do you think it has any implication to the possible protease action of receptors?

EDELMAN: I don't think that there is any evidence that the receptors act as proteases, and I think that point was covered in one of the talks. Bill Schrader did that. He explicitly explored that in the isolated A and B subunits and found no protease activity.

THOMPSON: We have looked at some of the glucocorticoid–receptor studies in which a large number of protease inhibitors have no effect on activation, but we never did the forward experiment on binding.

Discussants: I.S. EDELMAN, T.D. GELEHRTER, J. McGINNIS, G.C. MUELLER, E.B. THOMPSON, and J. STEVENS

Chapter 10

Studies on the Mechanism of Glucocorticoid Hormone Action

Lorin K. Johnson, Steven K. Nordeen, James L. Roberts, and John D. Baxter

I. Introduction

Extensive data are now available about many aspects of glucocorticoid hormone action. However, the fundamental mechanism through which these hormones act remains obscure. It is currently thought that most of the inhibitory and stimulatory actions of the glucocorticoid hormones are initiated by the binding of the steroid to specific glucocorticoid receptors (for review see Baxter and Rousseau 1979). Following an event termed "activation" (Higgins et al. 1973) the hormone–receptor complex then binds to the nucleus. This interaction is presumed, by mechanisms that are currently unknown, to modulate levels of specific mRNAs (Schultz et al. 1973; Ringold et al. 1975; Martial et al. 1977a; Nakanishi et al. 1977). The translation products of these mRNAs then mediate the glucocorticoid response.

It is likely that the diversity of glucocorticoid responses elicited in various cell types result from differences in cellular elements distal to the receptor. This notion derives from currently available data that the receptors for these hormones are identical in the various glucocorticoid target tissues. For instance, somatic cell fusion experiments show that the capability of receptors from a responsive cell to elicit a response is determined by the tissue source of the nuclei (Gehring and Thompson 1979). In addition, no tissue-specific differences have been observed in the receptor's binding affinity of various glucocorticoid analogs (Rousseau and Baxter 1979) or in the physicochemical characteristics of the receptor (Litwack and Singer 1972). Thus, nuclear elements dictated by the differentiation of the cell apparently determine the selection of genes to be influenced by the receptors once they bind to chromatin. It is to the nature of these general issues as well as the way that glucocorticoid receptors modify chromatin functions that our current studies are addressed.

II. Considerations

A. Specificity of Glucocorticoid Responses

The proteins whose synthesis in a given target cell is influenced by glucocorticoids is usually quite selective, encompassing but a small subset of the expressed genes. In both rat pituitary tumor cells (GH) (Martial et al. 1977b) and rat hepatoma cells (HTC) (Ivarie and O'Farrell 1978) the domain of the glucocorticoid response as determined by two-dimensional gel electrophoresis is estimated to encompass less than 1% of the detectable gene products.

The specificity of the glucocorticoid response has recently been examined at the level of mRNA in a line of cultured pituitary cells (GC), where growth hormone and its mRNA are stimulated by glucocorticoids. In Fig. 1 are the proteins synthesized by mRNA from control and dexamethasone-treated cells in a heterologous cell-free system (Martial et al. 1977a) after separation by gel electrophoresis. The activities of most of the mRNAs detectable in this manner are not affected under conditions in which growth hormone mRNA (Fig. 1) and a few other cellular mRNAs are clearly influenced by glucocorticoids (Fig. 1). The complexity of the changes in RNA in GC cells has also been analyzed by cDNA–RNA hybridization. The data in Fig. 2 show that induced cells do not contain a large number of poly(A$^+$)–RNA species that are absent in uninduced cells. Further, no mRNA species is induced that constitutes a large proportion of the total cellular mRNA. Thus, glucocorticoids only increase certain mRNAs and models for receptor function must ultimately account for such selectivity.

Furthermore, glucocorticoids are generally not the primary inducers of the genes that they influence. Although one exception to this general rule may be the induction of mouse mammary tumor virus mRNA (Stallcup, et al. 1978), the evidence to date indicates that in most systems glucocorticoids modulate the levels of specific mRNAs, which are also synthesized in their absence. Upon first consideration this appears to contrast with certain ac-

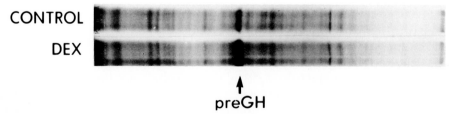

CONTROL

DEX

↑
preGH

Fig. 1. Specificity of the glucocorticoid response at the mRNA level in rat pituitary tumor cells (GC). Poly(A) plus RNA, isolated from cells grown in media containing 10 n*M* triiodothyronine and in the presence or absence of dexamethasone (1 μ*M*) for 50 h was translated in a heterologous (wheat germ) cell-free system as described (Martial et al 1977b). The [^{35}S]methionine-labeled products were then separated by SDS–polyacrylamide gel electrophoresis. (Martial et al. 1977b)

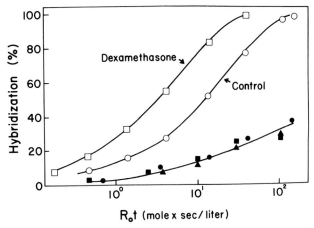

Fig. 2. Kinetics of hybridization of total cytoplasmic RNA from control glucocorticoid and thyroid hormone treated GC cells with cDNA prepared from total poly(A)–cellular RNA (filled symbols) or with cDNA highly enriched for sequences complimentary to growth hormone mRNA (open symbols). The cDNAs were hybridized with total cytoplasmic RNA from control (○) or dexamethasone treated (1 μM, 3 days) cells (□) (Martial et al. 1977a).

tions of the sex steroids, which closely approximate cases of de novo gene activation. Illustrations here are the avian oviduct (O'Malley and Means 1974) and liver (Shapiro et al. 1976) wherein estrogen is the primary inducer of ovalbumin and vitellogenin, respectively. Additionally, estrogens trigger cellular proliferation and differentiation in the immature oviduct and uterus (Palmiter et al. 1973). Thus, in these tissues the estrogenic domain, the number of regulated gene products, may be quite large. The primary receptor event may however, still be very similar for all steroid hormone receptors; the differentiation of sex steroid target tissues could simply program a more extensive response to the estrogen or androgen receptor than occurs in the glucocorticoid target cell.

B. Direction of Hormone Responses

Another key feature of glucocorticoid action that must also be taken into account is that of directionality. Glucocorticoid hormones exert varied and often contrasting effects on cellular metabolism and on the regulation of specific gene products. For instance, in cell types such as muscle, skin, adipose, lymphoid and connective tissue, glucocorticoids produce inhibitory influences on protein, fat, nucleic acid, and carbohydrate metabolism (Cahill 1971; Fain and Czech 1975; Fauci and Dale 1974). In liver the response is characteristically stimulatory (Exton 1972; Feigelson et al. 1971). In still other cells such as fibroblasts, both stimulatory and inhibitory effects on DNA synthesis have been described (Pratt and Aronow 1966; Thrash and Cunningham 1973).

Such bidirectional glucocorticoid responses are by no means limited to general effects on cellular metabolism. For example, in cultured rat pituitary cells (GH$_3$) two major gene products, growth hormone and prolactin, are oppositely regulated by corticosteroids. Growth hormone is stimulated, while prolactin is inhibited (Dannies and Tashjian 1973). In the HTC cells discussed above there are proteins whose rates of synthesis are inconsistently repressed by the steroid and others whose synthesis is always stimulated (Ivarie and O'Farrell 1978). Thus, the documented biological responses of glucocorticoid target cells range from effects on general cellular metabolism and morphology to influences on specific gene products; these may be stimulatory or inhibitory. The data to follow show that such directionality of glucocorticoid action can also be regulated by other hormones and by the metabolic state of the cell.

C. Regulation of the Cellular Responses to Glucocorticoids

A third consideration in modelling mechanisms of glucocorticoid action arises from the fact that the cellular environment plays a major role in modulating the program of hormone responses set during differentiation of the tar-

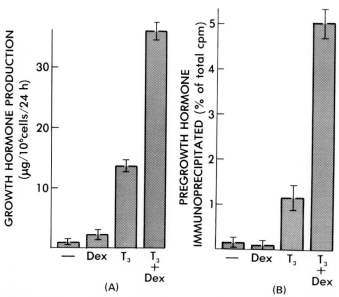

Fig. 3. Influence of thyroid hormone on glucocorticoid induction of growth hormone synthesis and growth hormone mRNA. GC cells grown in media containing serum from a thyroidectomized calf were induced for 48 h with either T$_3$ (10 n*M*), dexamethasone (1 μM), or both hormones together.
A Cultures were then pulsed with [^{35}S]methionine for 2 h and growth hormone secreted into the media quantitated by immunoprecipitation. **B** Poly(A) plus mRNA isolated from the induced cells was translated in a cell-free system (wheat germ) and the amount of pregrowth hormone quantitated by immunoprecipitation. (Martial et al. 1977b)

get cell. This includes both specific gene product responses such as the induction of tyrosine aminotransferase (TAT) and growth hormone and general metabolic responses such as the glucocorticoid stimulation of [^{3}H]thymidine incorporation by fibroblasts.

In primary cultures of hepatocytes, glucocorticoids are unable to induce tyrosine aminotransferase mRNA unless both glucagon and cyclic adenosine monophosphate are present in the culture media (Ernest and Feigelson 1979). Similarly, in the rat pituitary tumor cells discussed above, cultured in serum-containing media, dexamethasone is not an effective inducer of growth hormone mRNA unless thyroid hormone is also present (Fig. 3). In more recent studies with the use of a chemically defined serum substitute media, Schachter and co-workers (unpublished) have shown that glucocorticoids alone can, in fact, stimulate growth hormone synthesis, and that insulin can suppress this. However, the magnitude of the response is still increased synergistically by the presence of thyroid hormone.

In cultured fibroblasts, where thymidine incorporation is modulated by glucocorticoids, both fibroblast (FGF) and epidermal (EGF) growth factors can influence the response. Gospodarowicz and Moran (1974) using quiescent 3T3 cells observed that dexamethasone was unable to induce thymidine incorporation in the absence of serum. However, when FGF was included in the growth media, the glucocorticoid was 58% as effective as optimal concentrations of serum in promoting DNA synthesis. With human skin fibroblasts, Polansky and co-workers have shown that cortisol is a more potent and reproducible stimulus of [^{3}H]thymidine incorporation when EGF is present (Polansky et al. unpublished data, 1979). In these cells EGF has no effect on either the number of glucocorticoid receptors or on the binding affinity of a variety of agonists and antagonists. Such results suggest that the loss of glucocorticoid-mediated stimulation of thymidine incorporation in the absence of growth factors or serum results from their influence on cellular elements distal to the receptor itself.

The cellular milieu not only modifies the magnitude of the response to glucocorticoids but can, in fact, determine whether the response is stimulatory or inhibitory. This is illustrated by our recent studies with human skin fibroblasts, where the response was measured either in the presence of fresh culture media or in nutritionally deprived cultures (Johnson et al. 1979a). When the fibroblast cultures were about 75% confluent (late log phase of growth), half of the cultures received fresh media containing increasing concentrations of cortisol, whereas the remaining cultures received the glucocorticoid directly without the media change. Sixty-six hours later, the cultures were pulsed with [^{3}H]thymidine for 6 h and the acid precipitable radioactivity were determined. As is evident in Fig. 4 (Johnson et al. 1979a), the steroid increased thymidine incorporation up to 2.5-fold in the cultures that received fresh media. In contrast, thymidine incorporation in the cultures grown in depleted media was inhibited up to 80%. In both cases the steroid concentration producing a half-maximal response (C_{50}) was about 0.1 μM. Although in the experiment shown it appears that the stimulatory

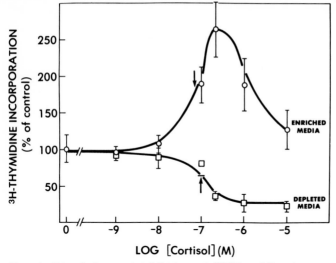

Fig. 4. Stimulation and inhibition of [³H]thymidine incorporation by cortisol in human skin fibroblasts. Fibroblast cultures, induced for 72 h with increasing dosages of cortisol under enriched and depleted culture conditions (see text), were pulsed with [³H]thymidine for 6 h, after which time the acid precipitable radioactivity was determined. (Johnson et al. 1979a)

steroid response is blunted at the higher steroid concentrations, a large stimulation even at $1–10 \ \mu M$ dexamethasone is consistently observed when the cells are cultured in the presence of epidermal growth factor (Polansky et al. unpublished data, 1979). Thus, in fibroblasts the nature of the glucocorticoid effect on thymidine incorporation can be profoundly modified when the metabolic status of the target cell is altered, here by media depletion or by growth factors.

The effect of media conditions on the killing of lymphoid cells by glucocorticoids has also been examined. In this experiment, mouse lymphoma (S49) cells were collected from suspension culture by centrifugation and resuspended in either fresh media or the same media from which they had been collected. One-half of each suspension received dexamethasone, and samples were then taken from the cultures at various times for counting of cells that exclude trypan blue. It is apparent from Fig. 5 that the use of fresh media at the time of induction significantly decreased the killing effect, whereas in no case did the steroid promote an increase in cell number above the control values. In cultures with fresh media, dexamethasone reduced the number of trypan blue-excluding cells to 50% of the value obtained from the control culture by 20 h. In marked contrast, dexamethasone killed over 90% of the cells in depleted media by 20 h. In addition, neither of the two glucocorticoid-resistant receptor-defective mutant cell lines (r⁻, Nᵗⁱ) were killed by dexamethasone treatment. In another lymphoid cell line (lymphosarcoma P1798) Nicholson and Young have shown that a receptor-containing variety that is glucocorticoid resistant in vivo becomes susceptible to the

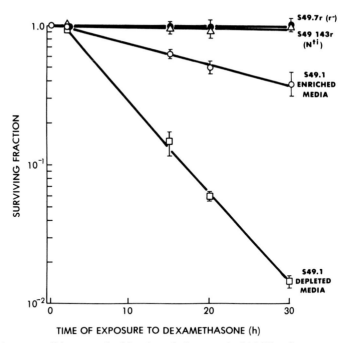

Fig. 5. Effect of culture conditions on the kinetics of glucocorticoid killing in mouse lymphoma cells. At various times after the addition of dexamethasone (1 μM) to enriched and depleted cultures (see text), duplicate aliquots of cells were removed and the number of viable cells determined by trypan blue exclusion. S49.1, wild type; S49.7r (r$^-$), glucocorticoid receptor minus mutant; S49 143r (N^{ti}), nuclear transfer increase glucocorticoid unresponsive mutant. (Johnson et al. 1979a)

nucleolytic effects of glucocorticoids when incubated in vitro (Nicholson and Young 1978). These authors have suggested that hormone resistance in vivo is the result of the development of hardier nuclear membranes, which is overcome in vitro presumably by the stress of incubation. Thus, the sensitivity of lymphoid cells to the killing actions of glucocorticoids can also be significantly influenced by the cellular environment.

In addition to these general effects of glucocorticoids on growth, we have examined the induction of a specific gene product by dexamethasone in enriched and depleted media. Figure 6 depicts the responses of rat pituitary tumor cells (GC) to the glucocorticoid dexamethasone, where either fresh media was added 4 h before the steroid or no fresh media had been added for 72 h. Confirming previous results (Martial et al. 1977a), dexamethasone increased growth hormone production in cultures containing enriched media without altering the general profile of cellular protein synthesis. The increase in growth hormone production was detected by immunoprecipitation of growth hormone secreted into the culture media and by SDS–polyacrylamide gel electrophoresis of newly synthesized proteins labeled during a 2-h pulse with [^{35}S]methionine. In contrast, dexamethasone did not induce growth hormone synthesis in GC cells under depleted conditions. In fact,

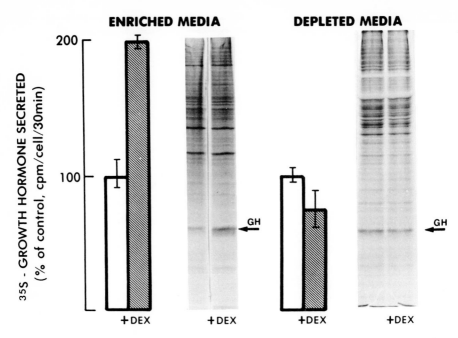

Fig. 6. Effect of cell culture on glucocorticoid induction of growth hormone synthesis by rat pituitary tumor cells. Seventy-two hours after plating the culture media was either removed and replaced with fresh media (enriched condition) or was not removed (depleted condition). Four hours later dexamethasone was added (1 μM). Forty-eight hours after addition of dexamethasone the cells were labeled by a 1-h pulse with [^{35}S]methionine, and the amount of growth hormone was analyzed by immunoprecipitation of proteins secreted into the culture media and by SDS–polyacrylamide gel electrophoresis of the total cellular protein as previously described. (Johnson et al. 1979a)

the steroid commonly decreased growth hormone production. In the experiment shown in Fig. 6 this decrease was about 20%. The synthesis of other proteins as analyzed by SDS–polyacrylamide gels was not detectably altered. These results indicate that the media conditions can also profoundly affect the induction of a specific gene product by glucocorticoids.

Preliminary studies using the rat pituitary cells and skin fibroblasts have shown that EGF, T_3, and insulin have no effect on: (1) the number of cytoplasmic glucocorticoid receptors; (2) the affinity of agonist and antagonist binding by the receptor; (3) the percentage of receptors that activate and translocate to the nucleus; (4) nuclear retention of receptor–hormone complexes. Although receptor hormone binding studies have not yet been completed in all cases, the data collected thus far argue for the participation of some step(s) subsequent to hormone–receptor association with nuclei and prior to the accumulation of mRNA. This step must be quite sensitive to the metabolic state of the cell. It is emphasized that such factors are capable not only of inhibiting or allowing the response to occur; they can, in fact, reverse its direction.

The following experiments on the nature of receptor–nuclear interactions may provide clues to possible mediators of glucocorticoid action at the nuclear level.   In the following sections a variety of experiments on the nature of glucocorticoid–receptor–nuclear interactions are also presented.   Finally, the data from these diverse approaches are then drawn upon to formulate a testable model for glucocorticoid–receptor funtion.

III. Nuclear Events in Glucocorticoid Action

A. Role of the Glucocorticoid Receptor in Positive and Negative Glucocorticoid Effects

Certain findings can be interpreted to suggest that at least some stimulating effects of glucocorticoids may be a rather direct consequence of the receptor interaction with chromatin.   For example, the glucocorticoid-induced increase in accumulation of tyrosine aminotransferase mRNA has been shown to occur even when protein synthesis is blocked (Peterkofsky and Tomkins 1968).   The synthesis of mouse mammary tumor virus RNA is also stimulated within minutes after exposure of cells to glucocorticoids, and this effect can likewise be observed even when protein synthesis is blocked (Stallcup et al. 1978).   Thus, in these cases there is no induction of proteins that in turn mediate the steroid-induced increase in mRNA.

How do receptor–steroid complexes elicit negative influences?   In rat thymus cells, which are irreversibly committed to inhibitory effects of glucocorticoids within a few minutes, indirect experiments with the use of inhibitors of macromolecular synthesis could be interpreted to mean that the major early influence of the receptor is not to directly inhibit the synthesis of specific mRNAs.   In this system the inhibitory actions of glucocorticoids on glucose uptake and cellular metabolism can be blocked by concurrent addition of cycloheximide, cordycepin or actinomycin D (Hallahan et al. 1973; Mosher et al. 1971; Young et al. 1974).   These data have lead to the hypothesis that the receptor stimulates the synthesis of an RNA whose translation product is inhibitory.   Unfortunately, direct confirmation has not been presented because identification of the protein or mRNA involved in such inhibitory actions has proved particularly difficult.

Investigations on the role of the receptor in the production of inhibitory glucocorticoid actions would be facilitated by a model system in which the effects on a specific mRNA involved in an inhibitory response could be examined.   With this thought, we have studied cultured mouse AtT20 cells that produce adrenocorticotrophic hormone (ACTH).   Like the case in vivo with steroid feedback inhibition on the pituitary, glucocorticoids inhibit ACTH production by the cells (Roberts et al. 1979).   This occurs because the glucocorticoid decreases the mRNA that codes for the common protein precursor to ACTH and β-lipotropin (ACTH–LPH mRNA) (Nakanishi et al. 1977; Roberts et al. 1979).

As shown in Fig. 7, the inhibition of this gene product by glucocorticoids

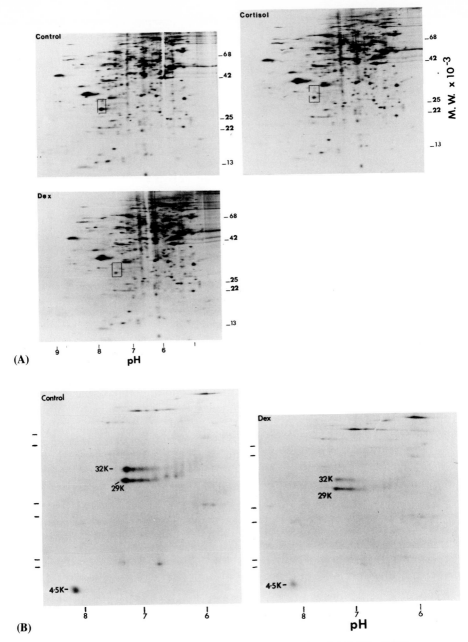

Fig. 7A,B. Influence of glucocorticoids on the synthesis of ACTH–βLPH precursor. Cells were pretreated for 24 h with either no hormones (control), cortisol (1 μM), or dexamethasone (1 μM). After labeling 1 h with [³⁵S]methionine, the cell proteins were solubilized directly (Ivarie and O'Farrell 1978) and fractionated by isoelectric focusing in the horizontal dimension, followed by SDS–polyacrylamide gel electrophoresis in the vertical dimension. **A** Total cellular proteins; **B** proteins precipitated with antiserum to ACTH–βLPH. (Roberts et al. 1979a)

Table 1. Effect of Cycloheximide on the Decrease in ACTH–Endorphin mRNA Caused by Dexamethasone: ACTH–Endorphin mRBA Levels[a] (untreated control, %)

	Control (%)	Dex (%)
Control (%)	100	79
Cycloheximide (%)	84	49

[a] Four identical 10-cm dish cultures of AtT-20 cells at ~60% confluency were used. Two of the cultures were given cycloheximide to 10^{-4} M. After 30 min, 1×10^{-6} M dexamethasone was added to one control and one cycloheximide-treated culture, and all four cultures were harvested 10 h later. Total cytoplasmic RNA was prepared, and the ACTH–endorphin mRNA levels were determined by cell-free translation (reticulocyte lysate) followed by immunoprecipitation.

occurs in a highly specific manner. In this experiment cells were treated with either cortisol or dexamethasone for 24 h; the proteins synthesized were then analyzed on two-dimensional gels. The major effect was to decrease the incorporation of [^{35}S]methionine into the 29 000- and 32 000-dalton precursor proteins to ACTH–LPH. The steroid did not effect the rate of synthesis of any of the other detectable proteins. Thus, as with the stimulatory actions of glucocorticoids on growth hormone mRNA production (see Fig. 1), the inhibitory actions of these hormones can also be quite selective.

In order to determine whether the inhibition of ACTH–LPH mRNA required the induction of a protein, which in turn inhibited the mRNA levels, cells were treated with cycloheximide in the presence and absence of the glucocorticoid dexamethasone. After 10 h the poly(A) plus RNA was isolated from the cells, translated in a cell-free protein-synthesizing system, and the amount of ACTH precursor quantitated by immunoprecipitation. As shown in Table 1 dexamethasone was still able to decrease the ACTH–LPH mRNA even though 98% of cellular protein synthesis had been blocked by cycloheximide. Therefore, these data indicate that glucocorticoids can elicit inhibitory as well as stimulatory effects on specific mRNAs; the inhibitory influences need not be mediated by other proteins induced by the steroid. Thus, the hypothesis should be considered that the receptor can inhibit mRNA levels through a process that is intimately associated with the interaction of the hormone–receptor complex and chromatin.

B. Studies on the Nature of Nuclear Acceptor Sites
For the Binding of Receptor—Glucocorticoid Complexes

The most popular model to explain steroid hormone action assumes that receptor–steroid complexes bind specifically at a limited number of sites on chromatin to regulate transcription of specific mRNAs. The site-specific localization of the glucocorticoid–receptor complex would in this case result from either specific sequences on the DNA that bind the complexes, or chromatin proteins located at specific sites that bind or influence the binding of receptor–steroid complexes. Such a model is an attractive analogy to the well-studied examples of prokaryotic gene regulation. It has also been assumed by many that such "relevant operator sites" for receptor action may be undetectable owing to an enormous background of nonspecific binding by the receptor in nuclei (Yamamoto and Alberts 1975) (again analogous to the prokaryotic models).

Whereas elements of such a formulation may turn out to be the case, the available data suggest that there are differences in the details and that these may be helpful in developing models. For example, when the quantity of nuclear, bound, receptor–glucocorticoid complexes is plotted as a function of the cytosol complexes, a constant proportion of the total (about 50%) is found in the nucleus (Baxter 1976). Thus, many of the cytosol receptors are not consumed by the nuclear binding, which increases linearly with more cytosol-bound receptors. Such a linear binding curve implies that if there were more receptors, there would be more nuclear binding, and therefore that the observed nuclear binding sites for receptor–glucocorticoids complexes are not filled. These sites may also be construed as being "nonspecific" because of this. Most have assumed that the "specific operators for steroid action" would be saturated at much lower concentrations of complex but that this would not be observed because it would be a small proportion of the total binding (Yamamoto and Alberts 1975).

To determine the receptor–dexamethasone complex occupancy of the actual "operators" mediating glucocorticoid hormone action, we plotted a biological response (TAT induction) as a function of nuclear binding (Fig. 8) or cytosol binding (not shown). This relationship is similarly linear in either case. Thus, it appears that the nuclear sites for receptor *actions* are not saturated in these cells. It is unlikely therefore, that there exist "operators" mediating glucocorticoid hormone action that display an affinity for binding receptors much higher than the nuclear binding that is observed. The data are more consistent with the idea that the receptors act through acceptors that are in excess and are unfilled by the receptors in the cell even when all receptors are bound by the hormone. Alternatively, the chance association of a receptor–steroid complex with one of the "right" acceptors is the event that triggers subsequent steps in steroid hormone action. Thus, in this system the receptors themselves and not the number of nuclear acceptor sites apparently are limiting.

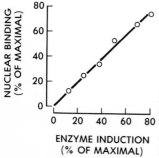

Fig. 8. Correlation between nuclear binding of dexamethasone and enzyme (tyrosine aminotransferase, TAT) induction in HTC cells. (Data from unpublished observations of J.D. Baxter, E. Bloom and D.T. Matulich.)

Fig. 9. Activated cytoplasmic receptors with bound [³H]dexamethasone were incubated with either nuclei or DNA–cellulose at the indicated ionic strengths and the bound receptor complexes determined as described previously (Baxter et al. 1972; Higgins et al. 1973).

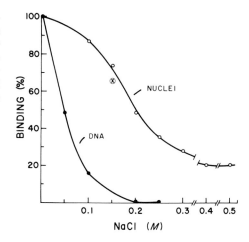

The chemical nature of the nuclear acceptor sites that bind receptor–steroid complexes has been studied by numerous investigators, and the general consensus is that DNA is in some manner involved in the nuclear localization of the receptor. The evidence for this includes the following. First, nuclei predigested with DNAse I lose all capacity to bind hormone-bound receptors (Baxter et al. 1972). Second, glucocorticoid–receptor complexes (like other steroid receptors) bind to DNA under cell-free conditions (Rousseau et al. 1975). Third, receptors from a glucocorticoid-resistant S49 lymphoma cell mutant, termed NT⁻ (nuclear transfer minus) show a lower affinity for DNA, whereas receptors from the S49 Nᵗⁱ (nuclear transfer increase) mutant have been shown to have a higher affinity for binding to DNA–cellulose in vitro (Yamamoto et al. 1974). Also, "activation" of the receptor–glucocorticoid complex by salt or heat treatment enhances binding of the complex to isolated DNA or to nuclei (Rousseau et al. 1975). Receptors do not bind to RNA although they fail to distinguish between DNA isolated from a variety of eukaryotes, prokaryotes, and bacteriophages (Rousseau et al. 1975).

Although DNA is likely to be involved in the nuclear localization of receptors, data from several laboratories as well as the experiments described below indicate the presence of other nuclear factors that participate in receptor binding. This is most evident upon examination of the ionic strength sensitivites of DNA and chromatin binding by the receptor. In Fig. 9 are the salt sensitivity curves of glucocorticoid receptor binding to both nuclei and DNA–cellulose. As is evident, the elution of receptor complexes from DNA occurs much more readily as the ionic strength increases than with nuclei. For DNA, half-maximal binding occurs at 0.05 *M* NaCl, whereas nuclei do not liberate 50% of the receptors until 0.20 *M* NaCl is added. Clearly, subnuclear chromosomal components not present in pure DNA affect the binding of receptors to nuclei.

Thus, the picture that has emerged from most studies to date concerning the nature of nuclear acceptor sites for glucocorticoid receptors is the following:

1) The nuclear acceptor sites for receptor action are probably in excess of the number of receptors that can translocate to the nucleus.

2) The binding affinity of receptors for the nuclear sites that mediate receptor action may be similar to that of the observed receptor binding by nuclei. (There is no evidence for hidden high affinity operator sites.).

3) The sites contain DNA although receptor binding is also influenced by other aspects of chromatin as well. This could involve effects on the conformation of the DNA or of chromatin superstructures.

Alternatively, nuclear binding of receptors could be influenced by specific nuclear proteins as suggested by work in systems responsive to sex steroids (Spelsberg et al. 1971).

Because of the possibility that the observed nuclear acceptor sites may be similar to those that actually mediate glucocorticoid hormone action, we began to examine the possibility that perhaps receptors were actually acting at the acceptors to which they were bound, but that most of the influences would not be transmitted into effects or specific genes. In this case, analogy with cAMP action in eukaryotes might be considered. Whereas the cAMP-dependent protein kinase phosphorylates numerous substrates, only a few of these result in effects on metabolic processes in the cell (Rubin and Rosen 1975). Thus, the question arose as to whether there were biochemically detectable effects of receptor–steroid complexes on chromatin that superficially might appear to be of a rather nonspecific nature, but might reflect a biologically important receptor function.

C. Glucocorticoids and Chromatin Structure

The extensive packing of DNA in chromatin (10 000 : 1 packing ratio in the supersolenoid (Bak et al. 1977) results in a significant fraction of the DNA template being totally inaccessible to a variety of chemical and enzymatic probes, including the RNA polymerase enzymes that transcribe the DNA in vivo. For example, in calf thymus chromatin only 10% of the possible DNA initiation sites for a nonspecific bacterial RNA polymerase are accessible to this probe (Cedar and Felsenfeld 1973). More recent experiments using DNAse I and micrococcal nuclease digestion have revealed that this more accessible fraction of chromatin is, in fact, enriched in transcribed sequences (Levy and Dixon 1978; Weintraub and Groudine 1974). Although such probes of chromatin structure do discriminate between expressed and repressed sequences, there is evidence that DNase I does not discriminate among sequences that are transcribed at diverse rates (Garel et al. 1977). Thus, chromatin superstructure may be a determinant of which DNA sequences are available for transcription, whereas other components [possibly soluble initiation factors (Spindler 1979)] dictate the rate at which given sequences will be transcribed by the endogenous RNA polymerases. These ideas are supported by experiments of Weintraub and Groudine (1979) showing that in mature chick red blood cells, the globin gene remains equally accessable to DNAse I digestion as in earlier developmental stages even though at maturity no globin mRNA is transcribed within the cell. In addition, cell-free transcription experiments of duck reticulocyte chromatin using either *Eschericia coli* RNA polymerase, sheep liver RNA polymerase II, or the endogenous chromatin-bound RNA polymerase have shown that the ratio of globin sequences transcribed were 1:4:50 000, respectively, when normalized per unit of catalyzing activity (Steggles et al. 1974). Therefore, the lack of fidelity of exogenous bacterial or eukaryotic polymerase [with the possible exception of RNA polymerase III from *Xenopus* (Parker and Roeder 1977)] renders this approach ineffective as a means of quantitating the transcription of specific eukaryotic genes. However, the above results do show that either exogeneously added bacterial or eukaryotic enzymes are still equally useful as nonspecific probes of DNA accessibility in chromatin. Because the precise molecular events mediating initiation, elongation, and chain termination by bacterial RNA polymerases are known in greater detail, we have used this probe to ask if glucocorticoid receptors exert effects on chromatin structure that can be detected by changes in overall template availability. Previous studies by O'Malley and co-workers had also demonstrated that estrogen and progestins were effective in producing detectable changes in chick oviduct chromatin with kinetics that paralleled receptor binding to nuclei (Schwartz et al. 1975; Tsai et al. 1976). It was therefore of interest to determine whether other classes of steroids (glucocorticoids) might produce similar responses and if such gross effects on chromatin could occur in a cell type where the biological response is limited to a small subset of the expressed genes.

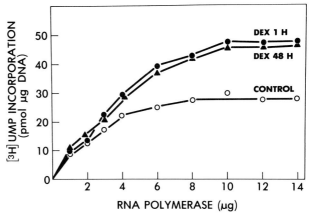

Fig. 10. Early effect of dexamethasone on GC cell chromatin template activity. Chromatin was isolated after treatment of the cells for the indicated times with 1 μM dexamethasone, and template activity was assayed under cell-free conditions with bacterial RNA polymerase as described in Johnson and Baxter (1978).

In Fig. 10 are the results obtained when cultured rat pituitary tumor cells are induced with dexamethasone and the chromatin subsequently isolated and used as a template in a cell-free transcription reaction with *E. coli* RNA polymerase. The surprising result is that in this cell, where the glucocorticoid response is quite specific and restricted to a small subset of the detectable gene products (Martial et al. 1977a), there is some gross change in chromatin within 1 h of hormone treatment that allows up to 40% more RNA to be transcribed by the nonspecific bacterial enzyme. However, because such standard transcription reactions reflect a combination of initiation, elongation, and termination events, it was not possible to determine which step in the reaction had been influenced by hormone treatment. To circumvent this difficulty we adopted the rifampicin challenge assay, developed for prokaryotic templates (Hinkle and Chamberlin 1970) and adopted for chromatin-templated reactions by Tsai et al. (1976) for use in our system. In this assay the template is first incubated with RNA polymerase to form enzyme–template preinitiation complexes. The enzymes bound in such complexes then are allowed to initiate RNA synthesis by the addition of nucleoside triphosphates. Rifampicin is added simultaneously to inhibit all RNA polymerase molecules that are not prepared to immediately initiate transcription. Reinitiation after one round of synthesis is also prevented by the drug. Thus, provided that the size of the RNA chains is known, the number of chains synthesized and hence the number of sites where RNA polymerase enzymes initiated can be estimated.

Figure 11A demonstrates that when these procedures are used, about 50% of the cell-free RNA synthesis is rifampicin resistant. The chromatin prepared from control and hormone-treated cells do not exhibit different sensitivities to the drug. This indicates that the effect of the hormone on cell-free transcription is not to alter the rate of initiation once the enzyme has formed the preinitiation complex; such a change would result in a shift in the rifampicin sensitivity curve (Chamberlin and Ring 1972).

Additional control experiments were performed to determine if the observed changes in chromatin template capacity resulted from initiation of RNA synthesis by the bacterial polymerase on the RNA in chromatin rather than on chromosomal DNA. Ribonucleic acid-templated RNA synthesis has been shown to occur in cell-free transcription systems with eukaryotic chromatin and *E. coli* RNA polymerase (Zasloff and Felsenfeld 1977). Because RNA-directed RNA synthesis is insensitive to actinomycin D, the cell-free reactions were tested for their sensitivity to this inhibitor. In all cases, control and dexamethasone induced chromatin displayed similar sensitivites to the drug. Figure 11B shows that progressive increases in actinomycin D markedly inhibit rifampicin-resistant uridine monophosphate UMP incorporation by chromatin from either control or dexamethasone-treated cultures. The inhibition at low concentrations ($0.25\mu g/ml$) is greater than 50%, whereas 90% of the RNA synthesis was inhibited by 5 $\mu g/ml$. Thus, under the assay conditions employed here, most if not all of the cell-free RNA synthesis is directed by the DNA template.

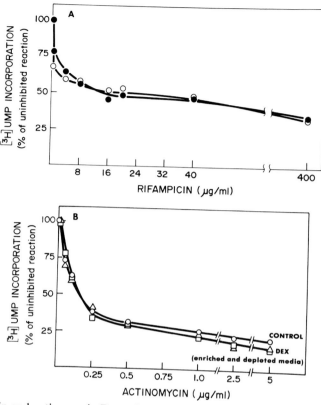

Fig. 11,A,B. Rifampinin and actinomycin D sensitivity of RNA synthesis templated by control and glucocorticoid induced GC chromatin. **A** Chromatin (1 μg) was preincubated with enzyme (10 μg) for 20 min followed by the addition of nucleoside triphosphates, heparin, and various concentrations of rifampicin. RNA synthesis was terminated after 15 min and measured as described (Johnson and Baxter 1978). **B** Chromatin and enzyme were preincubated with the indicated concentrations of actinomycin D and the reactions initiated with nucleoside triphosphates, heparin, and 40 μg/ml rifampicin. (Johnson et al. 1979a)

Other parameters of the cell-free transcription reactions were also investigated. These included the kinetics of RNA polymerase binding to chromatin and RNA chain elongation as well as the distribution of RNA chain sizes. As is evident in Fig. 12A, dexamethasone does not change the rate at which the polymerase binds to the template. Similarly, when the cell-free reactions are terminated at various times after initiation (Fig. 12B), it can be seen that elongation of RNA from rifampicin-resistant sites does not appear to be altered by hormone treatment. Finally, the size distribution of the RNA chains was analyzed to determine whether chromatin from glucocorticoid-treated cells was supporting the synthesis of larger RNA. Figure 12C shows the profile of the RNA synthesized from each template. Also shown is a plot of the reciprocal of the number average molecular weight of the RNA throughout the gradient. Two important points can be made from this

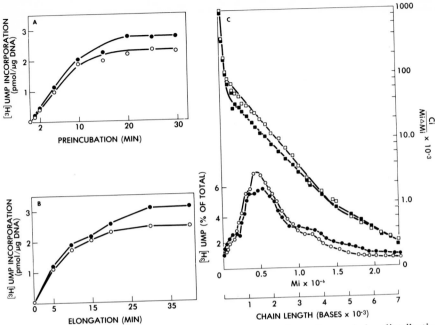

Fig. 12,A–C. Kinetics of RNA polymerase binding, elongation and size distribution of RNA synthesized on control and dexamethasone induced chromatin. **A** Chromatin (1 µg) from control (○) and 1 h induced cells (●) was preincubated with 1 µg RNA polymerase for the indicated time intervals at 37°C. The bound enzymes were then allowed to start RNA synthesis in the presence of rifampicin and heparin. In **B** The preincubation was for 20 min and the amount of UMP incorporated was determined at various intervals after initiation. In **C** Chromatin (10 µg) from control (○) and induced (●) cells was preincubated with 10 µg of *E. coli* RNA polymerase and then allowed to synthesize RNA in the presence of rifampicin and heparin. Sucrose gradient sedimentation was then performed on the products as described in Johnson and Baxter (1978).

experiment. First, the size distribution of the RNA transcripts is not altered by glucocorticoid treatment. Second, three discrete size classes of chains are present. The smallest is 75 bases and comprises 12% of the total bases. The second size class averages 901 bases and represents about 80% of the total, whereas the third class is about 2000 bases and comprises about 10% of the total bases.

Since the hormone-induced increase in cell-free transcription does not result from a change in the rate of chain elongation, RNA length, or the kinetics of polymerase binding to chromatin, it is likely that the steroid effect is on the number of initiation sites. Because the level of initiation could be affected both by the number of sites and by the affinity of the polymerase binding to such sites, we developed methodology to measure both parameters. To measure the affinity of RNA polymerase binding to rifampicin-resistant initiation sites, we quantitated the number of enzymes bound in initiation sites and the number of free enzymes measured with an excess of DNA after removal of the bound enzymes by centrifugation at various concentrations of added RNA polymerase (Johnson and Baxter 1978). A Scatchard plot of the data then reveals the affinity of RNA polymerase binding to chromatin initiation sites. Figure 13 depicts the results from this experiment. The data show one apparent class of binding sites on a thermodynamic basis with an apparent K_d (equilibrium dissociation constant) of 5 nM for chromatin preparations from control hormone-induced cells. However, although the hormone does not alter the affinity of the sites for binding RNA polymerase, there is an increase in the number of sites available for binding.

The glucocorticoid-induced increase in initiation sites was next confirmed by an independent method using γ-[^{32}P]GTP. The results of this experiment are summarized in Table 2 and agree with the data from Scatchard plots of the equilibrium binding curves. When the results of both experiments are expressed as initiation sites per cell, it can be seen that the effect of the hormone is to increase the number of polymerase binding and initiation sites by about 500 000 per cell. This value is an order of magnitude greater than the number of glucocorticoid receptors in these pituitary cells (50 000).

Although estrogen was similarly found to increase the number of initiation sites for a nonspecific structural probe such as bacterial polymerase in chick oviduct (Tsai et al. 1976) and rat uterus (Markaverich et al. 1978), the increase in the number of initiation sites per cell is much greater in the pitui-

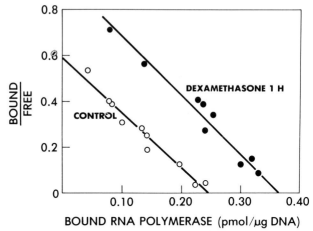

Fig. 13. Scatchard analysis of *E. coli* RNA polymerase binding to GC cell chromatin from control (○) and dexamethasone induced (●) cells (1 h). Following preincubation (20 min) of the chromatin with various enzyme concentrations the active enzymes bound at rifampicin-resistant sites were allowed to initiate RNA synthesis, while enzymes free in solution were assayed with an excess of calf thymus DNA after removal of the chromatin by centrifugation. The picomoles of UMP incorporated by both the bound and free enzymes were then converted into the number of RNA chains synthesized by dividing the total bases incorporated as a given size class by the number chain length of each size class. The average UMP content was assumed to be 25% of the incorporated bases. In the absence of reinitiation the number of chains synthesized is equal to the number of active RNA polymerase molecules bound. (Johnson and Baxter 1978)

Table 2. Effect of Dexamethasone on RNA Polymerase Initiation Sites in GC Cell Chromatin

	Initiation sites				
	Active enzyme per μg DNA (pmol)		Sites per cell[b]		
Chromatin source	Rifampicin assay	$[\gamma\text{-}^{32}\text{P}]$GTP assay	Rifampicin assay	$[\gamma\text{-}^{32}\text{P}]$GTP assay	$K_d{}^a$ (nM)
Control	0.18	0.11	1.2×10^6	0.86×10^6	5.12
Dexamethasone	0.32	0.17	1.8×10^6	1.33×10^6	5.30

[a] Apparent equilibrium dissociation constant determined from Scatchard plots of the equilibrium binding data (Fig. 13).
[b] Based on 13 pg DNA per GC cell.

tary cells.   In oviduct and uterus the number of new sites generated by a 1-h estrogen treatment was 25 000 per cell and 100 000 per cell, respectively. The larger increase in initiation sites in pituitary cells was also surprising in view of our previous studies, which demonstrated that the glucocorticoid response in GC cells at the level of transcription is restricted to a small subset of the translatable mRNAs.   These apparent contradictions led us to ask whether the observed effect on chromatin was indeed a receptor-mediated event and if glucocorticoids elicited a similar response in another target cell.   We also designed experiments to ask if such an effect was also present in isolated nuclei.   For these experiments the ACTH-producing mouse pituitary (AtT20) cell line was compared to the rat pituitary growth hormone producing cell.

Figure 14 shows that the stimulation of initiation sites in GC cells by dexa-

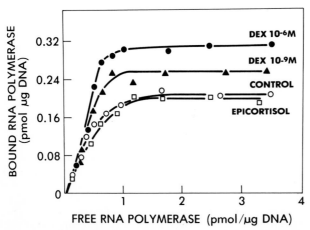

Fig. 14. Concentration and steroid structural requirements for influencing the capacity of chromatin to bind bacterial RNA polymerase.   Cells were induced with either 11α-hydroxycortisol (epicortisol) 1 μM or dexamethasone at concentrations of 1 μM and 1 nM.   After 1 h the chromatin was isolated and was assayed for the number of *E. coli* RNA polymerase initiation sites plotted on the ordinate as a function of the free RNA polymerase concentration as described in Johnson and Baxter (1978).

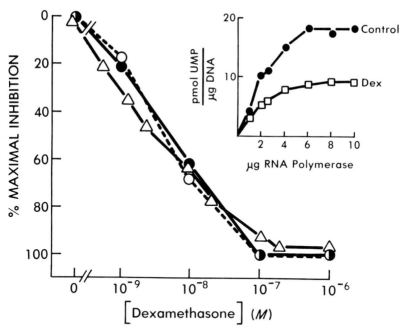

Fig. 15. Glucocorticoid dose response for the inhibition of RNA polymerase initiation sites (●, chromatin; △; nuclei) in isolated AtT20 cell nuclei and chromatin and
inhibition of ACTH–βLPH mRNA activity (○). For initiation site assays the AtT20
cells were induced for 1 h with various concentrations of dexamethasone and either
nuclei or chromatin was prepared and assayed as described in Johnson and Baxter
(1978). For the ACTH–βLPH mRNA assays poly(A) mRNA was isolated from cells
after 48 h of hormone treatment, translated in a reticulocyte cell-free system, and the
amount of ACTH–βLPH determined by immunoprecipitation as described in Roberts et al. (1979a).

methasone is proportional to the quantity of glucocorticoid–receptor complexes in the nucleus. The inactive steroid, epicortisol, which does not bind
to the receptor (Rousseau et al. 1973), failed to elicit the chromatin response,
while 1 nM dexamethasone produced a half-maximal response when compared with 1 μM dexamethasone. A similar experiment was next performed using the ACTH-producing mouse pituitary AtT20 cell line. If such
a nuclear effect reflects, in fact, some important biological response to glucocorticoids, one might expect this cell type to show an inhibition of chromatin-binding capacity for RNA polymerase. The results in Fig. 15 show
that when chromatin or nuclei from AtT20 cells are examined after a 1-h exposure of the cells to dexamethasone, a decrease in the number of initiation
sites is observed. Moreover, the dose–response curves for inhibition of
polymerase binding in either chromatin or nuclei totally paralleled the dose
response curve for inhibition of ACTH–LPH mRNA as assayed by cell-free
translation and immunoprecipitation. Thus, this nuclear response appears
to parallel the biological response to glucocorticoids. Stimulation of growth
hormone synthesis in GC cells is preceded by a stimulation of chromatin-

binding capacity for RNA polymerase, whereas inhibition of this chromatin function also preceds inhibition of ACTH–LPH by glucocorticoid in the AtT20 cells. In order to test these conclusions and to more closely examine the role of the glucocorticoid receptor in the nuclear response, we designed experiments to ask if the change in chromatin structure could also be influenced by manipulations of the cellular milieu known to influence the biological response to glucocorticoids. In these experiments we again employed the media depletion conditions to alter the glucocorticoid response.

D. Bidirectional Glucocorticoid Effects on Chromatin Structure

Figure 16A shows the results observed when chromatin isolated from GC cells is analyzed after a 1-h exposure of the cells to dexamethasone in fresh media, where growth hormone in production is increased by glucocorticoids, or depleted media, where growth hormone is not induced by the steroid (discussed earlier). In this experiment the number and affinity of rifampicin-resistant initiation sites for E. coli RNA polymerase were examined in each chromatin preparation by measuring transcription at several polymerase concentrations. Scatchard analyses of the equilibrium binding data show that the glucocorticoids produced an increase in the number of initiation sites in chromatin from the media-enriched cells as in earlier experiments (see Fig. 13). However, in media-depleted cells, the hormone rapidly induced a decrease in the number of initiation sites (Fig. 16B). In

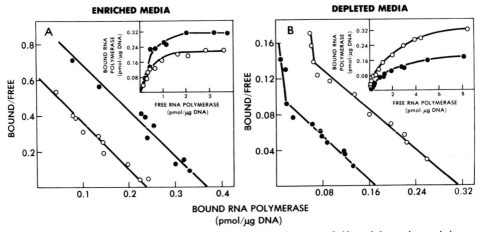

Fig. 16,A,B. Stimulation and inhibition by dexamethasone of rifampicin-resistant initiation sites for *E. coli* RNA polymerase in pituitary tumor cell chromatin. One hour after addition of dexamethasone (1 μM) to media-enriched or -depleted GC cells, chromatin was isolated and the number of rifampicin-resistant initiation sites determined from Scatchard plots of the equilibrium binding data (*insert figures*). The values for bound and free RNA polymerase were calculated as previously described (Johnson and Baxter 1978) by dividing the amount of UMP incorporated into RNA by the number average chain lengths of the transcripts. In A The chain lengths were 901 and 1034 nucleotides for control and dexamethasone treated chromatins, respectively. In **B** the chain lengths were 943 and 1981. (Johnson et al. 1979a)

Table 3. Effect of Cortisol on RNA Polymerase Initiation Sites in Human Skin Fibroblast Chromatin[a]

Chromatin source	UMP incorporated (pmol/μg DNA)	RNA chain length: Nucleotides[b]	Initiation sites: Sites/cell[c]
Enriched media			
Control	38	856	1.1×10^6
Cortisol (0.1 μM)	56	910	1.5×10^6
Depleted media			
Control	52	872	1.45×10^6
Cortisol (0.1 μM)	33	898	0.90×10^6

[a] Fibroblasts were induced for 1 h with cortisol. Chromatin was then isolated and the quantity of rifampicin-resistant initiation sites determined using 1 μg of chromatin and 10 μg of RNA polymerase as described in Johnson et al. (1979a).
[b] Determined from sucrose gradients as described (Johnson and Baxter, 1978).
[c] Based on 10 pg of DNA per cell.

both cases, the affinity (apparent equilibrium dissociation constant, K_d) of the sites for binding bacterial polymerase was not affected by the steroid. Under conditions favoring stimulation, a 1-h glucocorticoid treatment produced over 700 000 additional sites per cell, whereas hormone treatment of media-depleted cells resulted in over 802 000 fewer rifampicin-resistant sites per cell. Chromatin from media-depleted cells may have two classes of binding sites whose affinity differ by about an order of magnitude (as judged by the resolution of the curve of Fig. 16 into two straight lines), whereas only one apparent class was detected in chromatin from cells from media-enriched conditions. In the chromatin preparations from depleted cells the calculated values for the K_d were 3.2 nM and 22 nM, respectively, for the two classes, whereas the sites in chromatin from media enriched cells showed an apparent K_d of 5.2 nM.

The effect of a glucocorticoid on chromatin from cultured skin fibroblasts was also examined. As shown in Table 3, cortisol stimulated an increase of about 400 000 polymerase initiation sites per cell within 1 h in the fibroblast cultures in which tymidine uptake was stimulated. In contrast, the steroid rapidly decrease (by about 550 000 per cell, within 1 h) the number of sites under conditions where thymidine uptake is inhibited.

Dexamethasone also changed the number of initiation sites in mouse lymphoma (S49) cells (Fig. 17). With the lymphoma cells, the kinetics of the response was somewhat slower than in either the fibroblast or pituitary cell systems (2 h versus 1 h to reach a maximal effect). The direction of the influence of culture conditions on the hormone effect on initiation sites also paralleled the biological steroid effect. Cells in enriched media exhibited a much lower decrease (20%) in initiation sites than that produced by dexamethasone in cells in depleted media (75%). As was shown earlier (see Fig. 5), dexamethasone exerted a much greater cell-killing activity in cells in depleted media compared with those in fresh media. The apparent correlation between the direction of hormone effects on chromatin and the nature of the subsequent biological effect (stimulatory or inhibitory) even within the same

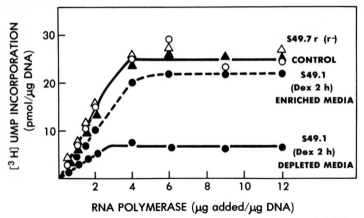

Fig. 17. Inhibition by dexamethason of rifampicin-resistant initiation sites in chromatin from mouse lymphoma cells.   Two hours after dexamethasone addition to media-
enriched or -depleted cultures, chromatin was isolated and titrated with *E. coli* RNA
polymerase.   Rifampicin-resistant UMP incorporation was then measured at each
polymerase concentration: control S49.1 wild type (○) plus dexamethasone (●); S49
7r, receptor minus control (△), plus dexamethasone (▲).   Number average chain
lengths were 842 and 898 nucleotides fdor dexamethasone treated chromatins and 901
for control chromatin.   (Johnson et al. 1979a)

cell type prompted further studies to relate the receptor−nucleus interaction
to the effects on chromatin structures as assayed with the RNA polymerase
probe.

E. Correlations between Receptor Binding and Glucocorticoid Effects on Chromatin Initiation Sites

To better assess the role of the glucocorticoid−receptor in the steroid effects
on chromatin, we examined two steroid-resistant lines derived from mouse
S49 lymphoma cells.   In one line (r⁻, receptor minus), there is very little or
no receptor-binding activity (Rosenau et al. 1972; Sibley and Tomkins
1974).   In the other line (Nt^i, nuclear transfer increased) the total steroid-
binding activity is normal, but a greater percentage of the receptor−steroid
complexes bind to the nucleus than in the parent steroid-sensitive cells (Yamamoto et al. 1974).

Dexamethasone did not induce any change in the number of initiation sites
in chromatin from mutant S49ʳ⁻ cells (Fig. 17).   This result strongly suggests
that the receptors are required for the effect.   Figure 18 shows the effect of
dexamethasone on cell viability, uridine uptake, and chromatin in wild-type
and Nt^i S49 cells in the presence of enriched media.   Under these conditions
the chromatin of the wild-type lymphoma line displayed a time-dependent,
hormone-induced inhibition of rifampicin-resistant initiation sites, which
was maximal by 24 h (Fig. 18A).   During this time period the glucocorticoid-
mediated killing of the cell approached 50%, and there was an 80% inhibition
by the steroid of the incorporation of uridine into acid precipitable material.

In contrast, dexamethasone did not kill the N^{ti} cells, nor did it decrease the chromatin binding of polymerase even after 30 h. In fact, in a number of experiments we have noted that after 2 h the number of initiation sites are commonly greater than the control level by about 25% when the cells are induced in enriched media. Parallelling the lack of an inhibitory response on chromatin and frequent transient stimulation of the number of poly-merase binding sites is a slight increase (about 30%) in uridine uptake.

A kinetic analysis of the decrease in initiation sites produced by dexa-methasone in S49 lymphoma was also performed, as shown in Fig. 19B. The levels of nuclear-bound dexamethasone were determined in parallel cul-tures. The inhibition of the initiation sites by the steroid lagged slightly be-hind glucocorticoid–receptor complex binding to the nucleus. Nuclear dex-amethasone binding reached a maximum by 1 h, whereas the rifampicin-resistant initiation sites were maximally inhibited by approxi-mately 2 h. A similar relationship was also noted with media-depleted GC cells (Fig. 19A) where the effect on chromatin closely followed accumulation of receptor–glucocorticoid complexes in the nucleus. In these cells the ki-netics of initiation site inhibition and of binding (maximal by 0.5 and 1 h, re-spectively) were somewhat faster than in the S49 cells.

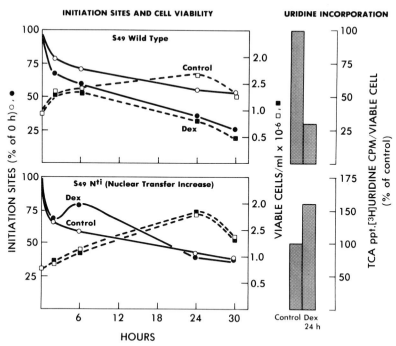

Fig. 18. Effect of glucocorticoids on cell viability, uridine incorporation and rifampi-cin-resistant initiation sites in wild type and mutant S49 cells. At various times after addition of dexamethasone to media enriched cultures, aliquots of cells were removed for chromatin isolation, and cell viability determination (trypan blue exclusion). After 24 h, samples were also removed for determination of acid-precipitable uridine incorporation. (Johnson et al. 1979a)

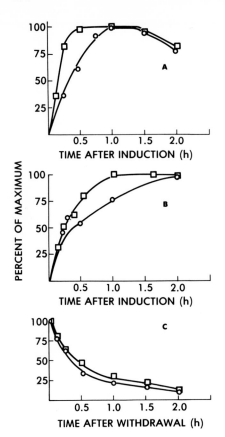

Kinetics of glucocorticoid receptor accumulation in GC and S49 cell nuclei and alteration of rifampicin-resistant initiation sites in isolated chromatin. **A** GC (depleted media; **B** S49.1 (depleted media; **C** GC (enriched media). In **A** and **B**, at various times after hormone addition to media-depleted GC and S49 cells, respectively, chromatin was isolated and the number of rifampicin-resistant initiation sites analyzed with saturating quantities of *E. coli* RNA polymerase. In parallel cultures the nuclear accumulation of [^{3}H]dexamethasone was determined at identical times. In **C**, 4 h after the addition of dexamethasone to media enriched GC cultures, the cells were washed 5 times with glucocorticoid-free media and incubated at 37°C. At the indicated time intervals chromatin was isolated from the cells and the initiation sites determined. In parallel cultures subjected to an identical experimental protocol, nuclear-bound [^{3}H]dexamethasone was assayed: (□) nuclear bound [^{3}H]dexamethasone; (○) rifampicin resistant initiation sites. Maximal nuclear binding was 2.3 fmol of dexamethasone per microgram of cellular DNA. (Johnson et al. 1979a)

Using the pituitary cell system with enriched culture conditions, it was also noted that the hormone-induced increase in initiation sites required the continued presence of the glucocorticoid–receptor complex in the nucleus. Figure 19C shows that once the glucocorticoid is removed from the induced cultures, bound hormone in the nucleus quickly drops. This is paralleled by a disappearance of the hormone-induced effect on chromatin. In both cases, a $t_{1/2}$ of about 15 min was required for the disappearance. In all cases the effect on chromatin reached a maximum well before the biological effects of the hormone. These temporal relations suggest that the hormone effect on initiation sites is an early step in the mechanism of glucocorticoid action.

IV. A General Model for Glucocorticoid—Receptor Function

The experiments described above have shown that functional glucocorticoid receptors when bound to the nucleus influence a process that alters chromatin structure. This can be readily assayed in either chromatin or nuclei (using bacterial RNA polymerase as a probe). Such results are similar to findings on the estrogen-responsive chick oviduct (Schwartz et al. 1975; Tsai et al. 1976) and rat uterus (Markaverich et al. 1978), as well as with results on androgen action presented here by Liao et al. (Chap 12). Thus these findings may represent an early event common to the action of all steroid hormones.

Before considering molecular models that can explain the observed nuclear effects, it is necessary to reiterate that the nature of the hormone-induced changes in chromatin measured as an altered number of initiation sites for added bacterial RNA polymerase is still unknown. Such sites do not represent sites where RNA synthesis is initiated in the intact cell (Johnson et al. 1979b). Rather, the evidence to date indicates that the assay itself reflects a structural conformation in chromatin that results from partially destabilized base pairing of DNA in localized areas. These ideas are derived from temperature dependence studies on the formation of rifampicin-resistant initiation sites performed by O'Malley and co-workers using chick oviduct chromatin and DNA. In these experiments the formation of rifampicin-resistant preinitiation complexes with RNA polymerase and chromatin is much less dependent on temperature than that with protein-free DNA (Tsai et al. 1975). In the latter case destabilization is accomplished by the σ subunit of $E.\ coli$ RNA polymerase and is strongly temperature dependent (Hinkle and Chamberlin 1970). Thus, intrinsic chromosomal proteins may serve to destabilize base-paired DNA in given regions of chromatin.

There are also no experiments to date which show that such hormone-induced changes in chromatin cause changes in the transcription of specific mRNAs. However, because these effects are tightly coupled on a kinetic and dose–response basis to the presence of receptors in the nucleus, the hypothesis must be considered that it in some manner reflects a biologically important function of the steroid–receptor complex. Therefore, the remainder of this discussion concerns possible mechanisms whereby the hormone–receptor complex could mediate such changes in chromatin structure, consistent with current knowledge of the glucocorticoid receptor and in particular the considerations mentioned in Sect. II.

Certain constraints are placed on possible mechanisms of receptor action by our data showing that receptors are able to influence the appearance or disappearance of a large excess of initiation sites for RNA polymerase — about 10 sites per receptor. This is not consistent with the most straightforward mechanism whereby each receptor influences one initiation site. Alternatives such as receptor binding repeatedly at multiple sites, thereby altering chromatin structure at many regions, or receptor binding causing stabilization or destabilization of the local packing of DNA over a region sufficiently large to affect the binding of many polymerase molecules cannot be excluded by the present data.

However, as mentioned previously, an important consideration is the considerable influence the cellular milieu has upon glucocorticoid action without demonstrably altering the hormone receptors themselves. It therefore must be hypothesized that the mechanisms of receptor function involve other intermediate factors, which influence how the receptors interact with nuclear acceptor sites or how the chromatin responds to receptors once they are bound to the acceptor sites. An attractive hypothesis is that the intermediate factors sensitive to the metabolic state of the cell are enzymatic reactions that catalyze the covalent modification of chromosomal proteins. Potential modification reactions could include phosphorylation (Kleinsmith et al. 1976; Teng et al. 1971), acetylation (Jackson et al. 1975; Marushige 1976), methylation (Allfrey 1971), and ADP ribosylation (Hayaishi and Veda 1977; Mullins et al. 1979), all of which are known to correlate with changes in gene activity and possibly chromatin conformation.

From the known selectivity of the glucocorticoid response at the RNA and protein level it is possible that most of the modifications of chromatin structure do not subsequently influence the expression of specific genes. This mechanism has precedents in the mechanism of action of two other gene activators, cAMP and butyrate, that operate via the covalent modification of proteins. In the case of cAMP there is strong evidence for the phosphorylation of numerous substrate proteins, most of which apparently do not result in physiological influences (Schwartz et al. 1975). With butyrate the parallel is even more striking. Butyrate treatment of certain hematopoetic cells induces globin synthesis, which is accompanied by the direct inhibition of histone deacetylation (Leder and Leder 1975; Riggs et al. 1977). Experiments by others using a number of cell types have shown that the influence of butyrate on histone deacetylation is quite extensive, occurring throughout the chromatin, and does not always influence gene activity (Cousens et al. 1979). In HeLa cells, the butyrate-induced increase in histone acetylation results in a structural change in chromatin detectable by increased kinetics of digestion of nuclei by DNase I but not micrococcal nuclease (Simpson 1978). This results because acetylation of nucleosome core histones (H_4, H_3) alters the structure of the nucleosome. The DNA within the nucleosome is accessible to DNase I but not to micrococcal nuclease, which only cuts the DNA between nucleosomes.

In the rat pituitary cells used for the current studies butyrate does cause an accumulation of acetylated core histones but has no effect on gene expression, as determined by two-dimensional gel analysis, or on the number of initiation sites for bacterial RNA polymerase (Ivarie R, Johnson LK, Baxter JD unpublished data). While the involvement of protein modifications in glucocorticoid action remains an attractive hypothesis, the butyrate data indicate that the mechanism may not involve acetylation of core histones. The experiments also suggest that the initiation site assay reflects changes in chromatin structure outside or of a higher order than the nucleosome.

The findings above show that certain protein modifications that demonstrably regulate gene activity can produce widespread alterations in chromatin structure (such as those we have observed with glucocorticoids) and still result in a specific biological result. Such a "catalytic" intermediate of glucocorticoid receptor function is also flexible enough to allow modulation of the hormone responses by the metabolic status of the cell.

The following working model for receptor function is consistent with all of

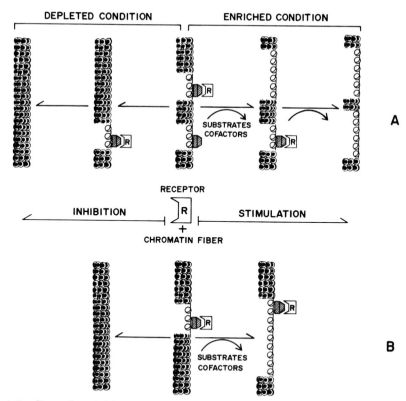

Fig. 20,A,B. General model for glucocorticoid–receptor-induced changes in chromatin structure.

A Multiple-site model; receptors bind repeatedly and influence the structure of chromatin at many sites. **B** Single-site model. Each receptor only binds to one site, but the resultant change in chromatin structure affects the binding of many (~ 10) RNA polymerases per receptor. In each case the change in chromatin is mediated by the covalent modification of chromosomal proteins. Thus, depending on the availability of reactants and products, the reaction can be driven forward (stimulation) or backward (inhibition).

the data to date. After binding the steroid the activated receptor complex translocates to the nucleus and binds extensively to acceptor sites, where it initiates a modification of chromosomal protein(s). The binding of receptors could drive the modification reaction(s) either forward or backward, depending on the availability of substrates or cofactors involved. Alternatively, each receptor could either contain an enzymatic activity or be a co-factor or subunit of an enzyme that catalyzes the modification of chromosomal proteins. As illustrated by the model shown in Fig. 20, the modification of chromosomal proteins might occur either as in Fig. 20A, at multiply spaced or as in Fig. 20B, at contiguous sites, thereby altering the structure of chromatin such that when analyzed in vitro the binding of many RNA polymerases are affected. Thus, if the reaction should proceed in one direction, there would be an increased binding of RNA polymerase, whereas if the reverse reaction occurs, there will be a decrease in the available binding sites within chromatin. Whereas the modifications may occur at numerous loci in chromatin, the majority of these must not subsequently influence the expression of specific genes. Thus, the differentiation of the target cell would determine whether such modifications will affect the expression of particular genes. Through such processes gross receptor-mediated changes in the capacity of chromatin to bind bacterial RNA polymerase could be observed in the face of a highly selective biological response.

V. Summary

Glucocorticoid hormones are known to exert many of their biological effects through the regulation of specific mRNAs. In order to formulate a testable hypothesis for the possible mechanisms involved, various aspects of glucocorticoid action have been considered. First, the specificity of the hormone response in cultured cells at the mRNA and protein level was shown to be quite selective, encompassing only a small percentage of the detectable gene products. Second, these specific responses in various cell types can be either stimulatory, as in the case of growth hormone production by rat pituitary tumor cells, or inhibitory, such as with ACTH–LPH synthesis in mouse pituitary tumor cells. Third, both the magnitude and direction of the glucocorticoid response in a given cell type can be modulated by other hormones, growth factors, and even media depletion without apparent influences on the glucocorticoid receptors or their nuclear translocation. Thus, some component in the pathway between hormone–receptor binding to nuclei and the observed change in mRNA is sensitive to the metabolic status of the cell.

Evidence from systems in which glucocorticoids stimulate specific RNAs (those coding for tyrosine aminotransferase and mouse mammary tumor virus) indicate that the increases are a rather direct consequence of the receptor interaction with chromatin; no induced protein intermediate is required. In contrast, studies with lymphoid cells have indicated that a protein(s) intermediate(s) is first induced by glucocorticoids, which then produces the observed inhibitory responses. However, using the ACTH-producing AtT20 cell line it was observed that glucocorticoids could directly inhibit mRNA levels of this protein even in the presence of cycloheximide. Thus, the possibility exists that both stimulatory and inhibitory glucocorticoid actions can result from a direct interaction between the receptor and chromatin.

An examination of the nature of the nuclear binding sites for receptors indicates that they are probably in excess of the number of receptors in the cell. In addition, the observed binding probably does not differ in affinity from that of the sites where the receptors initiate their actions (that is, no hidden high affinity "operator" sites have been detected). The sites also contain DNA although other subchromosomal components also influence the binding to chromatin or nuclei. This prompted experiments to determine if the receptors were in fact acting at the acceptor sites, with only a portion of these actions successfully transmitted into influences on specific genes.

Using a nonspecific bacterial RNA polymerase to probe gross changes in chromatin structure, major glucocorticoid-induced alterations were observed in the number of initiation sites for this enzyme in all cell types measured (~ 10 sites per receptor). The direction of the influence on chromatin frequently correlated with that of the biological response in the cell. In addition media depletion by the cells resulted in a more inhibitory (or less stimulatory) response by chromatin, while hormone induction in the presence of fresh culture media favored a more positive (or less inhibitory) response to the steroid. Both the positive and negative nuclear responses also displayed kinetics that lagged slightly behind receptor binding to chromatin ($\sim 20-30$ min). This could indicate an intermediate event between receptor binding and the resultant change in chromatin. Such an intermediate is implicated by findings with the N^{ti} S49 lymphoma cell mutant, which has a defective receptor that binds more avidly to nuclei yet is unable to modify chromatin structure. It is hypothesized that the receptors once bound to chromatin influence a process that reversibly modifies chromosomal proteins. The magnitude and direction of the reaction, either inhibitory or stimulatory, are determined by the metabolic state of the cell, whereas the selectivity of the response is a function of its differentiation.

Acknowledgments. The authors thank Ms. Susan Bromley for preparation of the manuscript. S. K. N. is supported by a Jane Coffin Childs Cancer Fund postdoctoral fellowship. J. D. B. is an Investigator of the Howard Hughes Medical Institute. This work was supported by NIH Grant AM 19997-02.

References

Allfrey VG (1971) In: Phillips DM (ed) Histones and Nucleohistones. Plenum Press, New York, p 241

Bak A, Zenthen J, Crick F (1977) Proc Natl Acad Sci USA 74: 1595–1599

Baxter JD (1976) Pharm Ther Biol 2: 605–659

Baxter JD, Rousseau GG (eds) (1979) Glucocorticoid Hormone Action. Springer-Verlag, Heidelberg

Baxter JD, Rousseau GG, Benson MC, Garcia RL, Ito J, Tomkins GM (1972) Proc Natl Acad Sci USA 69: 1892–1896

Cahill GF (1971) In: Christy N (ed) The Human Adrenal Cortex. Harper and Row, New York, pp 205–239

Cedar M, Felsenfeld G (1973) J Mol Biol 77: 237–242

Chamberlin M, Ring J (1972) J Mol Biol 70: 221–237

Cousens L, Gallwitz D, Alberts B (1979) J Biol Chem 254: 1716–1723

Dannies PS, Tashjian AH (1973) J Biol Chem 248: 6174–6179

Ernest JM, Feigelson P (1979) In: Baxter JD, Rousseau GG (eds) Glucocorticoid Hormone Action. Springer-Verlag, Heidelberg, pp 219–241

Exton JH (1972) Metabolism 21: 945–990

Fain JN, Czech MP (1975) In: Blaschko H, Sayers G, Smith A (eds) Handbook of Physiology: Endocrinology, The Adrenal Am. Physio. Soc., Washington, DC, pp 169–178

Fauci AS, Dale DC (1974) J Clin Invest 53: 250–256
Feigelson P, Yu F, Hanoune J (1971) In: Christy N (ed) The Human Adrenal Cortex. Harper and Row, New York, pp 257–272
Garel A, Zolan M, Axel R (1977) Proc Natl Acad Sci USA 76: 1682–1686
Gehring U, Thompson EB (1979) In: Baxter JD, Rousseau GG (eds) Glucocorticoid Hormone Action Springer-Verlag, Heidelberg, pp 399–421
Gospodarowicz D, Moran J (1974) Proc Natl Acad Sci USA 71: 4584–4589
Hallahan C, Young DA, Munck A (1973) J Biol Chem 248: 2922–2927
Hayaishi O, Ueda K (1977) Ann Rev Biochem 46: 95–116
Higgins SJ, Rousseau GG, Baxter JD, Tomkins GM (1973) J Biol Chem 248: 5866–5872
Hinkle D, Chamberlin M (1970) Cold Spring Harbor Symp Quant Biol 35: 65–72
Ivarie RD, O'Farrell PH (1978) Cell 13: 41–55
Jackson V, Shires A, Chalkley R, Granner D (1975) J Biol Chem 250: 4856–4863
Johnson LK, Baxter JD (1978) J Biol Chem 253: 1991–1997
Johnson LK, Lan NC, Baxter JD (1979a) J Biol Chem 254: 7785–7794
Johnson LK, Rousseau GG, Baxter JD (1979b) In: Baxter J, Rousseau G (eds) Glucocorticoid Hormone Action. Springer-Verlag, Heidelberg. pp 305–325
Kleinsmith LJ, Stein J, Stein G (1976) Proc Natl Acad Sci USA 73: 1174–1178
Leder A, Leder P (1975) Cell 5: 319–322
Levy B, Dixon G (1978) Proc Natl Acad Sci USA 76: 1682–1686
Litwack G, Singer S (1972) In: Litwack G (ed) Biochemical Actions of Hormones. Academic Press, New York, Vol II, pp 113–165
Markaverich BM, Clark JM, Hardin JW (1978) Biochemistry 17: 3146–3151
Martial JA, Baxter JD, Goodman HM, Seeburg PH (1977a) Proc Natl Acad Sci USA 74: 1816–1820
Martial JA, Seeburg PH, Guenzi D, Goodman HM, Baxter JD (1977b) Proc Natl Acad Sci USA 74: 4293–4295
Marushige K (1976) Proc Natl Acad Sci USA 73: 3997–3609
Mosher KM, Young DA, Munk A (1971) J Biol Chem 246: 654–659
Mullins DW, Giri CP, Smulson M (1979) Biochemistry 16: 507–518
Nakanishi S, Kita T, Taii S, Imura H, Numa S (1977) Proc Natl Acad Sci USA 76: 3283–3288
Nicholson ML, Young DA (1978) Cancer Res 38: 3673–3680
O'Malley BW, Means A (1974) Science 183: 610–620
Palmiter RD, Catlin GM, Cox RF (1973) Cell Diff. 2: 163–170
Parker CS, Roeder RG (1977) Proc Natl Acad Sci USA 74: 44–48
Peterkofsky B, Tomkins GM (1968) Proc Natl Acad Sci USA 60: 222–228
Pratt WB, Aronow L (1966) J Biol Chem 241: 5244–5251
Riggs MG, Whittacker RG, Neumann JR, Ingram VM (1977) Nature 268: 462–464
Ringold GM, Yamamoto KR, Tomkins GM, Bishop J, Varmus HE (1975) Cell 6: 299–305
Roberts JL, Budarf ML, Baxter JD, and Heibert E (1979)a Biochemistry 18: 4907–4914
Roberts JL, Budarf ML, Johnson LK, Allen R, Baxter JD, Herberg E (1979)b In: Sato, G and Ross R (eds) Hormones and Cell Culture: Cold Spring Harbor Symp Cell Proliferation. Cold Spring Harbor Laboratory, New York, Vol. II pp 827–841
Rosenau W, Baxter JD, Rousseau GG, Tomkins GM (1972) Nat New Biol 237: 20–24
Rousseau GG, Baxter JD (1979) In: Baxter JD, Rousseau GG (eds) Glucocorticoid Hormone Action. Springer-Verlag, Heidelberg, pp 49–77
Rousseau GG, Baxter JD, Higging SJ, Tomkins GM (1973) J Mol Biol 79: 539–554
Rousseau GG, Higgins SH, Baxter JD, Gelfand D, Tomkins GM (1975) J Biol Chem 250: 6015–6021
Rubin CS, Rosen OM (1975) Ann Rev Biochem 44: 831–887
Schultz G, Beato M, Feigelson P (1973) Proc Natl Acad Sci USA 70: 1218–1221

Schwartz, RJ, Tsai M-J, Tsai SY, O'Malley BW (1975) J Biol Chem 250: 5175–5182
Shapiro D, Baker H, Stitt D (1976) J Biol Chem 251: 3105–3111
Sibley CH, Tomkins GM (1974) Cell 2: 213–227
Simpson RT (1978) Cell 13: 691–699
Spelsberg TC, Steggles AW, O'Malley BW (1971) J Biol Chem 246: 4188–4197
Spindler S (1979) Biochemistry 18: 4042–4048
Stallcup MR, Ring J, Yamamoto KR (1978) Biochemistry 17: 1515–1521
Steggles AW, Wilson GN, Kantor JA, Piccinao DJ, Falvey AK, Anderson WF (1974) Proc Natl Acad Sci USA 71: 1219–1223
Teng CS, Teng CT, Allfrey VG (1971) J Biol Chem 246: 3597–3609
Thrash CR, Cunningham DD (1973) Nature 242: 399–401
Tsai M-J, Schwartz RJ, Tsai SY, O'Malley BW (1975) J Biol Chem 250: 5165–5174
Tsai M-J, Towle MC, Harris SE, O'Malley BW (1976) J Biol Chem 251: 1960–1968
Weintraub H, Groudine M (1974) Science 193: 848–858
Yamamoto KR, Alberts B (1975) Cell 4: 301–310
Yamamoto KR, Stampfer MR, Tomkins GM (1974) Proc Natl Acad Sci USA 71: 3901–3905
Young DA, Barnard T, Mendelsohn S, Giddings S (1974) Endo Res Commun 1: 63–72
Zasloff M, Felsenfeld G (1977) Biochemistry 23: 5135–5145

Discussion of the Paper Presented by J.D. Baxter

GOLDBERGER: I think that you made a proper disclaimer about the initiation site assay when you began. But, somehow I find it very disquieting to see the bacterial RNA polymerase initiation sites enumerated. The fact that you are using this as a probe for changes in the structure of chromatin is fine. I also like the fact that you have shown something interesting, but I certainly hope that no one would take home the idea that there are 50 000 new sites for initiating new messages. Besides, bacterial polymerases binding to eukaryotic DNA probably has nothing whatsoever to do with the initiation of messenger synthesis in eukaryotic cells.

BAXTER: I would certainly agree with you. We absolutely did not want to imply in any way that what we are looking at here is initiation of specific genes. Now it may be that these polymerases are binding somewhere near the loci where actual genes in vivo are initiated; we would not want to say whether or not that is true. Certainly the polymerase is not going to initiate with fidelity or specifity, or it doesn't look as if it is. We don't want to come to that kind of conclusion. The only conclusion I want to make from these data is that there is some change in the chromatin induced by the hormone, and interestingly the changes are correlated with the biological response.

GOLDBERGER: Could I just say that the RNA polymerase does bind to SV40 DNA very specifically and in the wrong place, so that specificity is not in itself the point. It is whether this is physiologically meaningful, and here I agree with you.

OMALLEY: I don't know why you need to get anymore upset over the use of initiation site assays than for DNase digestion experiments. Some people actually say that they measure the same things. I don't feel that this is the case since they do not show a direct correlation. In other words, if you give hormone, an expressed gene in a primary response becomes sensitive to DNase. If you take hormone away again, it still remains relatively sensitive to DNase even though the gene stops being expressed. In contrast, the initiation site assay follows the response. It increases with induction, regresses again when you take away the hormone, and gene expression ceases. They might be both looking at some aspect of chromatin structure, but certainly not the identical parameters.

BAXTER: I'd like to reiterate the DNase situation. In this case we are usually looking at rather minor effects on the expression of the gene. The ACTH message only goes down to 39% of the control. In many conditions using the growth hormone, although we can get a maximal of 250-fold induction if we really try, most of the time we are talking about three- to fourfold effects. Here in most cases we are clearly looking at genes that are expressed both in the absence and presence of the hormone, and so this probably really bears no relation to the DNase susceptibility that other people have examined.

CLARK: I don't know what using *E. coli* polymerase for probing chromatin means either, but to substantiate John Baxter's information, we have just published a paper describing our use of estradiol, estriol, and nafoxidine to demonstrate that initiation site activity parallels receptor binding, uterotropic function, and activation of RNA polymerases. So it follows that when the receptors are in the nucleus, they stimulate polymerase activity and initiation site activity, All these corollaries add up to changes that occur in chromatin.

EDELMAN: I am going to try to put the phenomenology into a slightly different context, because I'm not yet satisfied that the *E. coli* polymerase assay, in fact, measures a change in the native state of chromatin in the nucleus. You are well aware of the fact that when you prepare chromatin, you get enhanced template activity in comparison to normal cell and this is independent of hormone action. Now one cell type that we have studied this in carefully was the chicken erythrocyte, in which the nucleus is condensed. When you prepare the nuclei, you get very little sensitivity to *E. coli* polymerase. When you prepare the chromatin, you get a marked increase. In fact, now the chromatin has almost (this is independent of hormone) the same polymerase activity or template activity in response to *E. coli* polymerase as spleen or kidney chromatin from the rat. What is possibly happening is that in the nucleus there is a hormone-sensitive change all right, but the change has to do with something that is not at all related to the native state of chromatin or to the mechanism by which the chromatin is being regulated.

BAXTER: Everybody, I think, agrees that chromatin as you isolated it is different. There is no reason to think right now or to know that chromatin is identical once you isolate it from the cell. Also, I try to emphasize that we don't know what this change is, or even what it means, nor have we proved that this change is responsible for the hormone effect. I don't know the answer to that, nor do I want to make a conclusion on it. The thing is, though, that when you isolate the chromatin under these two conditions, the control is different. So the hormone has done something. What this is we don't know.

O'MALLEY: A final comment to Izzy's (Edelman's) comment. I disagree, and I want to clarify that issue about what you said about chromatin versus nuclei. It is true that normally if one is working with nuclei or with chromatin, one gets a lot more RNA made through chromatin, but that is a spurious result, depending on how you set up the experiment. Most people add a much larger molar excess of polymerase per unit DNA or gene DNA to chromatin than is actually in nuclei. In nuclei there are somewhere between 10 000 and 80 000 polymerases per haploid genome, whereas in chromatin the tendency is to have a lot of this enzyme, especially when *E. coli* enzyme is present, and per unit enzyme there is less RNA synthesis, whereas actually in the overall reaction there may be more. In fact, if you calculate the reaction, and if you adjust it on the basis of the polymerase-to-chromatin ratio, so far in every chromatin I've seen, the rate of transcription for DNA sequences is less in chromatin than in nuclei. In fact, that is one of the paradoxes that we are trying to figure out now— what is missing in chromatin to give us that enhanced rate of transcription one sees in nuclei. I think that what you said might vary, depending on the individual conditions of the experiment; it does not hold as a general rule for transcription in those two states. This, I think, has nothing to do with whether *E. coli* polymerase has any biological phenomena. That is just a specific.

EDELMAN: I think that the basic point is that there is no way at the moment in which you can relate defined and isolated chromatin to the native state under hormone conditions. You have to be very careful how you look at this phenomenon as an induction effect of some kind, or as a hormone-dependent effect. But the nature of such an effect translating into the statement that there is no question that the chromatin has changed, draws a conclusion that I think is not yet justified.

STEVENS: You didn't say anything about what you think might be present or missing from your enriched or depleted media in terms of being able to modulate that steroid response in the very impressive way that you have shown. [Have you tried to identify these factors?] I became aware of a report by a Japanese group about a substance that they claimed modifies TAT induction by glucocorticoids. They recently have shown, in a paper that I haven't read yet (so I am embarrassed) that it increases the sensitivity of the L5178Y cells to glucocorticoid-induced lymphocytolysis by a factor of about two. So I wonder whether you could tie this in with the kind of effect you described here.

BAXTER: I don't have very much information about either. It has been frustrating for us, because we have studied huge numbers of things, including: cyclic AMP, EGF, FGF, insulin, calcium, thyroid. Just about all of these agents do modulate the cellular sensitivity to glucocorticoids in several of these systems. But what we don't have at all is a clear picture, and so I'm not saying anything until it becomes clarified a little more.

THOMPSON: With respect to the factor that is not yet purified, one question of course is whether that would turn out to be identical to an already known growth factor. So far, its properties suggest that it may not be, but it is not at all clear. It could be a very interesting compound, obviously, if in fact it acts to potentiate existing steroids. I have communicated with the laboratory where this substance is being studied. They are just actively working very hard on trying to get total purity before proceeding with the physiological work.

STEVENS: Do they think that it is a carbohydrate-containing substance?

THOMPSON: They think it contains carbohydrate, but it's not pure carbohydrate. It probably contains protein as well.

MUELLER: Do you actually do any studies on the state of the receptors from your cells that were in the actively growing state and the nonactively growing state? Did you actually do things such as the kinetics of binding in competition and exchange phenomenon with receptors from cells in those two states?

BAXTER: We have not done any extensive studies. I have shown some data of receptors actually in both states for on and off rates. We do know that the number and the proportion in the nucleus and that roughly the kinetics of binding are identical. We have not studied the receptors in any more detail. You could expect a lot of this data by saying these cells somewhat modify the receptor. There are some rather long-winded arguments against that, but nothing tight. That is one of the viable possibilities.

MUELLER: It is the long windedness of the arguments to the situation that bothers me.

BAXTER: I think that is a state, when you have something, you do not know what it means.

Discussants: J.D. BAXTER, J.H. CLARK, I.S. EDELMAN, R.F. GOLDBERGER, G.C. MUELLER, B.W. O'MALLEY, J. STEVENS, and E.B. THOMPSON.

Androgens

Chapter 11

Enigmas in the Molecular Biology of Androgen Action

H. G. WILLIAMS-ASHMAN

"The secret of being tiresome is in telling everything"
Voltaire

It has been said that a valuable scientific hypothesis is not so much a cogent explanation of phenomena as a successful policy for research. For more than fifteen years it has become increasingly accepted that control of gene expression mechanisms is a salient feature of androgen action, and that somehow or other, androgens profoundly influence the transcription of specific genes in the nuclei of target cells (Mainwaring 1977; Williams-Ashman 1965; Williams-Ashman and Reddi 1972). Around the turn of this decade, specific androgen–receptor proteins were discovered in a number of responsive tissues. These receptors interact firmly with testosterone and even more tenaciously with 5α-dihydrotestosterone (DHT), a major metabolite of circulating testosterone in many but not all androgen-sensitive cells. The androgen–receptor complexes are readily translocated into cell nuclei and retained there in association with chromatin (Jensen et al. 1977; Liao, 1975; Liao and Fang 1969; Mainwaring, 1977; Williams-Ashman 1975; Williams-Ashman and Reddi 1971; Wilson and Gloyna 1970). However despite various claims that DHT–receptor complexes can directly stimulate RNA polymerase reactions in isolated nuclear preparations, the mechanisms by which these hormones affect RNA transcription and processing in the nucleus remain enigmatic. And there is increasing evidence hinting that not all of the biological effects of androgenic hormones are contingent upon their influencing intranuclear biosynthetic events. By way of introduction to the three contributions that follow, it may be helpful to highlight some unique features of mammalian androgen physiology that must be taken into account by any comprehensive theory of the molecular basis of action of these hormones.

I. Androgens and Male Sex Differentation in the Fetus

In eutherian mammals, in which the heterogametic sex appears invariably to be the male one, androgens secreted by the fetal testis at critical periods of embryonic life serve as crucial triggers for the differentiaton of male reproductive organs originating from the urogenital sinus, the urogenital tubercle, and the Wolffian duct. (A separate fetal testicular hormone—probably a glycoprotein—induces regression of the Müllerian duct during normal male sexual development.) Wilson (1978) has pioneered in the study of the biochemistry of androgen-induced differentiation of the prostate gland and seminal vesicles in the fetus. Testosterone is the only androgen formed in substantial amounts by the fetal testis prior to and at the time of male accessory gland differentiation. Conversion of testosterone into DHT is extensive in the urogenital sinus and external genital primordia just prior to their differentiation into male structures. In contrast, DHT formation in the Wolffian duct is not demonstrable until development of the seminal vesicles and epididymis is already well advanced in the embryo. Very few biochemical data are available concerning the role of androgen receptors and amplificatory gene expression processes in the androgen-induced fetal development of a phenotypically male primitive reproductive tract. However, studies on mutants in a number of species that exhibit partial or complete resistance to androgens accord well with the notions that specific androgen receptors must be present in the appropriate fetal tissues when the male phenotype is established, and that DHT formation from testosterone is involved in the differentiation of the prostate gland, penis, and scrotum, but not in that of the seminal vesicles and epididymis. There are indications that androgens may primarily affect the mesenchyme of the primitive indifferent urogenital tract and that mesenchymal–epithelial interactions are of critical importance in the formation of male genital structures (Cunha 1972). Whereas conversion of testosterone to both DHT and phenolic estrogens seems to be involved in the androgen-mediated virilization of the hypothalamus that occurs, depending on the species, in late fetal or early neonatal life, it is not known whether estrogen as well as DHT production is involved in male genital tract differentiation.

II. Role of Metabolites in the Biological Actions of Testosterone

Although DHT certainly seems to be a major "active form" of testosterone in the adult prostate and seminal vesicles, this is not the case with regard to some other androgen-responsive tissues in the adult male. Furthermore, the importance of 5α-androstan-3β, 17β-diol which can be formed from DHT in some tissues of the male reproductive tract (Roy et al. 1972) in the androgenic regulation of these organs remains obscure. There is evidence that the dog prostate uniquely contains an androgen receptor protein that readily combines with testosterone and 5α-androstan-3α, 17α-diol but not with DHT (Evans and Pierrepoint 1975), but whether this particular 17α-androstandiol is of central importance in the androgenic regulation of the canine prostate remains to be clarified. In any case, the clear-cut role of certain of its metabolites in the action of testosterone on some tissues stands in marked contrast to the overwhelming evidence that 17β-estradiol and progesterone act on female reproductive tissues without undergoing chemical transformations.

III. Use of Mutations in the Analysis of Androgen Action

Mutants in a number of mammalian species have proved useful in probing various features of androgen action. One set of mutations are those in which there exist partial or complete blocks to the biological actions of testosterone, because of defects in (1) androgen receptor proteins (for example, the X-linked *Tfm* mutations in certain rodents or some forms of the testicular feminization syndrome in man); (2) intranuclear actions of androgen–receptor complexes; or (3) inadequate operation of the steroid 5α-reductase that catalyzes the NADPH-dependent reduction of testosterone to DHT. These mutants present a picture of genotypic (XY) males with testes and no trace of ovarian tissues, but with varying degrees of phenotypic female characteristics. In some species, phenocopies of the testicular feminization mutants exhibiting defects in androgen receptors can be induced by treatment of male fetuses with antiandrogenic drugs such as cyproterone acetate (Elger et al. 1970).

Entirely different types of mutants have been successfully employed in the analysis of androgen regulation of specific enzyme synthesis. This is exemplified by the studies described in the contribution by Leslie Bullock in this volume, in which mutations of structural and regulatory genes concerned with β-glucuronidase synthesis have provided new insight into the control by testosterone of the formation of this enzyme in mouse kidney.

IV. Concluding Comment

Prediction of the course of research in any scientific discipline is always hazardous. Yet sometimes it may be easier to foresee what is unlikely to be accomplished in the near future than to anticipate discovery. In trying to get to the heart of the actions of androgens and other types of steroid hormones on gene expression processes, we must not lose sight of the fact that there remain many huge gaps in our understanding of the mechanisms of formation and intracellular accumulation of specific RNA molecules. Until much more is understood about the intranuclear processing of functional mRNA molecules from initial transcripts of their nuclear DNA genes, and also the fine details of mRNA translation by cytoplasmic polyribosomes, it would seem improbable that attempts to influence these processes by the direct addition of steroid hormone–receptor complexes to in vitro systems are going to be very fruitful.

References

Cunha GR (1972) Anat Rec 172:529

Elger W, Neumann F, Steinbeck H, Hahn JD (1970) In: Gibian H, Plotz EJ (eds) Mammalian Reproduction. Springer-Verlag, New York, pp 32–44

Evans CR, Pierrepoint CG (1975) J Endocrinol 64: 539

Jensen EV, Catt KJ, Gorsky J, Williams-Ashman HG (1977) In: Frontiers of Reproduction and Fertility Control. Greep RO, Koblinsky MA (eds) MIT Press, Cambridge, Mass, Chap. 25, p 245

Liao S (1975) Int Rev Cytol 41: 87

Liao S, Fang S (1969) Vitamins Horm. 27: 17

Mainwaring WIP (1977) The Mechanism of Action of Androgens. Springer-Verlag, New York

Roy AK, Robel P, Baulieu EE (1972) Endocrinology 92: 1216

Williams-Ashman, H.G. (1975) In: Greep RO and Astwood EB (eds) Handbook of Physiology, Section 7: Endocrinology Am Physiol Soc, Washington, DC, Chap. 24, p 473

Wilson JD (1978) Annu Rev Physiol 40: 279

Wilson JD, Gloyna RE (1970) Rec Prog Horm Res 26: 309

Chapter 12

Receptors and Factors Regulating Androgen Action in the Rat Ventral Prostate

R. A. HIIPAKKA, R. M. LOOR, AND S. LIAO

I. Introduction

In the rat ventral prostate, testosterone is rapidly reduced to 5α-dihydrotestosterone (DHT), which can be selectively retained in prostate cell nuclei (Anderson and Liao 1968; Bruchovsky and Wilson 1968). This nuclear retention, which is antagonized by antiandrogens (Fang and Liao 1969) and can be observed by cellular fractionation or by means of autoradiographic techniques (Sar et al. 1970), received immediate attention since, in this target tissue, (1) DHT was known to be more potent than testosterone as a growth promotor (see Liao and Fang 1969); (2) the earliest biochemical response known at the time was the enhancement of RNA synthesis in prostate cell nuclei (Liao et al. 1965); and (3) the existence of a testosterone receptor had not been clearly demonstrated. The suggestion that DHT rather than testosterone plays a major role in the target tissue was a surprise, for it was an accepted view that, in the female rat, the major circulating estrogen, 17β-estradiol, binds to a receptor protein in the uterus and presumably functions without metabolic conversion (Jensen et al. 1974).

During the last decade, receptors or receptor-like proteins that bind androgen specifically have been found in a variety of androgen target tissues. The majority of androgen-sensitive tissues appear to have receptor proteins that can bind both DHT and testosterone. Therefore, the relative metabolic activities of these two androgens and their relative affinities toward the cellular receptors become important factors in providing information as to which of the two androgens plays key roles in androgenic responses.

II. Androgen Receptors in Different Tissues

5α-Dihydrotestosterone appears to play a predominant role in binding to the receptor in the ventral prostate, seminal vesicles (Liao et al. 1971; Tveter and Unhjem 1969; Zakár and T'oth 1977), and epididymis (Blaquier and Calandra 1973; Hansson and Djøseland 1972; Tindall et al. 1972); it is formed readily from testosterone and binds to the prostate receptor more firmly than testosterone (Krieg and Voigt 1976; Liao et al. 1973a), and it is only slowly metabolized further to hydroxylated compounds (Attramadal et al. 1976). In the kidney (Bardin et al. 1975; Gehring et al. 1971; Gustafsson and Pousette 1975; Ritzen et al. 1972), uterus (Giannopoulos 1973; Jungblut et al. 1971), ovary (Louvet et al. 1975), and submaxillary gland (Dunn et al. 1973; Gustafsson and Pousette 1975) DHT is formed at moderate rates from testosterone, but is also metabolized rapidly to diols. Since the receptors in these tissues bind both testosterone and DHT, both androgens may be effective. In tissues such as the adult testis (Galena et al. 1974; Hansson et al. 1974), levator ani muscle (Jung and Baulieu 1972; Krieg et al. 1974, 1976), and thigh muscle (Dionne et al. 1977; Gustafsson and Pousette 1975; Kreig 1976), DHT is formed slowly or not at all. Therefore, testosterone binding by the receptor may be very important, regardless of the differences in the receptor affinities toward these androgens (Attramadal et al. 1976).

Testosterone or DHT receptor proteins have also been found in target tissues such as the sebaceous and preputial glands (Bullock and Bardin 1970; Eppenberger and Hsia 1972; Gustafsson and Pousette 1975; Takayasu and Adachi 1975); coagulating gland (Gustafsson and Pousette 1975); spermatozoa (Wester and Foote 1972); hair follicles (Fazekas and Sandor 1972); skin (Bonne et al. 1977; Collier et al. 1979; Evain et al. 1977); bone marrow (Valladares and Mingnell 1975); heart, gizzard, and lung (Dubé et al. 1974, 1977); chick magnum (Palmiter et al. 1973); cock's comb, and other head appendages (Dubé et al. 1974, 1977); specific areas of the brain, including the hypothalamus, pituitary, preoptic area, brain cortex, and the pineal gland (Armstrong and Villee 1977; Attardi and Ohno 1978; Fox 1975; Jouan et al. 1973; Kato 1975; Loras et al. 1974; Naess et al. 1975); the preen gland (Daniel et al. 1977); and certain androgen-sensitive tumors (Attramadal et al. 1977; Bruchovsky and Rennie, 1978; Bruchovsky et al. 1975; Mainwaring and Mangan 1973; Norris et al. 1974).

Liver has often been considered a relatively androgen-insensitive tissue. In the liver, however, androgen can enhance nuclear RNA synthesis (Tata 1966). In the rat liver production of α_{2u}-globulin and DHT-binding activity are dependent on the androgen level (see Roy et al., Chap. 14 in this book). Estradiol administered to adult male rats can result in the reduction of both activities. Roy et al. (1974) suggested that estradiol binds to a distinct site on the receptor, induces an allosteric effect, and prevents androgen binding at a distant site of the receptor.

In some of the target tissues, the biological responses to the two androgens are clearly discriminatory. For example, in some species, anovulatory sterility (Whalen and Luttge 1971a) and sexual behavior (Beyer et al. 1973; Whalen and Luttge 1971b) are affected by testosterone, but not by DHT. Testosterone is also much more effective than DHT in stimulating glandular secretions; it increases the height of the luminal epithelium in the rat uterus (Gonzales-Diddi et al. 1972) and the secretory output of fructose and citric acid in bull seminal vesicles (Mann et al. 1971). In such target tissues, the differential effects may result from the large differences in metabolic and binding activities; it is also possible, however, that identical or different receptors interact with the two androgens or with others, and form dissimilar complexes that can induce different biological responses.

The biological action of certain androgenic compounds may involve different types of receptors that interact with androsta(e)ne derivatives other than testosterone or DHT. For example, in the vagina, 3α-hydroxy- and 3-ketoandrostanes can stimulate the production of mucus by the superficial cells, whereas 3β-hydroxy steroids can affect deeper layers, and $3\beta,17\beta$-dihydroxy-androst-5-ene (Δ^5-diol) and estrogens can cause keratinization of the epithelium (Huggins et al. 1954). Receptor proteins for some of these hydroxylated androsta(e)nes have been found in the cell nuclei of the vagina and uterus (Shao et al. 1975). The Δ^5-diol has been shown to interfere with binding of estradiol and DHT to the receptors in human myometrial and mammary cancer tissues (Poortman et al. 1975). Receptor-like proteins that bind 5-β-DHT in bone marrow cells (Valladares and Mingnell 1975) or liver (Lane et al. 1975), and androstenedione in liver (Gustafsson et al. 1975), have also been described.

In the dog prostate, there is a suggestion that androgenic action may depend upon a receptor for $3\alpha,17\alpha$-dihydroxy-5α-androstane. Recently $3\alpha,17\beta$-dihydroxy-5α-androstane has been implicated in the induction of benign prostatic hyperplasia in dogs (Jacobi et al. 1978; Walsh and Wilson 1976), indicating the presence of a receptor for this diol. Certain unsaturated androgens can also be aromatized to estrogens (Beyer et al. 1973; Kato 1975; Naftolin and Ryan 1975) and, therefore, may bind to estrogen receptors. This process could be important in some biological responses related to certain brain functions.

It now appears that, in different tissues of the same animal, there are different receptor proteins for various steroids related to androgens. It is unlikely that a single gene is responsible for the synthesis of all of these receptor proteins. However, certain posttranslational alterations or association of other cellular molecules with the proteins may generate multiple forms of receptors that are individually responsible for the induction of specific biological responses.

5α-Dihydrotestosterone is known to induce growth of the rat uterus in vivo. Rochefort and his associates (Rochefort and Carcia 1976; Rochefort and Lignon 1974) found that DHT at high concentrations can promote the translocation of the estrogen receptor into cell nuclei in vitro and in vivo, and suggested that this nuclear retention of the estrogen receptor is responsible for the growth-promoting effect of DHT. Ruh and Ruh (1975) also showed that androgens were effective in the induction of the synthesis of a specific uterine protein (IP) in vitro, and that antiandrogens were incapable of inhibiting this effect. Schmidt et al. (1976), however, reported that high levels of androgens which elicited an increase in uterine weight were not capable of evoking any detectable translocation of estrogen receptor or inducing IP in rats. They proposed that the action of androgens on the uterus in vivo is probably not mediated directly through the estrogen–receptor system. The possibility that androgens act through binding to an estrogen receptor has also been discussed by Zava and McGuire (1978) in their study of the human breast cancer MCF7 cell line.

III. Androgen Receptors in Rat Ventral Prostate

In the rat ventral prostate, there are at least two proteins that can bind DHT (Fang and Liao 1971; Liao and Fang 1970). At low concentrations, DHT binds predominantly to a high-affinity ($K_a \sim 10^{11}\ M^{-1}$) and low-capacity protein (β-protein). The complex (complex II) formed can eventually be retained by nuclear chromatin (Fang and Liao 1971; Mainwaring and Peterken 1971; Steggles et al. 1971). 5α-Dihydrotestosterone can form a complex (complex I) with another low-affinity ($K_a \sim 10^7\ M^{-1}$) and high-capacity protein (α-protein). Whereas the high-affinity binds only active natural and synthetic androgens (Fang and Liao 1971; Liao et al. 1973a), the low-affinity protein also binds estradiol, but not glucocorticosteroids. The biological significance of the low-affinity binding protein is not clear; it does not bind to chromatin in vivo or in vitro. As will be described later, this protein inhibits the high-affinity DHT-binding protein from binding to the nuclear chromatin.

There is general agreement that complex II represents a form of the cellular androgen–receptor complex. The interaction of DHT with the receptor is effectively antagonized by typical antiandrogens such as cyproterone acetate (Fang and Liao 1969) and flutamide (Liao et al. 1974; Peets et al. 1974). As much as 30% of the cellular receptor protein is found in the cytoplasmic particulate (microsomal) fraction (Liao et al. 1975; Robel et al. 1974). Only a small quantity ($\sim$200 DHT-binding sites per nucleus) of the receptor is present in the cell nuclei of castrated rats. After injection of testosterone or DHT, the nuclear DHT–receptor content is increased to 2000–6000 DHT molecules per nucleus (Bruchovsky et al. 1975; Fang and Liao 1971; Liao and Fang 1970). In comparison, in the well-developed prostate of recently ($<$24 h) castrated rats there are about 10 000 DHT-binding receptor sites per cell; the number may be considerably smaller ($\sim$1000 sites) in rats castrated several weeks previously, or in aged rats (Shain et al. 1975).

Like other steroid–receptor complexes, DHT–receptor complexes have often been found to sediment as $7–12S$ and $3–5S$ units (Baulieu and Jung, 1970; Liao et al. 1975; Mainwaring 1969; Unhjem et al. 1969). The large form can be transformed to the small form by incubation at 30°C, or by raising the salt concentration to $0.4\ M$ KCl. Under some conditions, the small form can also be converted to the large form (Hu and Wang 1978; Liao et al. 1975). In $0.4\ M$ KCl or in $2\ M$ urea, the cytosol DHT–receptor complex ($3.8 \pm 0.3\ S$) sediments slightly faster than the nuclear complex ($2.9 \pm 0.3\ S$) (Liao et al. 1975; Tymoczko and Liao 1976). At $0.1\ M$ KCl and at a pH below 7.5, the $4\ S$ complex gradually aggregates to the $8\ S$ and larger forms. When the pH is raised to 9, the extent of aggregation is reduced, and the complex sediments at $8\ S$ and $3–4\ S$ forms. At pH 9.5, only a $7\ S$ form is clearly observed. The 8 *S* and the aggregated forms were not observed with hydroxyapatite-purified complex (Liao and Liang 1974), suggesting that other cellular materials are involved in the formation of these large complexes. With the nuclear complex, extensive aggregation is observed at pH 6.4; but at a pH above 8.3, most of the complex sediments as a 3 *S* form in 0.4 *M* KCl, or even in the absence of KCl (Liao et al. 1975).

The interaction of the androgen–receptor complex with divalent metal ions and nucleotides has been studied by gradient centrifugation. At 1–5 m*M*, $MnCl_2$, $MgCl_2$, or $CaCl_2$ produced no significant effect in a medium containing 0.4 *M* KCl, but $CoCl_2$ facilitated aggregation of the 3.8 *S* form. With $ZnCl_2$, a shift in the sedimentation coefficient from 3.8 *S* to 4.5 *S* was observed after incubation at 0°C for 20 min, but there was no change in the total amount of [^{3}H]DHT bound to the receptor. At 20°C, the radioactive peak broadened (5 ± 2 *S*), and eventually a considerable amount of the radioactive androgen dissociated from the receptor (Liao et al. 1975).

We also found that ATP and GTP can interact with and stabilize the DHT –receptor complex. Such an effect is observed most clearly when a freshly prepared receptor protein is incubated with 1–5 m*M* of these nucleotides at 20°C for 20 min. Under these conditions the nucleotides appear to retard the release of DHT from the receptor as well as the transformation to the nuclear 3*S* form. CTP, UTP, ADP, and GDP tend to be less effective or do not affect the sedimentation property of the androgen–receptor complex (Liao et al. 1975).

The biological significance of the receptor interaction with divalent ions and nucleotides is not clear. As was reviewed elsewhere (Liao 1977; Liao et al. 1975), metal ions have been shown to affect the receptors for estrogens, mineral corticosteroids, and progesterone. In some cases, this could result from an indirect effect, such as Ca^{2+} activation of a protease that may transform the estradiol–receptor complex (8*S* and 5*S*) to a form (4.5*S*) that does not aggregate. The interaction of the progesterone–receptor complex of the hen oviduct with ATP has also been described by Toft et al. (1976). Other investigators have speculated that the "activation" of the glucocorticoid receptor in mouse fibroblasts or thymus cells may be dependent on an energy supply system, or on ATP (Bell and Munck 1973; Munck et al. 1972). Experimental evidence has been presented to show that glucocorticoid binding to the receptor may be dependent on a prior phosphorylation of the receptor protein (Nielsen et al. 1977).

The correlation between the androgenicity and the receptor-binding affinity of many steroids tested is excellent (Liao et al. 1973a). With regard to receptor binding, the bulkiness and flatness of the steroid molecule, especially at the ring A area, appear to play a more important role than the detailed electronic structure of the steroid nucleus. Not only are the steroids with an A–B cis structure, such as the inactive (in rat prostate) 5β-isomer of DHT, not bound by the prostate receptor, but other relatively flat steroids with rings A–B in the trans form also differ in their receptor-binding affinity according to the bulkiness in the ring A–B area.

Testosterone binds to the receptor less firmly than DHT, apparently owing to the more rapid rate of dissociation of testosterone from the receptor (Wilson and French 1976). Since other Δ^4-3-ketosteroids can bind tightly to the receptor, the difference in dissociation rates may result mainly from the steric property at the ring A–B area, rather than from the presence of the Δ^4 double bond. This is best shown by the fact that potent androgens like $7\alpha,17\alpha$-dimethyl-19-nortestosterone (DMNT), 2-oxa-17α-methyl-17β-hydroxyestra-4,9,11-triene-3-one, and 17α-methyl-17β-hydroxyestra-4,9,11-triene-3-one are capable of binding to the androgen receptor more tightly than DHT (Liao et al. 1973a). Since they have conjugated double bonds that extend from rings A and B to C, the estratrienes are indeed very flat molecules. The removal of the angular methyl group between ring A and ring B also makes the area less bulky and possibly facilitates tight binding of these androgens to the receptor. This view is also supported by the fact that A-nor-17β-acetoxyestra-4,9,11-trien-3-one, with five carbons in ring A, is a very potent androgen, whereas A-homotestosterone, as well as A-homo-DHT, with seven carbons in ring A, is virtually inactive (cf. Liao et al. 1973a).

If a methyl group is substituted on the C-7α position on 19-nortestosterone, the receptor-binding affinity increases significantly (Fig. 1). It may be that the receptor protein has a specific (M) site for binding the 7α-methyl group, and that both the binding affinity and the androgenicity are enhanced manyfold. Alternatively, the 7α-methyl group may strain the steroid nuclear structure and make it bind more tightly to the receptor protein.

A unique structural aspect of natural steroid hormones is that they contain an oxygen function at the C-3 position. Whether this group is absolutely required for androgen action is not clear, for many synthetic steroids without this oxygen group have shown to be androgenically active (Liao and Fang 1969). These deoxy steroids generally show a very low binding affinity for the prostate androgen receptor and have relatively low androgenicity compared with that of natural androgens. It is possible that some of the 3-deoxy androgens are oxygenated and become androgenic, or that various 3-oxygenated androgens have different affinities toward different receptors and exhibit diverse biologic responses.

With the exception of the possible involvement of 17α-hydroxylated androgens in the dog prostate (Evans and Pierrepoint 1975), the 17β-hydroxy group appears to be needed for the high-affinity binding of androgens to the receptor and for androgen action. It is not clear, however, whether the 17β-hydroxy group is needed only for the formation of a tight androgen–receptor complex, or for the triggering process itself. Perhaps the triggering action at a functional site involves a specific interaction of a cellular molecule with the hydroxy group of the steroid still bound to the receptor protein. This may weaken the androgen–receptor interaction and result in a structural reorientation of the complex. Whether the 17β-hydroxy group is required for the triggering action can be tested by the use of 7α-methyl-19-nor-17-deoxyandrostenes. The 7α-methyl group in these compounds could bind to the M

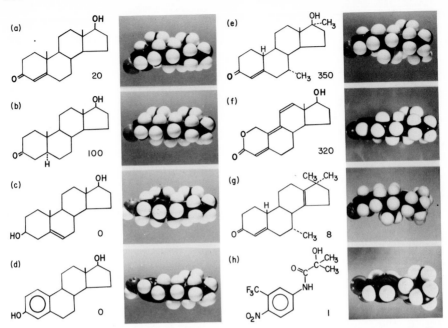

Fig. 1. CPK models of androgens, antiandrogen, and estradiol: Corey–Pauling atomic models with Koltun connections were used to construct (**a**) testosterone, (**b**) DHT (5α-dihydrotestosterone), (**c**) 3β,17β-dihydroxyandrost-5-ene, (**d**) 17β-estradiol (**e**) 7α,17α-dimethyl-19-nortestosterone, (**f**) 2-oxa-17β-hydroxyestra-4,9,11-trien-3-one, (**g**) 7α,17,17-trimethylgona-4,13-dien-3-one, and (**h**) hydroxylated flutamide (a nonsteroidal antiandrogen). The numbers under the formulas indicate their relative affinities toward β-protein of rat ventral prostate, using DHT as 100 (Liao et al. 1979).

site on the receptor and provide the necessary binding affinity even in the absence of the 17β-hydroxyl group, and may allow us to evaluate whether the hydroxy group is absolutely needed for inducing the biological response.

The way in which a receptor protein recognizes the androgen structures suggests that, during interaction, the androgens are being "enveloped" in the hydrophobic cavity of the protein (Liao et al. 1973a). The localization of steroid-binding sites well inside the receptor proteins may be responsible for the very slow rates of association or dissociation of steroids from these proteins at low temperatures, for the very high affinity constants, for the acceleration of the rates of exchange of unbound steroid with bound steroid by freezing and thawing, and for the fact that the receptor-bound androgens are rather stable in ethanol (30%) and in detergents (2% Triton X-100 or deoxycholate).

Additional support for the hypothesis of steroid enveloping came from our study on the capacity of antisteroid antibodies to interact with steroids bound to various proteins (Castañeda and Liao 1975). Antibodies against DHT and testosterone were found to be effective in removing steroids bound to nonreceptor proteins of the blood and prostate, since these proteins (steroid-metabolizing enzymes or blood steroid-binding globulin) generally recognize only a portion of the steroid molecule, and since the rates of dissociation of the steroids from these nonreceptor proteins are much higher than that from the receptor protein. The steroids bound far inside the receptor protein have very low rates of dissociation and, therefore, are not readily removed by the antibody. The concept of steroid enveloping supports the suggestion that the receptor protein, and not the specific functional groups on steroids, participates in the key event that leads to steroid hormone action.

Purification of an androgen receptor has not been very successful (Hu et al. 1978; Mainwaring and Irving 1973). Among the difficulties encountered are instability of the receptor, loss of the steroid ligand from the receptor protein being purified, and the limited quantities of tissues available for purification. Although the crude cytosol androgen receptors isolated from the seminal vesicles, testis, epididymis, kidney, uterus, ovaries, brain, and other organs appear to have similar physicochemical properties, such as sedimentation patterns, thermolability, and electrophoretic mobility, it would be premature to conclude that all of the receptors are identical or have the same genetic origin.

IV. Nuclear Retention of the Androgen Receptor and Its Control

That nuclear androgen–receptor complex originates in the cytosol fraction of the prostate cells was first demonstrated by Fang et al. (1969), who showed that the selective retention of DHT by isolated nuclei requires the presence of a prostate cytosol fraction. Autoradiographic studies (Sar et al. 1970) and the cell-free experiments (Fang and Liao 1971; Mainwaring and Irving 1973) showed that the process is also temperature dependent, occurring readily at 20°C, but not at 0°C. This finding was confirmed by Mainwaring, who found, in addition, that the androgen–receptor complex undergoes a change in the isoelectric point from 5.8 to 6.5 (Mainwaring and Irving 1973). By gradient centrifugation, it was possible to show that the cytosol complex (3.8S) was altered to a nuclear form (3.0S) during the transformation of the receptor (Liao et al. 1975). Whether this process involves a change in the conformation of the receptor, a proteolytic action, or a bimolecular reaction has not yet been studied in detail.

The nuclear retention of the DHT–receptor complex appears to be tissue specific. In cell-free systems, cell nuclei or nuclear chromatin isolated from tissues that are less sensitive to androgens than the prostate have low capacities for retaining the prostate DHT–receptor complex. Since the nuclear receptor-binding activity is reduced if the nuclear preparation is heated to temperatures above 50°C, we have suggested that the receptor-binding activity requires heat-labile acceptor protein factors (Fang and Liao 1971). In most prostate nuclear preparations that we have studied, there are 2000–3000 acceptor sites per cell nucleus; this number may be higher ($\sim$10 000) in hyperplastic tissues and lower (~1000) in rats castrated many weeks previously.

Tymoczko and Liao (1971) first attributed the acceptor activity to a non-histone protein in the prostate nuclear chromatin. In the presence of purified DNA, the acceptor protein can form a nucleoprotein complex that can specifically bind the prostate DHT–receptor complex. Many properties of the acceptor activity of the nuclear proteins (but not of histones) are very similar to those of isolated prostate nuclei (Tymoczko 1973). They have identical heat denaturation curves, and the DHT–receptor complex bound can be released in 0.4 *M* KCl medium. The heat-labile factors are not dialyzable; they can be fractionated by ammonium sulfate and ethanol or precipitated from the solution by acid (pH 4.5). Equivalent liver nuclear protein preparations appear to contain a much smaller amount of acceptor-protein-like material, providing a reason for the low receptor-binding capacity of liver nuclei (Tymoczko and Liao 1971).

Klyzsejko-Stefanowicz and co-workers (1976) studied chromatin acceptor molecules by sequentially removing urea-soluble chromosomal nonhistone proteins, histones, and DNA-associated nonhistone proteins from the chromatin of the rat ventral prostate and testis. The prostate [³H]DHT–receptor complex was found to bind much more readily to the partially deproteinized chromatin, which still contains the DNA-binding nonhistone proteins, than to purified DNA. The receptor-binding activity of rat DNA was enhanced significantly by the addition of the DNA-associated nonhistone protein of the prostate or testis, but not that of the liver. The tissue specificity agrees with the results of the study described in the preceding sections. The DNA-associated nonhistone proteins are not phosphorylated, but phosphorylation of chromatin increases the receptor-binding capability of prostatic or testicular chromatin.

Using covalent linkage of nuclear components to Sepharose, Puca et al. (1974) studied the interaction of nuclear protein with the cytosol estrogen–receptor complex of the uterus and suggested that the nuclear acceptor molecule or molecules may be basic proteins. Mainwaring et al. (1976), who used the same technique, found that certain nonhistone basic proteins in the prostate nuclei, when bound to Sepharose, can retain the prostate cytosol [³H]DHT–receptor complex. Androgens present in free form or bound to the sex steroid-binding globulin were not retained. The acceptor activity was higher in the nuclear preparations from the prostate than in those from the liver, spleen, or other tissues that are relatively insensitive to androgens. When various radioactive steroids were incubated with prostate cytosol as the source of steroid–receptor complexes, [³H]DHT was about five times more active than labeled testosterone. Dexamethasone, progesterone, and androsterone were essentially inactive.

In the cell nuclei, the function of the steroid–receptor complex may depend on its interaction with different forms of acceptor molecules. To explore such a possibility, we (Liang and Liao 1974; Liao et al. 1973b) also studied the interaction of the DHT–receptor complex with ribonucleoprotein (RNP) particles extracted from prostate cell nuclei. The ternary complex sedimented as heterogeneous components (60–80S) and could be analyzed by gradient centrifugation. The receptor-binding sites on the RNP particles could be saturated with excess DHT–receptor, indicating that a limited number of binding sites exist. The apparent association constant calculated from such experiments is of the order of 10^{10} M^{-1}. Under saturating conditions with respect to the DHT–receptor complex, less than 5% of the isolated RNP particles can bind to the steroid–receptor complex. It is possible that only those RNP with heat-labile acceptor factors can associate with the DHT–receptor complex. Experiments with excess RNP demonstrated that only about 30–50% of the total DHT–receptor complex was capable of binding to the RNP particles, indicating that only certain forms of the steroid-receptor complex can interact with RNP.

The cytoplasmic polysome or 80S monosome forms of ribosomes do not bind the DHT–receptor complex, although the 40S and 60S subunit particles prepared from them can bind this complex readily (Liang and Liao 1974; Liao et al. 1975). Radioactive steroids alone do not associate with any of the nuclear RNP or ribosomal particles. Preliminary experiments suggest that naked receptor, unattached to an androgen, does not bind to the particles.

Wang (1978) reported that, when sequential extraction with 0.35 M NaCl and 2.0 M NaCl is used, the selective acceptor activity is also associated with the 2 M NaCl insoluble residual fraction of the prostate chromatin, which contains DNA bound to nonhistone proteins. Nyberg and Wang (1976) fractionated the androgen-labeled prostate chromatin after a single injection of radioactive testosterone into castrated rats. The quantity of radioactive androgen associated with a salt-soluble nonhistone protein fraction was high within the first hour, but declined rapidly during the second hour. The changes in the radioactivity of the salt-insoluble nonhistone protein and the DNA–histone complex fractions, however, exhibited the opposite pattern.

These studies and those described in the previous section suggest that the androgen–receptor complex interacts with various nuclear components, possibly at different stages. The possibility that the steroid–receptor complexes provide structural requirements not only for the formation, but also for the processing or functioning, or both, of certain RNAs in the target cells (Liao and Fang 1969; Liao et al. 1973b) is worth pursuing.

The control mechanism involved in the binding and release of the steroid–receptor complex from chromatin has not been studied in detail. The nuclear receptor-bound DHT may be metabolized and released from the receptor (Nozu and Tamaoki 1975), causing dissociation of the receptor protein from the nuclear acceptor site. As described above, we have also considered the possibility that the receptor complex associates with nuclear RNA or RNP particles and returns to the cytoplasm (Liao et al. 1973b). Klyzsejko-Stefanowicz et al. (1976) also reported that the ability of the prostate chromatin to bind the DHT–receptor complex can be enhanced by an in vitro phosphorylation of the chromatin. They suggested that phosphorylation may allow the acceptor protein to become more readily accessible to the receptor complex.

In 1971, we reported that a steroid-binding protein (α-protein) in the cytosol fraction can inhibit the DHT–receptor complex from binding to the nuclear chromatin (Fang and Liao 1971). Such a protein may also play a key role in the regulation of receptor binding to the chromatin. Other investigators have reported the presence of similar inhibitors in the cytosol of rat uterus (Chamness et al. 1974), chick oviduct (Buller et al. 1975), rat liver (Milgrom and Atger 1975), and rat hepatoma cells (Simons et al. 1976).

Bovine serum albumin, however, was also inhibitory in some of these systems; therefore, the inhibition has not been regarded as a highly specific phenomenon. With the prostate system, bovine serum albumin is totally inactive (Shyr and Liao 1978).

We have purified the prostate inhibitor extensively and found that the inhibitory activity is associated with the highly purified steroid-binding α-protein. This protein is a major one in the cytosol and apparently is identical to the prostatic binding protein studied recently by Heyns et al. (1976, 1977) and the prostatein reported by Lea et al. (1977). It is also found in the prostate fluid and apparently is a secretory protein. The protein has a subunit structure; the inhibitory activity is associated with the small (MW 8–9000) subunit (Chen et al. 1979).

The [³H]DHT–receptor complex of the ventral prostate binds much better to the cell nuclei of the prostate and testis than to the nuclei of the liver, thymus, or spleen. The extent of inhibition by the prostate inhibitor, however, was about the same, regardless of the source of cell nuclei (Table 1). Inhibitory activity was also found with the cytosol preparations of the rat testis, rat kidney, and beef kidney. The cytosol preparations of the rat liver and beef liver, however, showed very little inhibitory activity.

The inhibitor does not cause permanent damage to the prostate cell nuclei; these nuclei, incubated with the inhibitor and then washed, bound the radio-

Table 1. Tissue Specificity of the Nuclear Retention of the Androgen–Receptor Complex and Its Inhibition[a]

Source of nuclei	³H–Receptor complex retained		
	Control (cpm)	+ Inhibitor (cpm)	Inhibition (%)
Ventral prostate	4 823	2 480	49
Testis	4 173	2 210	47
Liver	1 828	850	54
Kidney	1 044	520	50
Thymus	328	181	45
Spleen	286	185	37

[a] The capability of the cell nuclei from various rat organs to retain the radioactive androgen–receptor complex was determined in the absence (control) or presence of 4.2 mg of the prostate inhibitor under the standard assay procedure. The specific radioactivity of [1,2,6,7-³H]5α-dihydrotestosterone used was 120 Ci/mmol. The radioactivity retained by the nuclei was expressed in cpm/100 μg DNA (Shyr and Liao 1978).

active receptor complex and responded to the inhibitor to the same extent as did the nuclei not previously incubated with the inhibitor (Shyr and Liao 1978). The inhibitor does not appear to act on the nuclear membrane and prevent the receptor complex from entering the nucleus. Intact prostate nuclei bound the same amount of the receptor complex and showed the same response to the inhibitor as did nuclei that had been washed with a nonionic detergent for removal of nuclear membranes prior to the binding assay. By comparing the DHT–receptor complex incubated with the inhibitor for different lengths of time prior to the assay, we also eliminated the possibility that the inhibitor inactivates the receptor complex irreversibly through a temperature- and time-dependent process.

The inhibition is reversible; removal of the inhibitor from the incubation medium completely restores the capability of the DHT–receptor complex and the nuclear chromatin to interact and bind to each other. One of the most interesting properties of the protein factor is that it can also release the DHT–receptor complex that is already bound to the chromatin. This process is temperature dependent, proceeding more readily at 20°C than at 0°C (Shyr and Liao 1979).

The inhibitor may act by altering the conformation of the androgen–receptor complex. Alternatively, the inhibitory protein may modify the chromatin from a receptor-binding status to one not favorable for such binding. One can conjecture that the steroid–receptor complex is also a chromatin modifier that is needed to maintain the genomic structure in the form required for orderly induction of cellular functions.

V. Early Effects of Androgen on Specific Proteins

The importance of RNA synthesis in androgen action was first indicated clearly in the early 1960s (Hancock et al., 1962; Liao 1965; Liao and Williams-Ashman 1962). The effect of androgen on nuclear RNA-synthesizing activity is very rapid, occurring within 1 h after the androgen reaches the prostate cells (Liao et al. 1965). The increase is most pronounced in the nucleolar regions, where the precursor of ribosomal RNA is synthesized (Liao and Lin 1967; Liao and Stumpf 1968; Liao et al. 1966). A clear increase in the quantity of certain mRNAs may be detected only many hours later, but one cannot exclude the possibility that the production of specific mRNAs is enhanced at a very early stage of androgen action.

Many steroid hormones can selectively increase the production of specific proteins or enzymes in their target cells. In several carefully studied examples (see Chap. 1 by O'Malley and co-workers in this book), the increased induction of these cellular proteins is well correlated with an increase in the cellular content of mRNAs for the proteins, suggesting that the hormones act at the transcription stage. As described by Roy et al. (Chapter 14 in this book), the hepatic level of mRNA for α_{2u}-globulin in rats is induced and controlled by androgen and other hormones. Androgen-stimulated increases in the prostate proteins such as aldolase (Mainwaring et al. 1974) and major proteins (Parker and Mainwaring 1977; Parker and Scrace 1978; Parker et al. 1978), including α-protein, a prostate-binding protein (Heyns et al. 1977), have also been shown to follow an increase in mRNA content for these proteins. Such findings are in accord with the proposal that the synthesis of these mRNAs is affected by the steroid–receptor complex; many of these androgen effects, however, are detectable only hours or days after androgen has been administered to experimental animals. For analysis of the very first event in androgen action, it would be highly desirable to carry out such a study with an mRNA or a protein that can be induced very rapidly (within 1 h) by androgen.

Motivated by the knowledge that steroid–receptor complexes may bind to nuclear RNP and to the subunit particles of ribosomes, but not to the 80S monosomes or polysomes, we also explored whether androgen and receptor have a direct effect on the initiation of protein synthesis (Liang and Liao 1975; Liang et al. 1977). We studied the effect of castration and androgen injection on the capacity of the prostate cytosol proteins to support binding of [^{35}S]methionyl-tRNA$_i$ (the IF activity) to the prostate ribosomal particles and found that the cytosol IF activity is reduced considerably within hours after castration. This loss is prevented by intraperitoneal injection of a relatively large dose (milligram quantity) of DHT or testosterone, which actually enhances the activity to a level higher than that seen in normal animals. Such enhancement can be seen within 1 h after androgens are administered to castrated rats.

The androgen effect was observed clearly within 30 min after the castrated rats had been injected intravenously with a low dose (10–20 μg) of DHT. The effects of castration and androgen injection could be seen when the IF activities were compared by measurement of the capacity of the cytosol proteins to support the formation of [^{35}S]Met-tRNA$_i$· eukaryotic initiation factor (eIF-2) complex. This rapid effect is rather insensitive to actinomycin D and cycloheximide, but can be suppressed by cyproterone acetate, an antiandrogen that inhibits the receptor binding of androgens (Liang et al. 1977). Our observation agrees with that of Ichii et al. (1974), who also suggested that the long-term (over several days) effect of androgen on protein synthesis may result from enhancement of the activity of the initiation factor (and, to some extent, the elongation factor) involved in protein synthesis in the rat prostate.

Polyamines are among the cellular factors that may affect the initiation factor activity mentioned above (Hung et al. 1976; Rannels et al. 1976). Therefore, we investigated polyamine-binding proteins in the rat ventral prostate. Our study (Liang et al. 1978) unexpectedly revealed that the cytosol fraction of the rat ventral prostate contains an acidic protein that can bind spermine selectively. The relative binding affinities of various aliphatic amines for the protein are spermine > thermine ≫ spermidine > putrescine > 1,10-diaminodecane, cadaverine, and 1,12-diaminododecane. The binding protein has an isoelectric point at pH 4.3 and a sedimentation coefficient of about 3S. Polyamines bind to this protein noncovalently, without prior metabolism. The ligand–protein interaction is reversible. The spermine–protein complex dissociates readily in media with high ionic strength or with high concentrations of divalent cations, indicating the ionic nature of the interaction. At low ionic strength, the apparent affinity constant for spermine binding to the protein is estimated to be $10^7 \ M^{-1}$. Since the binding activity can be abolished by intestinal alkaline phosphatase and reactivated by heart protein kinase in the presence of ATP, the binding protein appears to be a phosphoprotein. The spermine-binding activity in the cytosol fraction is reduced considerably after rats are castrated, but this effect can be reversed by injection of testosterone or 5α-dihydrotestosterone into the castrated animals. The action is very rapid and can be seen reproducibly within 10–30 min after androgen is administered (Fig. 2). Our recent experiments show that this androgen effect is sensitive to cycloheximide and to actinomycin D.

Some small, but probably important, changes in the nuclear proteins may be induced by androgen. For example, Anderson and his associates (Anderson et al. 1973; Kadohama and Anderson 1976) as well as Ahmed and Wilson (1975) reported that certain nuclear proteins could be selectively labeled with radioactive amino acids or phosphate within the initial hours after rats had been treated with androgen. Using radioactive actinomycin D, Mainwaring and Jones (1975) were able to show that androgen can enhance

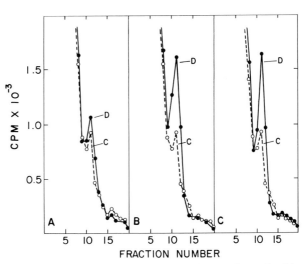

Fig. 2. Effect of androgen on spermine-binding activity assayed by polyacrylamide gel electrophoresis. Rats were castrated 42 h before they were killed. They were injected with 5α-dihydrotestosterone (D, 6 mg/rat) **A** 15min; **B** 30 min; **C** 60 min before their ventral prostates were removed. The control groups (C) received only the solvent carrier. Prostate cytosol proteins fractionated by ammonium sulfate were used for the assay. The concentration of [^{14}C]spermine in the incubation mixture was 180 μM. The quantity of prostate proteins in each assay tube was 1.2 mg (G. Mezzetti et al. 1979).

high-affinity binding of the antibiotic in vivo. This finding is in accord with the observation that the early effect of androgen on RNA synthesis occurs at the chromatin region, which is highly sensitive to very low concentrations of actinomycin D (Liao and Lin 1967; Liao et al. 1966).

VI. Concluding Remarks

Whereas the slow androgen responses (hours to days) may provide important information for an understanding of the physiological aspects of the androgen-sensitive organs, the elucidation of the primary trigger mechanism in androgen action may not be possible without a better knowledge of the biochemical responses that take place during the first hour after androgen enters into the target cells. The rapid responses described above indicate that some steroid hormones have multiple receptors or receptor-binding sites that function separately in different cellular loci. Thus, steroid hormones may control factors that have multiple roles in transcription, translation, and other processes to assure well-coordinated and efficient regulation of gene expression in the target cells.

The androgen-sensitive polyamine-binding protein may play a major role in the intracellular communication of the hormonal message, since the protein (and the polyamine-binding activity) activated by androgens may translocate to specific cellular organelles (chromatin, ribosomes, or membranes) so that they can fulfill their structural or functional roles in triggering the hormonal responses. It may be worthwhile to explore whether such a mechanism is responsible for the enhancement in vivo of the activity of prostate nuclear protein kinase by androgen, and by the addition of polyamine to the preparation of cell nuclei (Ahmed et al. 1979). Since the polyamine-binding protein that we have investigated appears to be phosphorylated, and the androgen has been shown to affect the pattern of protein phosphorylation in the rat ventral prostate (Ahmed and Ishida, 1971; Liu and Greengard 1976), detailed investigations in this area may also reveal the ways in which other peptide hormones, cyclic nucleotides, phosphorylated proteins, polyamines, and acidic proteins are related to steroid hormone action in the target cells.

Androgen has been shown to enhance the template activity of prostate chromatin in vivo, possibly in selective chromatin regions (Couch and Anderson 1973a,b; Liao and Lin 1967; Loor et al. 1977; Mainwaring and Jones, 1975; Mangan et al. 1968; Thomas et al. 1977). Changes in the physical properties of the rat prostate nuclear chromatin, such as the thermal denaturation pattern, circular dichroism spectrum, and the polylysine interaction by androgen, have also been observed (Loor et al. 1977).

As in studies of other steroid receptors (see Jensen et al. 1974; Liao 1975), crude androgen–receptor preparations have been found to enhance RNA synthesis by prostate nuclear chromatin (Davies and Griffiths 1974; Mainwaring and Jones 1975), or DNA (Hu et al. 1975). Although these observations support the general view that the androgen–receptor complex functions by interacting with nonhistone proteins to cause local modification and decondensation of chromatin, the precise way in which this is accomplished may not be understood until the roles of histones and nonhistones in supporting chromatin structure and function are clarified.

The α-protein inhibition of steroid–receptor complex binding to chromatin may play a key role in the control of prostate growth. As the prostate grows in response in androgen stimulation, inhibitory protein may accumulate in the cell, and the interaction of the receptor and the inhibitor with chromatin may reach an equilibrium state. A defect or loss of the inhibitory factor in such a balancing mechanism may cause the prostate to grow abnormally.

Acknowledgments. Research carried out in this laboratory was supported by Grants AM-09461 and CA-09183 from the U.S. National Institutes of Health and by Grant BC-151 from the American Cancer Society, Inc.

References

Ahmed K, Ishida H (1971) Mol Pharmocol 7: 323
Ahmed K, Wilson MJ (1975) J Biol Chem 250: 2370
Ahmed K, Wilson MJ, Goueli SA, Williams-Ashman HG (1979) Biochem J in press.
Anderson KM, Liao S (1968) Nature 219: 277
Anderson KM, Slavik M, Evans AK, Couch RM (1973) Exp Cell Res 77: 143
Armstrong EG Jr, Villee CA (1977) J Steroid Biochem 8: 285
Attardi B, Ohno S (1978) Endocrinology 103: 760
Attramadal A, Weddington SC, Naess O, Djøseland O, Hansson V (1976) Prog Clin
    Biol Res 6: 189–203
Attramadal A, Naess O, Haug E, Hansson V, Purvis K (1977) Acta Endocrinol 86:
    288
Bardin W, Bullock LP, Mowszowicz I (1975) Meth. Enzymol 39: part D, 454
Baulieu EE, Jung L (1970) Biochem Biophys Res Commun 38: 599
Bell PA, Munck A (1973) Biochem J 136: 97
Beyer C, Larsson K, Pérez-Palacios G, Morali G (1973) Horm Behav 4: 99
Blaquier JA, Calandra RS (1973) Endocrinology 93: 51
Bonne C, Saurat JH, Chivot M, Lehuchet D, Raynaud JP (1977) Brit J Dermatol 97:
    501
Bruchovsky N, Rennie PS (1978) Cell 13: 273
Bruchovsky N, Wilson JD (1968) J Biol Chem 243: 2012
Bruchovsky N, Sutherland DJA, Meakin JW, Minesita T (1975) Biochim Biophys
    Acta 381: 61
Buller RE, Schrader WT, O'Malley BW (1975) J Biol Chem 250: 809
Bullock L, Bardin CW (1970) J Clin Endocrinol Metab 31: 113
Castañeda E, Liao S (1975) J Biol Chem 250: 883
Chamness GG, Jennings AW, McGuire WL (1974) Biochemistry 13: 327
Chen C, Hiipakka RA, Liao S (1979) J Steroid Biochem 11: 401
Collier ME, Griffin JE, Wilson JD (1979) Endocrinology 103: 1499
Couch RM, Anderson KM (1973a) Biochem Biophys Res Commun 50: 478
Couch RM, Anderson KM (1973b) Biochemistry 12: 3114
Daniel JY, Vignon F, Assenmacher I, Rochefort M (1977) Steroid 30: 703
Davies P, Griffiths K (1974) Biochem J 140: 565
Dionne FT, Dubé JY, Tremblay RR (1977) Can J Biochem 55: 995
Dubé JY, Tremblay RR (1974) Endocrinology 95: 1105

Dubé JY, Chapdelain P, Tremblay R (1977) Can J Biochem 55: 263
Dunn JF, Goldstein JL, Wilson JD (1973) J Biol Chem 248: 7819
Eppenberger U, Hsia SL (1972) J Biol Chem 247: 5463
Evain D, Savage NO, Binet E (1977) J Clin Endocrinol Metab 45: 363
Evans CR, Pierrepoint CG (1975) J Endocrinol 64: 539
Fang S, Liao S (1969) Mol Pharmacol 5: 420
Fang S, Liao S (1971) J Biol Chem 246: 16
Fang S, Anderson KM, Liao S (1969) J Biol Chem 244: 6584
Fazekas AG, Sandor T (1972) 4th Int Congr Endocrinol (abstr), 80, Excerpta Med.,
    Amsterdam.
Fox TO (1975) Proc Natl Acad Sci US 72: 4303
Galena HJ, Pillai AK, Terner C (1974) J Endocrinol 63: 223
Gehring U, Tomkins GM, Ohno S (1971) Nature-New Biol 232: 106
Giannopoulos G (1973) J Biol Chem 248: 1004
Gonzalez-Diddi M, Komisaruk G, Beyer C (1972) Endocrinology 91: 1130
Gustafsson J-A, Pousette A (1975) Biochemistry 14: 3094
Gustafsson J-A, Pousette A, Stenberg A, Wrange O (1975) Biochemistry 14: 3942
Hancock RL, Zelis RG, Shaw M, Williams-Ashman HG (1962) Biochim Biophys
    Acta 55: 257
Hansson V, Djøseland O (1972) Acta Endocrinol 71: 614
Hansson V, Trygstad O, French FS, McLean WS, Smith AA, Tindal DJ, Weddington
    SC, Petrusz P, Nayfeh SN, Ritzen EM (1974) Nature 250: 387
Heyns W, Verhoeven G, DeMoor P (1976) J Steroid Biochem 7: 987
Heyns W, Peeters B, Mous J (1977) Biochem Biophys Res Commun 77: 1492
Hu AL, Wang TY (1978) J Steroid Biochem 9:53
Hu AL, Loor RM, Wang TY (1975): Biochem Biophys Res Commun 65: 1327
Hu AL, Loor RM, Chamberlin L, Wang TY (1978) Arch Biochem Biophys 185: 134
Huggins C, Jensen EV, Cleveland AS (1954) J Exp Med 100: 225
Hung SC, Liang T, Gluesing LM, Liao S (1976) J Steroid Biochem 7: 1001
Ichii S, Izawa M, Murakami N (1974) Endocrinol Japan 21: 267
Jacobi GH, Moore RJ, Wilson JD (1978) Endocrinology 102: 1748
Jensen EV, Mohla S, Gorell TA, DeSombre ER (1974) Vitam Horm 32: 89
Jouan P, Samperez S, Thieulant ML (1973) J Steroid Biochem 4: 65
Jung I, Baulieu EE (1972) Nature-New Biol 237: 24
Jungblut PW, Hughes SF, Gorlich L, Gowers U, Wagner RK (1971) Hoppe-Seyler's
    Z Physiol Chem 352: 1603
Kadohama N, Anderson KM (1976) Exp Cell Res 99: 135
Kato J (1975) J Steroid Biochem 6: 979
Klyzejko-Stefanowicz L, Chiu JF, Tsai YH, Hnilica LS (1976) Proc Natl Acad Sci US
    73: 1954
Krieg M (1976) Steroid 28: 261
Krieg M, Voigt KD (1976) J. Steroid Biochem 7: 1005
Krieg M, Szalay R, Voigt KD (1974) J Steroid Biochem 5: 453
Krieg M, Dennis M, Voigt KD (1976) J Endocrinol 70: 379
Lane SE, Gidari AS, Levere RD (1975) J Biol Chem 250: 8209
Lea OA, Petrusz P, French FS (1977) Endocrinology 100 (Suppl. 7): 165 (abstr)
Liang T, Liao S (1974) J Biol Chem 249: 4671
Liang T, Liao S (1975) Proc Natl Acad Sci US 72: 706
Liang T, Castañeda E, Liao S (1977) J Biol Chem 252: 5692
Liang T, Mezzetti G, Chen C, Liao S (1978) Biochim Biophys Acta 542: 430
Liao S (1965) J Biol Chem 240: 1236
Liao S (1975) Intl Rev Cytol 41: 87
Liao S (1977) In: Litwack G. (ed) Biochemical action of hormones. p. 351, Academic
    Press, New York, Vol 4, p 351
Liao S, Fang S (1969) Vitam Horm 27: 17

Liao S, Fang S (1970) In: Griffiths K, Pierrepoint CG (eds) Some aspects of aetiology and biochemistry of prostate cancer. Alpha Omega Alpha Publ., Cardiff, Wales, p 105

Liao S, Liang T (1974) In: McKerns KW (ed) Hormones and cancer. Academic Press, New York, pp 229–260

Liao S, Lin AH (1967) Proc Natl Acad Sci US 57: 379

Liao S, Stumpf WE (1968) Endocrinology 83: 629

Liao S, Williams-Ashman HG (1962) Proc Natl Acad Sci US 48: 1956

Liao S, Leininger KR, Sagher D, Barton RW (1965) Endocrinology 77: 763

Liao S, Barton RW, Lin AH (1966) Proc Natl Acad Sci US 55: 1593

Liao S, Tymoczko JL, Liang T, Anderson KM, Fang S (1971) Adv Biosci 7: 155

Liao S, Liang T, Fang S, Castañeda E, Shao T-C (1973a) J Biol Chem 248: 6154

Liao S, Liang T, Tymoczko JL (1973b) Nature-New Biol 241: 211

Liao S, Howell DK, Chang TM (1974) Endocrinology 94: 1205

Liao S, Tymoczko JL, Castañeda E, Liang T (1975) Vitam Horm 33: 297

Liao S, Mezzetti G, Chen C (1979) In: Busch H (ed) The cell nucleus. Academic Press, New York, Vol VII 201–227

Liu AY-C, Greengard P (1976) Proc Natl Acad Sci US 73: 568

Loor RM, Hu AL, Wang TY (1977) Biochim Biophys Acta 477: 312

Loras B, Genot A, Monbon M, Beucher F, Reboud JP, Bertrand J (1974) J Steroid Biochem 5: 425

Louvet JP, Harman SM, Schreiber JR, Ross GT (1975) Endocrinology 97: 368

Mainwaring WIP (1969) J Endocrinol 44: 323

Mainwaring WIP, Irving R (1973) Biochem J 134: 113

Mainwaring WIP, Jones DM (1975) J Steroid Biochem 6: 475

Mainwaring WIP, Mangan FR (1973) J Endocrinol 59: 121

Mainwaring WIP, Peterken BM (1971) Biochem J 125: 285

Mainwaring WIP, Mangan FR, Irving RA, Jones DA (1974) Biochem J 144: 413

Mainwaring WIP, Symes EK, Higgins SJ (1976) Biochem J 156: 129

Mangan FR, Neal GE, Williams DC (1968) Arch Biochem Biophys 124: 27

Mann T, Rowson LEA, Baronos S, Karagiannidis A (1971) J Endocrinol 51: 707

Mezzetti G, Loor R, Liao S (1979) Biochem J 184: 431.

Milgrom E, Atger M (1975) J Steroid Biochem 6: 487

Munck A, Wira C, Young DA, Mosher KM, Hallaham C, Bell PA (1972) J Steroid Biochem 3: 567

Naess O, Hansson V, Djøseland O, Attramadal A (1975) Endocrinology 97: 1355

Naftolin F, Ryan KJ (1975) J Steroid Biochem 6: 993

Nielsen CJ, Sando JJ, Pratt WB (1977) Proc Natl Acad Sci US 74: 1398

Norris JS, Gorski J, Kohler PO (1974) Nature 248: 422

Nozu K, Tamaoki B (1975) J Steroid Biochem 6: 487

Nyberg LM, Wang TY (1976) J Steroid Biochem 7: 263

Palmiter RD, Catlin GH, Cox RF (1973) Cell Diff 2: 163

Parker MG, Mainwaring WIP (1977) Cell 12: 401

Parker MG, Scrace GT (1978) Eur J Biochem 85: 399

Parker MG, Scrace GT, Mainwaring WIP (1978) Biochem J 170: 115

Peets EA, Henson MF, Neri R (1974) Endocrinology 94: 532

Poortman J, Prenen JAC, Schwarz F, Thijssen JHH (1975) J Clin Endocrinol Metab 40: 373

Puca GA, Sica V, Nola E (1974) Proc Natl Acad Sci US 71: 979

Rannels DE, Pegg AE, Rannels SR (1976) Biochem Biophys Res Commun 72: 1481

Ritzen EM, Nayfeh SN, French FS, Aronin PA (1972) Endocrinology 91: 116

Robel P, Blondeau JP, Baulieu EE (1974) Biochim Biophys Acta 373: 1

Rochefort H, Garcia M (1976) Steroids 28: 549

Rochefort H, Lignon F (1974) Eur J Biochem 48: 503

Roy AK, Milin BS, McMinn DM (1974) Biochim Biophys Acta 354: 213

Ruh T, Ruh MF (1975) Endocrinology 97: 1144
Sar M, Liao S, Stumpf WE (1970) Endocrinology 86: 1008
Schmidt WN, Sadler MA, Katzenellenbogen BS (1976) Endocrinology 98: 704
Shain SA, Boesel KW, Axelrod LR (1975) Arch Biochem Biophys 167: 247
Shao T-C, Castañeda E, Rosenfield RL, Liao S (1975) J Biol Chem 250: 3095
Shyr CI, Liao S (1978) Proc Natl Acad Sci US 75: 5969
Simons SS, Martinez HM, Garcea RL, Baxter JD, Tomkins GM (1976) J Biol Chem
    251: 334
Steggles AW, Spelsberg TC, Glasser SR, O'Malley BW (1971) Proc Natl Acad Sci US
    68: 1479
Takayasu S, Adachi K (1975) Endocrinology 96: 525
Tata JR (1966) Progr Nucleic Acid Res 5: 191
Thomas P, Davies P, Griffiths K (1977) Biochem J 166: 189
Tindall DJ, French FS, Nayfeh SN (1972) Biochem Biophys Res Commun 49: 1391
Toft D, Lohmar P, Miller J, Moudgil V (1976) J Steroid Biochem 7: 1053
Tveter KJ, Unhjem O (1969) Endocrinology 84: 963
Tymoczko JL (1973) PhD dissertation: "Factors Responsible for the Nuclear Reten-
    tion of Androgen Receptor of Rat Ventral Prostate" University of Chicago
Tymoczko JL, Liao S (1971) Biochim Biophys Acta 252: 607
Tymoczko JL, Liao S (1976) J Reprod Fertility, Suppl 24: 147
Unhjem O, Tveter KJ, Askvagg A (1969) Acta Endocrinol 62: 153
Valladares L, Mingnell J (1975) Steroids 25: 13
Walsh PC, Wilson JD (1976) J Clin Invest 57: 1093
Wang TY (1978) Biochim Biophys Acta 518: 81
Wester RC, Foote RH (1972) Proc Soc Exp Biol Med 141: 26
Whalen RE, Luttge WG (1971a) Endocrinology 89: 1320
Whalen RE, Luttge WG (1971b) Horm Behav 2: 117
Wilson EM, French FS (1976) J Biol Chem 251: 5620
Zakár T, Toth M (1977) Steroids 30: 751
Zava DT, McGuire WL (1978) Endocrinology 103: 624

Discussion of the Paper Presented by S. Liao

THOMPSON: What were the conditions of the nuclear binding assay? Were those linear nonsaturable nuclear binding assays? In glucocorticoid systems, at least, and
also in estrogen binding systems, there has been considerable discussion about
whether, in fact, there was apparent saturability of the nuclei, or whether that was
due to some exogenous factor. Is this the factor that makes pseudosaturation in the
prostate nuclei?

LIAO: I guess you could call this pseudosaturation. Somehow it is pretty easy for
the people who are working in the androgen field to show the so-called tissue specificity. The experiments were done under conditions where we see this specificity.
The inhibitor action is not reversed by the addition of excess androgen receptor complex. The portion that we are analyzing is the one not released by 0.1 M KCl but can
be released by 0.4 M KCl.

THOMPSON: As you are adding more receptors, and since it is not yet technically
possible to add pure receptor, are you adding more α-protein at the same time?

LIAO: The α-protein goes into the different fractions. The receptor that we used was
obtained at 35% ammonium sulfate. The α-protein comes at 55–75% saturation.

THOMPSON: So, the receptor preparation that you obtain free of α-protein will, if you
add more, show saturation?

LIAO: There may be some contamination, but you can separate them.

THOMPSON: Partially purified receptor preparation, added in increasing quantities,
will appear to saturate the nuclei. Is that correct?

LIAO: Yes, the curve will be hyperbolic. We wish we could use pure receptor to carry out this type of experiment.

BAXTER: I am having trouble understanding this relative testosterone–dihydrotestosterone action. Do you have any binding protein that recognizes dihydrotestosterone and does not bind at all to testosterone?

LIAO: No, it is a relative thing. As far as I can remember, all of the so-called "androgen receptor proteins" have been shown to bind both steroids, but with different affinities. Some of the receptor protein, especially one from kidney, seems to bind DHT and testosterone equally well. In vagina there is a protein that can bind androstenediols but not DHT or testosterone.

BAXTER: Do you think, then, that the qualitative differences that are seen in vivo, and also the striking differences in androgen sensitivity during development are dose–response phenomena? Or do you think that there are really actions of dihydrotestosterone, for example, that will not be elicited by testosterone?

LIAO: We do not really know this. But in the prostate, if there is a mechanism to keep testosterone at high concentrations, testosterone can probably carry out many actions normally done by DHT.

BAXTER: So, you would say that in vivo all of the things, for example, that happen in 5α-reductase syndrome would probably occur if there were enough testosterone around early in the development.

LIAO: It may be so. But as I mentioned, not all the actions of testosterone can be mimicked by DHT.

WILLIAMS-ASHMAN: I think there are many complexities to this. Of course, one of them is the effect on hypothalamus and sex behavior, where you may need testosterone both to be aromatized to estrogen and to go to dihydrotesterone. Of course, in vivo you have this additional problem of penetration of exogonous steroid.

CLARK: Jack Gorski and I did some crazy experiments showing that estrogen receptors are bound to glass and all sorts of things. They would bind to this rug if you put them on the rug. They bind very well. In fact, they bind with a dissociation constant of about 10^{-10} to 10^{-9} M. If you do competitive displacement of that glass binding by using cold complexes, they will compete that binding. I should say that the nonlabeled complexes will displace labeled complexes from glass. This displacement is not competitive; however, without the careful evaluation of saturation curves it will appear to be competitive. Therefore, in any nuclear binding experiment one needs to be able to define that which is specifically bound; otherwise, you are studying receptor binding to any surface. Did you do displaceable binding in those experiments that you performed?

LIAO: I do agree that estrogen receptor complex or androgen receptor complex will bind to many things in a nonspecific manner. We haven't done the displacement experiment as you suggested. As I stated previously, I don't know the reason why, but, apparently the androgen receptor complex seems to be able to recognize the different nuclei, that is, the nuclei from different tissues. The same thing applies to the progesterone receptor complex, in which, as Dr. O'Malley has shown, there is specificity in nuclear binding.

CLARK: We can do the same experiment and get the same results, but until one does the competitive displacement, the results are very ambiguous because you can always argue that spleen nuclei are not contaminated in the same way that uterine or prostate nuclei are.

LIAO: There is a lot of nonspecific binding of the androgen–receptor complex to the nuclei. The portion that we have analyzed is the portion that cannot be washed out by 0.1 M KCl but can be washed out by 0.4 M KCl. That fraction seems to be rather specific. If you take nuclei from spleen or other less-sensitive tissue, there you don't see the binding to the extent you can see with the prostate nuclei. I don't think that I can explain the specificity, but the fact is, there is a specificity and we are working under the conditions in which we can show this specificity.

CLARK: Yes, that makes it more interesting.

O'MALLEY: What is the status of your prostate androgen–receptor in terms of binding to nuclear RNA?

LIAO: We haven't done very much with that. About 1 or 2% of the total RNP particles we isolated from prostate nuclei will bind the androgen–receptor complex; the rest of them will not. That is why we thought there may be specificity. Other than that we don't have any specific message we can play with. Also we don't have a purified receptor.

O'MALLEY: How did you isolate those RNP particles?

LIAO: We used Wang's method in which you dissolve the chromatin in 2 *M* NaCl and then precipitate DNA by dialysis and then get the RNP particle out, which stays in the supernatant.

STEVENS: I'd like to go back to α-protein that you described, which seems to inhibit the nuclear binding of the androgen receptor. I believe you showed that the α-protein did not cause a reduction in steroid binding to the receptor itself. But, I wonder whether you looked into the possibility that there may be proteolytic activity in your preparation, perhaps an integral part of it, in fact, that could cause cleavage of the receptor into a meroreceptor form or something larger than that which would retain the steroid binding according to the type of assay you used as measurement. But then your receptor would have lost the nuclear binding capability, perhaps.

LIAO: No, the experiment was done by incubating the DHT–receptor complex with or without the inhibitor at the different lengths of time. The nuclei were added afterward and there was no change in the percentage of inhibition. So α-protein did not increase degradation of the receptor or reduce the ability of the receptor complex to bind to the nuclei.

STEVENS: I am not sure that I am 100% clear. If you test incubate your receptor with the inhibitor (forget about the nuclei now) and then you put the receptor preparation on sucrose gradient or some gel column, is the size of the receptor the same before and after it encountered the inhibitor?

LIAO: If you mix the DHT–receptor complex with the inhibitor for about 20 min and compare that fraction with the one that was not incubated in the absence of the inhibitor (but inhibitor was added after this incubation), there is no difference in their ability to bind to nuclei. So, during the 20 min of incubation the inhibitor did not cause the inactivation of the nuclear binding activity of the receptor complex.

STEVENS: About the size, I would like to know whether, after you expose the receptors to the inhibitors, the receptors change in size in any measurable way.

LIAO: Does not appear to be.

STEVENS: You could get exactly that result according to the way I understand Mary Sherman's work. According to how you measure the binding, you could still have a 20 000 molecular weight polypeptide with the receptor attached, which would not have the same binding properties. My question, is whether your inhibitor could be converting the native DHT receptor to something smaller that still retains steroid bound to it in a specific way.

LIAO: We did not see the change in sedimentation coefficient of the receptor complex. We have purified α-protein to homogeneity. The protein has no protease or esterase activity but retains the inhibitory activity (Chen et al. 1979).

WILLIAMS-ASHMAN: What about the heat lability of this effect? Is this α-protein effect heat stable?

LIAO: It is quite stable to heat.

WILLIAMS-ASHMAN: So it would be unlikely to be any enzymatic reaction. However, there are some heat-stable enzymes.

Discussants: J.D. BAXTER, J.N. CLARK, S. LIAO, B.W. O'MALLEY, E.B. THOMPSON, and J. STEVENS.

Chapter 13

The Mechanism of Androgen and Progestin Action on Mouse Kidney

L. P. Bᴜʟʟᴏᴄᴋ, T. R. Bʀᴏᴡɴ*, ᴀɴᴅ
C. W. Bᴀʀᴅɪɴ**

I. Introduction

Most studies on the mechanism of androgen action have employed tissues of
the reproductive tract, primarily the rat prostate. From these, and other investigations, the following concept of the mechanism of androgen action has
developed (Mainwaring 1977). Testosterone, the predominant plasma androgen, enters the target organ, where it is metabolized by 5α-reductase to
5α-dihydrotestosterone, which is, in turn, bound to a specific, soluble, cytoplasmic receptor protein. The dihydrotestosterone–receptor complex is
transferred to the nucleus, where it interacts with chromatin. This results in
the initiation of a series of molecular events, which include increases in
chromatin template and RNA polymerase activities and increased synthesis
of specific RNAs, proteins, and DNA.

While these studies have provided a great deal of useful information, they
also have inherent difficulties. The metabolism of testosterone to 5α-dihy-
drotestosterone, and its subsequent inactivation via reduction to the 5α-an-
drostanediols, make steady-state kinetic studies of steroid–receptor interactions difficult. Androgens stimulate cell division in reproductive tissues
such that most responses are evaluated in a changing cell population. To
avoid these disadvantages, we have been using the mouse kidney in our
studies of androgen action and its modulation by progestins. This contribution summarizes our investigations into the molecular interactions of androgens and progestins within the target cell and some of the cell's earliest responses.

* Current address: Division of Pediatric Endocrinology, Department of Pediatrics,
Johns Hopkins Hospital, Baltimore, Md.
** Current address: Director, Center for Biomedical Research, The Population
Council, The Rockefeller University, 64th St. at York Ave., New York, N.Y.

II. Metabolism of Androgens

In tissues of the reproductive tract, such as prostate and seminal vesicle, and epidermal derivatives, such as skin and preputial gland, testosterone is metabolized to 5α-dihydrotestosterone, which, by reason of its uptake and nuclear retention, is considered to be the "active" androgen (Mainwaring 1977). However, dihydrotestosterone may also be rapidly metabolized to the less active androstanediols. In contrast, in kidney there was little metabolism of testosterone to 5α-reduced products owing to the low 5α-reductase activity in this tissue (Mowszowicz and Bardin 1974). If, however, dihydrotestosterone was the substrate, it was rapidly metabolized to the androstanediols by high levels of 3-ketosteroid reductase. The activity of this latter enzyme in kidney cytosol was sufficient to reduce physiological concentrations of dihydrotestosterone at 0°C without the addition of cofactor. As a result, steady-state concentrations of dihydrotestosterone cannot be maintained in vitro for prolonged periods of time. This problem can be avoided, however, if testosterone is used.

III. Nuclear Uptake of Androgens

The absence of 5α-reductase activity in mouse kidney suggested that testosterone might be the active intranuclear androgen in this tissue. The results from a series of in vivo experiments supported this hypothesis (Bullock and Bardin 1974, 1975). When [³H]testosterone was administered to castrated male mice, testosterone was the major steroid recovered from renal cytoplasm and nuclei (Fig. 1A). Similar results were obtained with [³H]androstenedione. In contrast, after the administration of [³H]dihydrotestosterone, [³H]dihydrotestosterone and the [³H]5α-androstanediols were the predominant steroids in cytoplasm, whereas almost all nuclear radioactivity was dihydrotestosterone (Fig. 1B). The concomitant administration of a 100-fold excess of unlabeled testosterone, dihydrotestosterone, or cyproterone acetate with either [³H]androgen reduced nuclear radioactivity by 85 –95%. [³H]Dihydrotestosterone was also the major intranuclear steroid isolated following the administration of tritiated 5α-androstane-3α,17β-diol and its [³H]3β isomer (Fig. 1C, D). Similar results were obtained when normal female mice were used. In contrast, in kidney from androgen-insensitive Tfm/Y mice, there was no evidence of nuclear uptake of any of these steroids. Thus, in contrast to prostate, normal mouse kidney can concentrate either testosterone or 5α-dihydrotestosterone. However, since testosterone is the predominent androgen that is normally present in blood, it is considered to be the biological effector of androgen action in this organ. Subsequent studies by other investigators have indicated that testosterone is also the intranuclear or active androgen in a number of other tissues, including the brain, pituitary, submaxillary gland, muscle, and testis (Mainwaring 1977).

As concerns the ontogeny of the active androgen, the findings of Wilson

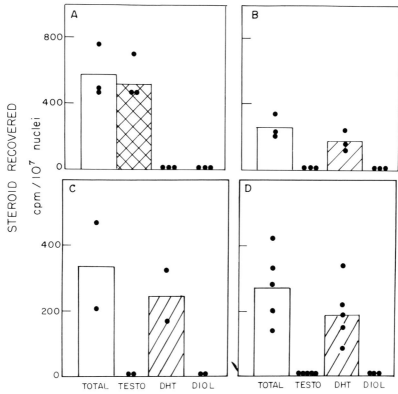

Fig. 1,A–D. [³H]Androgens isolated from kidney nuclei of functionally hepatectomized, castrated young adult male mice 60 min after the intravenous administration of 45 μCi of the [³H]steroid indicated: **A** [³H]testosterone; **B** [³H]dihydrotestosterone; **C** [³H]5α-androstan-3α,17β-diol; **D** [³H]5α-androstan-3β,17β-diol. Nuclear radioactivity was analyzed for total methylene chloride extractable radioactivity (TOTAL) as well as testosterone (TESTO), dihydrotestosterone (DHT), and androstanediols (DIOL). Note the absence of DHT following [³H]testosterone and the absence of TESTO following the administration of the other [³H]steroids. Each point represents the results from the pooled tissue of three animals.

and his colleagues concerning the "active" androgen in Wolffian ducts are of interest. During the initial stages of differentiation, 5α-reductase activity is not present in Wolffian ducts in rabbit (Wilson and Lasnitski 1971) and human fetuses (Siiteri and Wilson 1974), and testosterone is the intranuclear androgen (Wilson 1973). 5α-Reductase first appears after organogenesis, at which time dihydrotestosterone is synthesized and retained by nuclei in these tissues. It was thus proposed that testosterone is the active androgen in Wolffian ducts during differentiation and dihydrotestosterone is the active androgen during development and adult life (Wilson 1973). In contrast, in the urogenital sinus and genital tubercle of the rabbit and human, 5α-reductase activity is present in the fetus, and dihydrotestosterone is the "active" androgen during differentiation of prostate and external genitalia as well as in later life (Siiteri and Wilson 1974; Wilson 1973; Wilson and Lasnitsky 1971).

IV. Nuclear Uptake of Progestins

The ability of some progestins to mimic, potentiate, or inhibit androgen action has been recognized (Bullock and Bardin 1977b; Mainwaring 1977). These effects have been termed androgenic, synandrogenic, and antiandrogenic, respectively. There was little known, however, about the mechanism by which these effects were mediated. To learn more about this aspect of steroid action we chose to study medroxyprogesterone acetate (MPA) and its potential metabolite, 6α-methylprogesterone (6MP), since they were known to have androgenic, antiandrogenic, and synandrogenic effects on mouse kidney (Gupta et al. 1978; Mowszowicz et al. 1974). [H^3]Medroxyprogesterone acetate or [^{3}H]6MP were synthesized (Brown et al. 1979; Runic et al. 1976) for use in a series of in vivo and in vitro studies. Following the intravenous administration of [^{3}H]MPA, radioactivity accumulated in nuclei of kidney, submaxillary gland, and seminal vesicle (Table 1) (Bullock et al. 1978). Medroxyprogesterone acetate, rather than a metabolite, was the only nuclear steroid detected. The high nuclear uptake in seminal vesicle may reflect the higher proportions of steroid responsive cells in these organs relative to the other.

To demonstrate the low capacity and specificity of nuclear uptake of [^{3}H]MPA, the unlabeled steroid was administered subcutaneously owing to its poor solubility in the aqueous solution used for IV injection. Nuclear accumulation of [^{3}H]MPA was blocked by excess MPA or dihydrotestosterone in all these tissues. Conversely, excess MPA decreased nuclear uptake of ^{3}H-labeled androgen, regardless of whether the predominant steroid was testosterone, as in the kidney and submaxillary gland, or dihydrotestosterone, as in seminal vesicles. [^{3}H]Medroxyprogesterone acetate was not retained in nuclei of Tfm/Y mice. To determine if specific nuclear uptake of MPA occurred in androgen target organs of other species, similar studies were performed in rats (Bullock et al. 1978). Specific nuclear uptake of [^{3}H]MPA was demonstrated in nuclei from prostate and seminal vesicle. Medroxyprogesterone acetate also inhibited the nuclear retention of androgens in these organs. These results suggest that the androgenic actions of MPA are associated with specific in vivo binding to androgen receptors in various tissues and species. This hypothesis was strengthened by studies of the in vitro binding of progestins in mouse kidney cytosol, discussed below.

Similar studies were done to evaluate nuclear uptake of 6MP in kidney and prostate-seminal vesicles (Brown et al. 1979). [^{3}H]6α-Methylprogesterone and [^{3}H]testosterone were first administered subcutaneously to evaluate steroid absorption from subcutaneous sites and to determine the relationship between plasma and nuclear steroid concentrations. Plasma concentrations of radioactivity rose steadily during the 15–120-min of study with [^{3}H]testosterone administration producing tenfold higher values than did [^{3}H]6MP. An even greater difference was present in prostate–seminal vesicle nuclei, where concentrations of radioactivity following [^{3}H]testosterone were 100-fold greater than that after [^{3}H]6MP. In contrast, in kidney nuclei, the con-

Table 1. Nuclear Uptake of [³H]Steroids in Murine Androgen Target Tissues[a]

[³H]Steroid (route)	Competitor[b]	n[c]	Nuclear steroid isolated	Kidney	Submaxillary gland	Seminal vesicle
Medroxyprogesterone acetate (SC)	0	(3)	Medroxyprogesterone acetate	46[d]	91	710
	Medroxyprogesterone acetate (SC-50)	(3)	Medroxyprogesterone acetate	17	35	120
Medroxyprogesterone acetate (IV)	0	(6)	Medroxyprogesterone acetate	124	392	1338
	Dihydrotestosterone (IV-50)	(4)	Medroxyprogesterone acetate	32	39	96
Testosterone (IV)	0	(4)	Testosterone	253	959	2465
			Dihydrotestosterone	15	74	1921
	Medroxyprogesterone acetate	(3)	Testosterone	136	524	1337
			Dihydrotestosterone	14	37	833

[a] Castrate, functionally eviscerated male BALB/c mice were given 60 μCi of [³H]steroid $\pm$ unlabeled steroid in 0.25 ml of 6% ethanol in saline by the route shown. The competitors, when given by the same route, were included in the same solution. Animals were killed 30 min later, and tissues from 3 mice were pooled and nuclei obtained. Nuclear steroids were extracted into organic solvents and purified by thin-layer chromatography.

[b] The route of administration of competitor is shown in parentheses along with the ratio of competitor: [³H]steroid.

[c] n = number of experiments.

[d] Mean cpm/10⁷ nuclei.

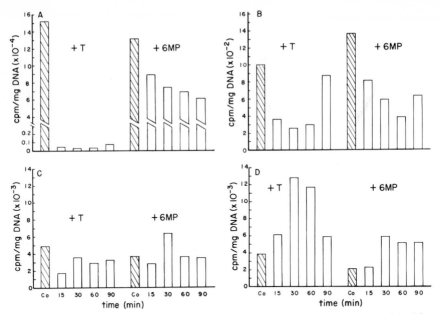

Fig. 2. Uptake of radioactivity by prostate–seminal vesicle or kidney nuclei of functionally hepatectomized, castrated young adult male mice: **A** [³H]testosterone, prostate–seminal vesicle; **B** [³H]testosterone, kidney; **C** [³H]6α-methylprogesterone, prostate–seminal vesicle: **D** [³H]6α-methylprogesterone, kidney. Pretreated animals (open bars) received a 500-fold excess of testosterone (+ T) or 6α-methylprogesterone (+6MP) SC at $t = 15, 30, 60,$ or 90 min prior to the injection of radioactivity. At $t = 0$ these and control (Co) animals received 50 μCi IV of [³H]testosterone (**A, B**) or [³H]6α-methylprogesterone (**C, D**). All animals were killed 30 min later, nuclei prepared, and radioactivity extracted with ethanol. The height of the bars represents the value obtained from tissue pooled from 3 animals. Note the different scales on the ordinate.

centration of radioactivity was similar for both steroids. The marked differences in the kinetics of steroid uptake in kidney and prostate–seminal vesicles imply that different factors regulate cell and/or nuclear uptake in these two tissues.

To determine the specificity of nuclear uptake, [³H]steroids were given IV 15–90 min after the administration of a 500-fold excess of testosterone or 6MP (Fig. 2) (Brown et al. 1979). In prostate–seminal vesicles the androgen and progestin reduced nuclear uptake from [³H]testosterone by 99 and 50%, respectively (Fig. 2A). Nuclear uptake in kidney was similarly reduced to background levels by testosterone, while 6MP was somewhat less active (Fig. 2B). This effectiveness is not readily obvious from Fig. 2 owing to the different scales used for depicting nuclear uptake in the two tissues. These results are consistent with 6MP binding to the androgen receptor. Similar studies were done with 6MP as the [³H]steroid (Fig. 2C, D). In prostate–seminal vesicles, nuclear uptake was significantly lower than that achieved with [³H]testosterone, and only testosterone had a slightly inhibitory effect. In kidney, however, not only did [³H]6MP result in greater concentrations of nuclear radioactivity than did [³H]testosterone, but both testosterone and 6MP increased the uptake. While the comparative differences in nuclear uptake in each tissue, resulting from the IV administration of the [³H]steroids alone, could have been predicted from the previous subcutaneous studies, the enhancement of nuclear [³H]6MP uptake by the unlabeled steroids was unexpected.

To characterize the nature of the nuclear radioactivity, plasma and nuclei from both organs were extracted with petroleum ether and subjected to thin-layer chromatography. The predominant steroid in plasma was [³H]6MP, although several more polar metabolites were also present. Kidney nuclei contained [³H]6MP and a single polar metabolite, which was thought to be similar to one in plasma. Derivative formation and thin-layer and paper chromatography were used to tentatively identify this compound as the 20α-hydroxy metabolite of 6MP. In contrast, in prostate nuclei only the parent compound, 6MP, was recovered.

There are thus marked differences in the metabolism and nuclear uptake of testosterone and 6MP in mouse kidney and prostate–seminal vesicles. The molecular basis for these differences is not yet understood.

V. Receptor Binding of Androgens and Progestins

The ability of mouse kidney to act as a target organ for androgen action and the retention of androgens by renal nuclei suggested the presence of an androgen receptor in this tissue. The in vivo competition studies described above suggested that testosterone and dihydrotestosterone shared a common binding site on this protein. This was confirmed when mouse kidney cytosol was incubated with [³H]androgens and analyzed by a variety of procedures. The binder had the physicochemical characteristics (Table 2) (Bullock et al. 1975) and the steroid specificity (Bullock and Bardin 1974) similar to the androgen receptor in the rat prostate (Mainwaring 1977). There was no consistent evidence for an androgen receptor in kidney cytosol from Tfm/Y mice. This, combined with the absence of intranuclear androgen uptake in these animals, was evidence that the absence of an effective andro-

Table 2. Physicochemical Properties of the Cytoplasmic Androgen Receptor from Mouse Kidney and Rat Prostate[a]

Parameter	Mouse kidney	Rat prostate
K_d (nM)	1.3	2–4
Molecular weight (daltons)	270 000	276 000
Sedimentation coefficient (S)	7.9	8
Stoke's radius (Å)	82	84
Frictional rate (f/f_0)	1.98	1.98
Isoelectric point	4.8	58
DNA binding	Yes	Yes
Heat stability	Poor	Poor

[a] Data from Bullock and Bardin (1974), Bullock et al. (1975); and Mainwaring (1977).

Table 3. Relative Binding Affinities and Biological Activities of Testosterone and Progestin in Mouse Kidney[a]

| | Relative binding affinity | | Biological activity | | |
	Cytosol	Nuclei	Androgenic	Anti-androgenic	Syn-androgenic
Testosterone	100	100	++++	−	−
6α-Methylprogesterone	36	10	++	+++	+++
Cyproterone acetate	30	2	−	+++	+++
Medroxyprogesterone acetate	8	10	+++	−	++
Progesterone	7	4	−	−	−

[a] Data from Bardin et al. (1978a) and Bullock et al. (1978).

gen receptor was the molecular basis for the defect of androgen sensitivity. This concept has been confirmed and extended by a number of other investigators (Bullock and Bardin 1977a).

The involvement of the androgen receptor, rather than one for progesterone, as the mechanism by which progestins mimic androgen action was suggested by the inability of progestins to elicit androgenic effects in androgen-insensitive Tfm/Y mice that lack the androgen receptor (Mowszowicz et al. 1974), and the ability of the progestin, cyproterone acetate, to block androgen action via competitive binding to the androgen receptor (Mainwaring 1977). This hypothesis was confirmed by in vitro studies (Bullock et al. 1978). Kidney cytosol was incubated with [³H]testosterone or [³H]MPA in the presence or absence of nonradioactive steroids and aliquots sedimented through sucrose gradients. [³H]Testosterone and [³H]MPA were both bound to high-affinity, low-capacity macromolecules that cosedimented in the 8S region of the gradient. Binding was displaced by nonradioactive MPA or testosterone in a dose-dependent fashion and by 100-fold excesses of several other progestins and estradiol, but not by dexamethasone. There was no detectable binding of [³H]testosterone or [³H]MPA to 8S components in kidney cytosol from Tfm/Y mice.

The competition by various progestins for [³H]testosterone binding in mouse kidney cytosol was studied further by adsorption of steroid receptor complexes to DEAE−cellulose filters (Bullock et al. 1978). The relative binding affinity of each steroid at 50% [³H]testosterone displacement was determined (Table 3). All the progestins studied effectively competed with testosterone for binding. However, the calculated affinities did not correlate directly with the magnitude or direction (agonist or antagonist) of the effect of progestin on an androgen-responsive renal enzyme, β-glucuronidase. For example, progesterone and MPA were calculated to have approximately the same relative binding affinities despite quite different androgenic potencies. These results were compared with those from an assay designed to evaluate the nuclear binding of these progestins (see Table 3) (Bardin et al. 1978a). Minced mouse kidney was incubated with [³H]testosterone with or without various concentrations of nonradioactive steroids, and the specific nuclear uptake of [³H]testosterone was quantitated. Relative binding affinities were calculated as for the cytoplasmic assay. There were differences in the relative binding affinities as determined by the two assays.

These data indicate that progestins affect androgen action via binding to the androgen receptor. However, their various actions (androgenic, synandrogenic, or antiandrogenic) cannot be predicted solely from their affinities for androgen receptors as assayed in kidney cytosol or nuclei. Obviously, other factors are involved in determining the ultimate biological response.

VI. Stimulation of RNA Polymerases by Androgens and Progestins

Steroid hormones are thought to initiate cellular responses through the interaction of the steroid–receptor complex with chromatin (Chan and O'Malley 1976; Mainwaring 1977). Changes in transcriptional events, either an increase in chromatin template activity or activation of DNA-dependent RNA polymerases, result in an increase in specific RNAs and subsequent protein synthesis. A series of studies were designed to learn more about this aspect of androgen action on mouse kidney. Ribonucleic acid polymerase I and II activities were evaluated in isolated nuclei following subcutaneous testosterone (1 mg) administration to normal female mice (Fig. 3) (Janne et al. 1976a). The activity of both kidney RNA polymerases increased within 15 min. Ribonucleic acid polymerase II reached peak activity within 1 h, whereas polymerase I activity was not maximum until 2 h, by which time polymerase II activity had returned to near control levels. Polymerase I activity declined less rapidly and was still slightly above control values 4 h after treatment. Similar patterns of responses to estrogen hormones have been reported previously (Chan and O'Malley 1976; Janne et al. 1976b).

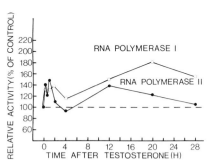

Fig. 3. Effect of testosterone on RNA polymerase activities in mouse kidney nuclei. Young adult female mice (10 per point) were treated with a single SC dose of testosterone (1 mg) and killed at the times indicated. Control animals were given sesame oil. Kidneys from each group were pooled, nuclei isolated, and RNA polymerase activity measured using [^{14}C]UTP incorporation and difference incubation conditions. Each point represents the mean of two separate experiments. The changes in RNA polymerase activity are expressed as percentage of control (= 100), which were 39 and 369 pmol UMP incorporated per mg DNA for RNA polymerases I and II, respectively. (Janne O, Bullock LP, Bardin CW, Jacob ST (1976) Biochim Biophys Acta 418: 330–343)

The patterns of polymerase activity were compared with nuclear uptake of [³H]testosterone following its subcutaneous administration (Janne et al. 1976a). Uptake was maximum within 30 min and remained relatively constant over the next 2 h. By 4 h, 40% of the peak value remained, and radioactivity was not detectable after 6 h. Thus, the fall in RNA polymerase II activity (1–2 h) occurred at a time when there was still a significant amount of testosterone in kidney nuclei. There was a second increase in polymerase I and II activities, which peaked at 20 and 12 h, respectively. While the early increases in the two activities were approximately equal, the secondary response was greater for polymerase I than for polymerase II. In contrast, testosterone had no effect on polymerase activity in kidneys of androgen insensitive Tfm/Y mice.

Further studies were done to determine if these changes in polymerase activities, as assayed in whole nuclei, were associated with changes in chromatin template activity or in the activities or concentrations of the polymerases themselves. Changes in template capacity induced by a single subcutaneous dose of testosterone were measured using mouse kidney chromatin and RNA polymerase II purified from pig kidney (Janne et al. 1976a). Changes in template capacity followed a pattern similar to that seen for polymerase I activity. A second dose of testosterone, given at 4 h when template activity had returned to control values, had little effect (Janne et al. 1976a). The template capacity of renal chromatin from Tfm/Y mice was not changed by androgen treatment. To evaluate the activity of the polymerases themselves, they were solubilized and assayed using calf thymus DNA as template (Lin et al. 1978). Testosterone increased the activities of RNA polymerase II and III within 4 h. In contrast, RNA polymerase I activity was unaltered. Testosterone had no effect on the activity of any of the solubilized polymerases in Tfm/Y mice. In summary, similar to what has been reported for other steroid-responsive systems, androgens act via the steroid–receptor complex to produce early effects on transcriptional events in mouse kidney.

In view of the androgenic effects of progestins on mouse kidney, it was of interest to determine whether MPA could also alter transcription in this organ. A single subcutaneous injection of MPA increased RNA polymerase I and II activities as assayed in isolated mouse kidney nuclei (Fig. 4) (Lin et al. 1978). The activities of both enzymes attained peak values by 4 h and declined to control values by 12 h after progestin stimulation. A second-peak of activity was observed for polymerase I at 20–28 h although there was no further increase in polymerase II activity. While a small dose of MPA (1 mg) had little effect on chromatin template activity, a larger dose (10 mg) increased chromatin capacity within 2 h for both polymerases I and II (Lin et al. 1978). Medroxyprogesterone also increased the individual activities of solubilized RNA polymerases I, II, and III (Lin et al. 1978). In contrast to these results in normal animals, MPA had no effect on any of these transcriptional events in kidneys from Tfm/Y mice.

Thus, although testosterone and MPA both induced changes in polymerase and chromatin template activities, the pattern of these changes was

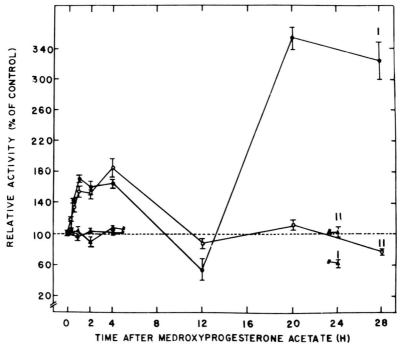

Fig. 4. Effect of medroxyprogesterone acetate on RNA polymerase activities in mouse kidney nuclei. Young adult female and androgen-insensitive (Tfm/Y) mice (6 per point) were treated with a single SC dose of medroxyprogesterone acetate (10 mg) and killed at times indicated. Kidneys from each group were pooled, nuclei isolated, and RNA polymerase activities measured as in Fig. 3. Each point represents the mean value of 3 separate experiments ±SE. RNA polymerase I (●) and II (○) activities from normal female mice and RNA polymerase I (▲) and II (△) activities from Tfm/Y mice were expressed as percent of the control (= 100). The mean values of RNA polymerase I and II activities for untreated normal mice were 112 and 322 pmol of UMP incorporated per mg of DNA, whereas those for untreated Tfm/Y mice were 233 and 305 pmol of UMP incorporated per mg per DNA, respectively. (Lin Y-C, Bullock LP, Bardin CW, Jacob ST (1978) Biochemistry 17: 4833–4838)

not always similar. Testosterone stimulated chromatin template activity earlier and in smaller doses than did MPA. Medroxyprogesterone acetate did not stimulate RNA polymerase II activity in isolated nuclei at later time points as did testosterone. While MPA stimulated an increase in extractable RNA polymerases I, II, and III, only the latter two polymerases were stimulated by testosterone.

VII. Summary

The mechanism of androgen action on mouse kidney has been shown to be similar to that of the reproductive tract. In kidney, androgens are bound to a soluble receptor with physicochemical characteristics similar to those of the androgen receptor in rat prostate. The steroid–receptor complex is translocated to the nucleus, where it stimulates an increase in chromatin template and RNA polymerase activities in a manner similar to that described for other steroid-responsive systems. Progestins, which have androgenic effects on mouse kidney, also act via the androgen receptor and stimulate early changes in transcriptional events.

Many of the problems inherent to studies of androgen action on the reproductive tract are not encountered when kidney is used. Since the kidney is not dependent on androgens for differentiation, studies may employ males, females, and androgen-insensitive animals. Whereas testosterone is the "active" androgen, owing to the absence of 5α-reductase, problems with metabolism can be avoided and steady-state binding kinetics can be achieved. Androgens induce renal hypertrophy but not hyperplasia, thus providing a stable cell population for study (Paigen et al. 1975). Androgens also increase the activity of a number of well-characterized renal enzymes (Bardin et al. 1978b). The genetic control of the androgenic response of one of these enzymes, β-glucuronidase, has been studied extensively (Paigen et al. 1975). In addition, other hormones such as progestins and pituitary hormones (Bullock and Bardin 1979; Swank et al. 1977) have been shown to modify the response of the kidney to androgens and provide a basis for further study.

Acknowledgments. The authors gratefully acknowledge the collaborate efforts of Drs. C. Gupta, S. Jacob, O. Janne, Y.C. Lin and I. Mowszowicz while in our laboratory. This paper is supported in part by NIH Contract Nol-2-2730 and NIH Research Cancer Development Award No. K04-HD00137 to Leslie P. Bullock.

References

Bardin CW, Brown TR, Mills NC, Gupta C, Bullock LP (1978a) Biol Reprod 18: 74–83
Bardin CW, Bullock LP, Mills NC, Lin Y-C, Jacob ST (1978b) In: O'Malley BW (ed) Receptors and Hormone Action. Academic Press, New York. Vol II, Chap 4 pp. 83–103
Brown TR, Bullock LP, Bardin CW (1979) In: Leavitt W (ed) Hormone receptor systems. Plenum, New York, pp 269–280
Bullock LP, Bardin CW (1974) Endocrinology 94: 746–756
Bullock LP, Bardin CW (1975) Steroids 25: 107–119
Bullock LP, Bardin CW (1977a) In: Martini L, and Motta M (eds) Androgens and antiandrogens. Raven Press, New York, pp 91–103
Bullock LP, Bardin CW (1977b) Ann NY Acad Sci 286: 321–330
Bullock LP, Bardin CW (1980) In: DiCarlo F (ed.) Pharmacological modulation of steroid action. Raven Press, New York, pp 145–158
Bullock LP, Mainwaring WIP, Bardin CW (1975) Endocr Res Commun 2: 25–45
Bullock LP, Bardin CW, Sherman MR (1978) Endocrinology 103: 1768–1782
Chan L, O'Malley BW (1976) New Engl J Med 294: 1372–1381
Gupta C, Bullock LP, Bardin CW (1978) Endocrinology 120: 736–744
Janne O, Bullock LP, Bardin CW, Jacob ST (1976a) Biochim Biophys Acta 418: 330–343
Janne O, Kontula K, and Vihko R (1976b) Acta Obstet Gynecol Scand (Suppl) 51: 29–45
Lin Y-C, Bullock LP, Bardin CW, Jacob ST (1978) Biochemistry 17: 4833–4838

Mainwaring WIP (1977) Monogr Endocrinol 10,

Mowszowicz I, Bardin CW (1974) Steroids 23: 793–807

Mowszowicz I, Bieber DE, Chung KW, Bullock LP, Bardin CW (1974) Endocrinology 95: 1589–1599

Paigen K, Swank RT, Tomino S, Ganschow RE (1975) J Cell Physiol 85: 379–392

Runic S, Miljkovic M, Bogumil RJ, Nahrwold D, Bardin CW (1976) Endocrinology 99: 108–113

Siiteri PK, Wilson JD (1974) J Clin Endocrinol Metab 38: 113–125

Swank RT, Davey R, Joyce L, Reid P, Macey MR (1977) Endocrinology 100: 473–480

Wilson JD (1973) Endocrinology 92: 1192–1199

Wilson JD, Lasnitski I (1971) Endocrinology 89: 659–668

Discussion of the Paper Presented by L.P. Bullock

EDELMAN: I just wanted to say a few words about receptor theory, because I was fascinated by the results that Dr. Bullock had shown, and yet, I beg your indulgence if I give a somewhat long introduction to what I'm going to say. There are some things about the theory of receptors that are worth bearing in mind. The first rule is that if a ligand binds to a receptor, then it must have some action. The only thing that is excluded is no action. Now, what are the nature of those actions? One possibility is that you get an agonist effect, another possibility is an antagonist effect, and the third possibility is that you get both. If you have complete depletion of the receptor, and the action is put on this scale, then an agonist will give 100% response. If you start at this level, an antagonist will give 0% response. Now a partial inducer, which was a phenomenon that was really first put on the map with respect to steroid hormone action by Samuels and Tomkins, will produce part of an effect.

Furthermore, the rule states that if you induce fully, you must have the sum of the agonist and antagonist effects equal to 100%. In other words, if you have a partial inducer that induces 30% at the maximum level, at the highest possible concentration, then it must be a 70% inhibitor if you fully induce. Now there are pseudocases in which there is no action and yet there is occupancy. That is a circumstance where you are already induced to the same level as the partial inducer. For example, if you have a background of 30% effect, then the addition of the partial inducer will appear to be to have absolutely no action. So to ascertain that you have no action, you must work from two standpoints; you must test both in the complete absence of the effect and the presence of 100% effect. All of the cases that have been carefully examined today, including the mineralocorticoid actions, fit this model.

Now how can one deal with this and the synandrogenic effect? One possibility is that the synandrogenic effect, in fact, is postreceptor action, and that is clearly a possibility, particularly because the action is being looked at over a period of days. So one could induce the synthesis of components that have a role in the process, and have nothing to do with the receptor. But, in fact, it could also be a function of receptor action in the following way. You can start with the question; if the mode which has been most widely developed, particularly by Sherman, namely the allosteric equilibrium model in which you have an active and inactive form, and you have the ligand which can bind to one, the other, or both. If the ligand binds to one, it stabilizes the active form. If it binds to the other, it stabilizes the inactive form. The ratio of these will determine the effect. The effect is proportionate only to the number of active forms that are present. But there is another element involved, and that is what is happening at the gene level. A paper was published recently by Cush and Farmin in which we studied nuclear occupancy, the response as a function of nuclear occupancy. In this case it happens to be active sodium transport, in response to aldosterone. This response is to a single receptor (type 1 receptors), and what we have found is a highly cooperative response. This could be modeled as being a consequence of the fact that, at the level of the genome, it took three complexes to bind simultaneously to produce the effect. This is a straight probabilistic effect. What you get here is apparently occupying 50% of the nuclear binding sites for this steroid-receptive complex, and you get virtually no effect. Then you get a very steep response, and 100% of the response corresponds precisely to 100% of nuclear occupancy in vivo. That implies that there is a cluster number. Now the cluster number need not be the same for various responses. Consequently, one can get responses at cluster numbers of 1, 2, 3, and 4. The shape of this curve then would be different, depending on the cluster number.

Now, what could be happening in the synandrogenic response is the following. You start with the case of alcohol dehydrogenase and the arginase response. The cluster number is smaller. Consequently, at the level of testosterone, which occupies 90% of the receptors on straight probabilistic interpretation, you have already saturated the response of arginase and alcohol dehydrogenase. If cyproterone acetate (CA) is promoting a higher yield of the active form of the receptor bound to the ligand, and if the cluster numbers for the polymerase and β-glucuronidase response is higher, then you will see the enhancing effect, although the steroid, that is, CA itself will have no response. This is just one possible way in which one can fit the existing, at the moment, most sensible model into this kind of special phenomenology.

There is another very well developed case, which should probably be given a name of "discoordinate induction." We are all familiar with the concept of coordinate induction. But equally important is discoordinate induction, because it means that with the same pool of active receptors, you can differentially modulate the kind of response you get, depending on this kind of requirement for the action. I was just going to say that there is another very well-documented example of this in some work that Tom Gelehrter has done.

BULLOCK: It just points out the fact that has been mentioned so often in this conference, which is that we just need to know a lot more about gene activation.

CLARK: I would like to make a comment on Dr. Edelman's comment. I would like to add one more complication to the scheme, and that is what we call differential cell stimulation. In the uterus triphenylethylenes stimulate the epithelium of the uterus in a purely agonistic way, that is, they act as estrogens. They cause the epithelium to grow, to undergo mitosis, hypertrophy, the whole business, whereas in the stroma and the myometrium of the uterus there is very little effect, and the triphenylethylenes may even be pure antagonists under those circumstances. In any heterogeneous systems like the uterus, or the kidney, where there is a possibility of differential cell effects, these effects must be ascertained before those kinds of calculations can be made. If only you could study one cell type, then you could make those calculations.

MUELLER: Just one comment about reconstitution. I think that you should put in some somatomedins, because actually they are probably tied to at least carbohydrate-metabolizing type systems. Without somatomedins and growth hormone, under pituitary depressions, you certainly wouldn't have much performance. Therefore, your supplementation probably is not quite right. But the other thing that is very interesting to me is this. Did you run a full range of concentrations of testosterone against the syndrogenic effects of the cyproterone?

BULLOCK: If I understand your question, I think that what you are asking is this: Is it the dose or the ratio that causes the effect? Yes, we used a constant dose of cyproterone acetate (0.2 mg), and range that against various doses of testosterone. We saw this enhancing effect only when testosterone and cyproterone acetate were given in about equal amounts.

MUELLER: Did you do any direct androgen-binding experiments under the same kind of conditions?

BULLOCK: I have tried androgen-binding experiments in cytosol and with a constant amount of testosterone and a wide, variety ranging from maybe a ratio of 0.1 cyproterone acetate up to 100-fold excess, and I was never able to show any enhancement of androgen binding under these conditions. But, when you think about it, maybe I was a little naive to expect this effect. Of, course, if it had come out well, everybody would have been happy, but cytosol and broken cell systems are too different from in vivo conditions. Also, when we didn't see synadrogenic effects with other end points, so many other end points, you might expect that androgens binding per se was not a primary factor.

MUELLER: I think that one of the interesting features here is that, of course, you are using a higher ratio of those steroid entities to potential receptors. And again, I think that one of the things we have encountered in the estrogen–receptor business is that there are other kinds of competitive effects indicating that steroids don't all work at the same spot. This is particularly true with delfoxidine. For example, I don't know about androgen receptors, but I would suggest that you go back and take a close look at that because, there are other components present in these cytosols that determine this kind of effect. You must be encountering such effects, and this is the interesting feature, since you are getting apparently one kind of genetic expression effect, and not getting some of the others. This may be a clue as to what the actual next contact points are, or the next things that the receptor actually does modulate. I think that this is the interesting feature.

LIAO: Have you seen whether flutamide will affect the testosterone effect and the cyproterone acetate effect?

BULLOCK: Flutamide will work as a classic antiandrogen, that is, it only inhibits. The synandrogenic effect seems to be specific for these progestins.

CLARK: Did I understand you to say that cyproterone acetate is a good progestational agent?

BULLOCK: Yes.

CLARK: Then that would predict that cyproterone acetate, by way of its progestational activity, would cause reduction in cytoplasmic estrogen receptors, and thereby it should be an antiestrogen and as well as an antiandrogen. Is that true?

WILLIAMS-ASHMAN: No, that is not true. Neumann went into that very carefully. Cyproterone has no antiestrogenic activity, but has strong progestational activity. That is the acetate. The free alcohol cyproterone is a feeble antiprogestational agent, but quite a good androgen antagonist, at least, in the dog and in the cat.

CLARK: That is very curious, because progesterone will clearly lower the estrogen receptor in the rat uterus, and at least some of us think that is the way it antagonizes the estrogen action. If cyproterone doesn't do that, and is still a good progestational agent, then it's a very interesting compound. Has anyone ever looked into the kidney to see what effects it has on estrogen receptors?

BULLOCK: Not to my knowledge. One of the things about the (mouse) kidney is that it contains both estrogen and androgen receptors. I have not looked extensively at estrogen receptors except just to show that they were there. The concentration of androgen receptors does not seem to be particularly sensitive to the sex hormonal milieu, that is, about the same number seem to be present in females, and in intact and castrated males. They are not affected by hypophysectomy. So it's completely different from the dependent prostate.

Discussants: L.P. BULLOCK, J.H. CLARK, I.S. EDELMAN, S. LIAO, G.C. MUELLER, and H.G. WILLIAMS-ASHMAN

Chapter 14

Hormone-Dependent Expression of α_{2u} Globulin Gene in Rat Liver

ARUN K. ROY, BANDANA CHATTERJEE, AND
AMRUT K. DESHPANDE*

I. α_{2u} Globulin and its Regulation by Various Hormones

α_{2u}-Globulin, the principal urinary protein of the adult male rat, is synthesized and secreted by the hepatic parenchymal cells (Roy and Neuhaus 1966; Roy and Raber 1972; Roy et al. 1966). The protein has a molecular weight of 21 000 daltons, a sedimentation coefficient of 2.2S (S_{w20}), and it consists of a single polypeptide chain containing 186 amino acid residues (Lane and Neuhaus 1972; Roy et al. 1966). As estimated from liver perfusion studies, α_{2u}-globulin is synthesized in the adult male rat at a rate of 100–150 μg/g liver/h (Roy and Neuhaus 1966). A survey of hepatic and urinary α_{2u}-globulin in male and female rats of different age groups showed that this protein is absent in immature and senescent male and normal female rats of all ages (Roy and Neuhaus 1967; Roy 1973a; Roy et al. 1974). Androgen treatment of spayed female rats led to the induction of α_{2u}-globulin and estrogen treatment of mature male rats completely inhibited α_{2u}-globulin synthesis (Roy and Neuhaus 1967).

Hypophysectomy, which is known to cause multiple hormone deficiency, resulted in complete inhibition of α_{2u}-globulin synthesis, and these animals did not respond to androgen administration (Kumar et al. 1969). This observation led to the investigation of the role of various growth and developmental hormones in the regulation of androgen-dependent synthesis of α_{2u}-globulin. Exploration of the effect of various hormonal combinations on the synthesis of α_{2u}-globulin in hypophysectomized male rats showed that the complete reversal of the effect of hypophysectomy requires simultaneous treatment with testosterone, corticosterone, thyroxine, and growth hormone. Drastic reduction in the hepatic synthesis of α_{2u}-globulin in male rats after adrenalectomy, thyroidectomy, and in diabetes was also observed (Irwin et al. 1971; Roy 1973b; Roy and Leonard 1973). Supplementation of the specific hormones in the above-mentioned endocrine deficiency states reversed the inhibition of α_{2u}-globulin synthesis.

Immature male rats do not synthesize α_{2u}-globulin, and the prepubertal

* Present address: Center for Biomedical Research, The Population Council, The Rockefeller University, New York, N.Y.

animals of both sexes were found to be androgen insensitive to α_{2u}-globulin induction.   Androgen treatment of maturing male and female rats showed that androgenic induction of α_{2u}-globulin can only be achieved after puberty.   The development of hepatic androgen sensitivity at the time of puberty is independent of the endocrine status of the animal during prepubertal life and possibly regulated by age-dependent developmental signals (Roy 1973a).   Senescent male rats were found to lose their ability to synthesize α_{2u}-globulin (Roy et al. 1974).   Besides the prepubertal and senescent states of androgen insensitivity, pseudohermaphroditic male (tfm) rats are also incapable of synthesizing α_{2u}-globulin (Milin and Roy 1973).   The above observations of age-dependent and genetic restrictions of androgen action and its multihormonal modulation have greatly enhanced the potential importance of the α_{2u}-model for exploring the biochemical and molecular mechanisms of hormone action.

II. Hepatic Androgen Binding Protein and its Role in the Androgen-Dependent Synthesis of α_{2u}-Globulin.

Studies in our laboratory have shown the presence of a 3.5S androgen-binding protein in the hepatic cytosol of rats that are capable of synthesizing α_{2u}-globulin under androgen stimulation.   This "cytosol androgen receptor" has a high affinity and specificity for androgenic as well as estrogenic steroids (Fig. 1).   Hepatic androgen insensitivity in prepubertal rats, in senescent rats, and in Tfm rats was found to result from the absence of cytosol androgen receptor (Milin and Roy 1973; Roy et al. 1974).   Spayed female rats, which normally have very low concentrations of this receptor protein, showed marked enhancement of the receptor activity and androgen responsiveness after androgen treatment.   On the other hand, treatment of mature male rats with estrogenic hormones resulted in the gradual decrease and ultimately complete loss of the androgen receptor activity, leading to the temporary androgen insensitivity of these animals (Milin and Roy 1973; Roy, 1977; Roy et al. 1974; Roy et al. 1975).

III. Bifunctional Nature of the Hepatic Androgen Binding Protein and Both Androgenic and Estrogenic Induction of α_{2u}-Globulin in Primary Response

Detailed analysis of the primary response and early events in the hormonal regulation of α_{2u}-globulin has been facilitated by the development of a radioimmunoassay procedure for this protein that is sensitive in the picogram range (Roy 1977).   Using this sensitive assay procedure, it has been observed that a single injection of either 5α-dihydrotestosterone (DHT) or 17β-estradiol-(estradiol) in spayed female rats leads to the induction of α_{2u}-globu-

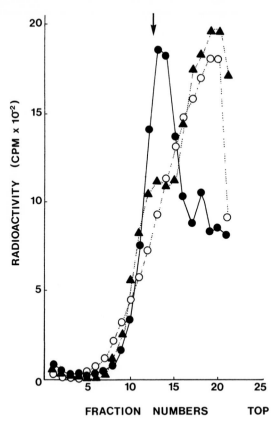

Fig. 1. Inhibition of receptor uptake of $5\alpha[1,2\text{-}^3H_2]$dihydrotestosterone by unlabeled 5α-dihydrotestosterone and estradiol in vitro. The cytosol preparations were incubated simultaneously with 400 n*M* $5\alpha[1,2\text{-}^3H_2]$dihydrotestosterone and 12 μ*M* unlabeled steroids and then subjected to sucrose density gradient analysis. (●) 400 n*M* $5\alpha[1,2\text{-}^3H_2]$dihydrotestosterone alone; (▲) 400 n*M* labeled + 12μ*M* unlabeled 5α-dihydrotestosterone; (○) 400 nM $5\alpha[1,2\text{-}^3H_2[$dihydrotestosterone + 12 μ*M* estradiol. (Roy AK, Milin BS, McMinn DM (1974) Biochim Biophys Acta 354: 213)

lin, reaching peak levels around 6 h after hormone administration. Daily pretreatments of these animals with DHT increased the sensitivity of subsequent androgenic response, whereas pretreatments with estradiol led to a gradual decrease and ultimate loss of the estrogenic induction of α_{2u}-globulin (Fig. 2). Our earlier findings concerning the dual affinity of the hepatic androgen receptor, along with these observations, indicate that binding of either the androgen or the estrogen to receptor protein may lead to α_{2u}-globulin induction. However, it seems that only the androgen is capable of receptor replenishment and receptor induction, whereas the estrogen may inhibit receptor synthesis (Roy et al. 1974; Roy 1977). Evidence for both androgen and estrogen action through one receptor system has also been reported in several other tissues (Jung-Testas et al. 1976; Ruh et al. 1975; Schmidt et al. 1976; Zava and McGuire 1978). Both androgenic and estro-

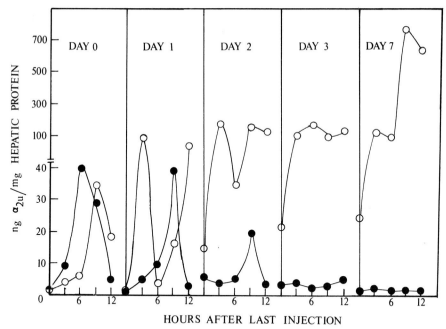

Fig. 2. Effect of daily pretreatments with estradiol or dihydrotestosterone on the hepatic levels of α_{2u}-globulin in castrated female rats after single final injection of the same hormone. The animals in day 0 did not receive any pretreatment and received only single injections of either estradiol (●) or dihydrotestosterone (○). Animals in days 1, 2, 3, and 7 received 1, 2, 3 and 7 daily pretreatments with either estradiol (●) or dihydrotestosterone (○) followed 24 h later by a single injection of the same hormone. Amounts per injection were 5 μg estradiol/100 g body wt and 30 μg dihydro-testosterone/100 g body wt. The animals were sacrificed at various times (hours) after the final injection. (Roy AK (1977) Eur J Biochem 73: 537)

genic regulation of α_{2u}-globulin through a bifunctional hormone receptor can be explained by the model described in Fig. 3.

IV. Properties of α_{2u}-mRNA and the Role of Terminal Poly (A) Residues in its Translational Stability in *Xenopus* Oocytes

Studies on the messenger RNA for α_{2u}-globulin have shown that this protein is coded by a poly(A)-containing mRNA. Fractionation of the hepatic mRNA in either sodium dodecylsulfate sucrose gradient or acid urea agarose gel electrophoresis followed by extraction, translation, and product identification show that α_{2u}-mRNA migrates as a band in the 14S fraction (Fig. 4). The above sedimentation coefficient would correspond to a polynucleotide containing 1200–1400 nucleotide residues. Based on its amino acid composition, it is estimated that α_{2u}-globulin containing 186 amino acids would require an mRNA with only 558 nucleotides to adequately account for its coding function. Even after considering the 180 poly(A) residues at the 3'

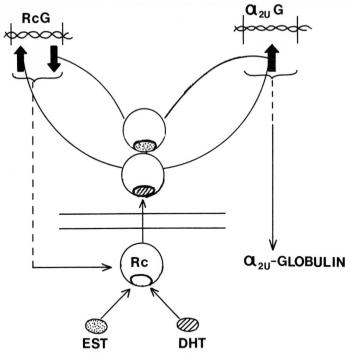

Fig. 3. Schematic model showing the action of both androgen and estrogen through a single receptor system. The androgen receptor (Rc) can accept either dihydrotestosterone (DHT) or estradiol (EST). Interaction of the androgen (DHT) with the receptor (Rc) may lead to derepression (↑) of both the α_{2u} gene (α_{2u}G) and the receptor gene (RcG). On the other hand, interaction of the estrogen (EST) with the receptor (Rc) may lead to derepression (↑) of α_{2u} gene (α_{2u}G) and repression (↓) of the receptor gene (RcG)

terminal of α_{2u}-mRNA (Deshpande et al. 1979), approximately half of the nucleotide residues in the α_{2u}-mRNA seem to be involved in the noncoding function. In addition to the terminal poly(A) residues, large amounts of other noncoding nucleotides in several other eukaryotic mRNAs such as ovalbumin (Woo et al. 1976), hemoglobin (Morrison and Lingrel 1976), and α_{A2}-crystallin (Berns et al. 1974) have been reported. Although the functional significance of these noncoding sequences is currently unknown, it is possible that these sequences may provide functional stability to the mRNA.

One multinational group of investigators has studied the functional stability of globin and histone mRNA in *Xenopus* oocytes after enzymatic deadenylation and adenylation of the 3' terminal. Based on these investigations it was concluded that the relative length of the poly(A) segment at the 3' end of the mRNA determines its functional stability (Huez et al. 1974, 1977; Marbaix et al. 1975). These results, however, do not explain the great differences in the stabilities of the hormonally inducible mRNAs (like α_{2u}-mRNA) in the multipotent cells and of the constitutive mRNAs (like globin mRNA) in the terminally differentiated cells, both having a similar degree of

Fig. 4. Preparative urea–agarose gel electrophoresis of the mRNA for α_{2u}-globulin. Total male rat liver mRNA was electrophoretically separated on urea–agarose gel and the mRNA obtained from different gel fractions were assayed for α_{2u}-mRNA activity in wheat germ cell-free translational system. α_{2u}-mRNA (dotted bars) was found to migrate as a 14S band

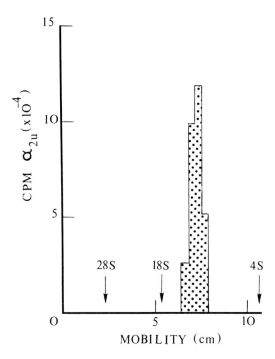

polyadenylation. Since the above-mentioned experiments on the role of poly(A) residues in mRNA stability were performed with two constitutive mRNAs, we have conducted similar experiments with the hormonally inducible α_{2u}-mRNA. One important difference in the design of our experiment from that of Huez and co-workers is the avoidance of the enzymatic treatment of the RNA. In order to maintain nearly physiological condition, we have fractionated normally occurring hepatic α_{2u}-mRNA having a different degree of polyadenylation by elution of the mRNA from the oligo(dT)–cellulose column with a KCl gradient (Deshpande et al. 1979). The functional stabilities of α_{2u}mRNA containing 40 and 175 average terminal poly(A) residues, respectively, in the *Xenopus* oocytes are shown in Fig. 5. After reaching maximum rate of translation, α_{2u}-mRNAs with short and long poly(A) terminals were found to decay with almost identical kinetics, both showing a half-life of about 50 h. The relatively short functional half-life of the hormonally inducible α_{2u}-mRNA in *Xenopus* oocytes is in sharp contrast to the almost unlimited stability of globin mRNA in the same system (Gurdon et al. 1972). Moreover, the lack of difference in the rate of decay of α_{2u}-mRNA with short and long poly(A) terminal, tends to contradict the validity of the universal correlation between terminal poly(A) size and mRNA stability and may suggest that besides poly(A) other structural features of the mRNA, such as the noncoding sequences and the secondary structure (Van et al. 1977), may contribute to mRNA stability in vivo. Therefore, it seems that hormonal regulation of mRNA stability through minor structural modifications may be a plausible possibility.

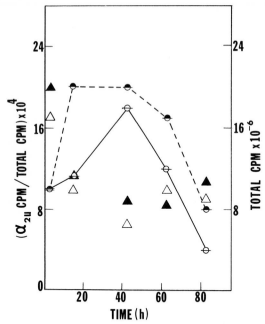

Fig. 5. Translational stability of hepatic mRNA containing approximately 40 and 175 terminal poly(A) residues in *Xenopus* oocytes. Three hundred oocytes were injected with the hepatic mRNA (20 ng/oocyte) and were cultured in nonradioactive medium. At different time periods, 40 oocytes from each group were removed and placed in [³H]leucine-containing medium for 5 h. At the end of this labeling period the oocytes were used for product analysis. The ratio of radioactivity in α_{2u}-globulin to total trichloroacetic acid insoluble protein in oocytes injected with mRNA containing 40 3'-terminal poly(A) residues $\ominus$ and 175 poly(A) residues $\oplus$ are shown against the time of in vitro culture after mRNA injection. The radioactivity in total trichloroacetic acid insoluble protein within the cytoplasmic extract of 40 oocytes injected with poly(A)-poor ($\triangle$) and poly(A)-rich ($\blacktriangle$) RNA is presented to indicate the continued protein-synthesizing activity of the oocytes throughout the period of culture. (Deshpande AK, Chatterjee B, Roy AK (1979) J Biol Chem 254: 8937)

V. Androgenic Induction and Estrogenic Repression of α_{2u}-mRNA

Unlike the scanty evidence for hormonal regulation of mRNA stability, a great body of data shows that both steroid and nonsteroid hormones regulate gene expression through changes in the cellular concentrations of the hormonally inducible mRNA species. The coordinated physiological role of the androgenic hormone and the hepatic androgen receptor in the expression of the α_{2u} gene through changes in the hepatic concentration of α_{2u}-mRNA can be demonstrated by monitoring levels of these parameters in maturing male rats (Fig. 6). The results of such studies show that in the prepubertal animals (<6 weeks old) there is no detectable level of hepatic α_{2u}-globulin and its corresponding mRNA, and in addition no urinary α_{2u}-globulin could be detected. Maturation at puberty is associated with the appearance and

GEL FRACTIONS

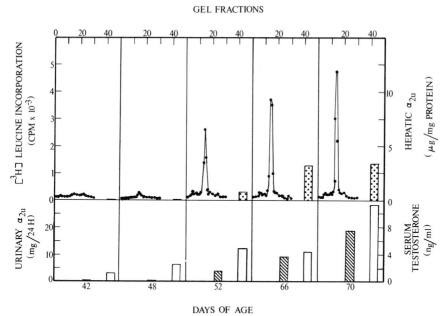

Fig. 6. Relationship among the levels of serum testosterone, hepatic messenger RNA for α_{2u}-globulin, and both hepatic and urinary levels of α_{2u}-globulin in the maturing male rats. Data in each vertical frame represent average values from same three animals. Equal amounts of the hepatic tissue from each of these animals were blended together for the extraction of RNA. Other values represent the mean of three separate assays. The messenger RNA activity for α_{2u}-globulin (●) is represented as the pattern of radioactivity on SDS–polyacrylamide gel electrophoresis of α_{2u}–anti-α_{2u} immunoprecipitate obtained from 2×10^6 cpm of released peptide chains synthesized in vitro under the direction of hepatic mRNA. Key to other symbols: dotted bars, hepatic concentration of α_{2u}-globulin; hatched bars, 24 h urinary output of α_{2u}-globulin; open bars, serum testosterone. (Roy AK, Schiop MJ, Dowbenko, DJ (1976) FEBS Lett 70: 137)

subsequent rise in the levels of hepatic α_{2u}-mRNA and both hepatic and urinary α_{2u}-globulin. These changes are again correlated with the rising levels of serum androgen and hepatic androgen receptor activity (Roy et al. 1974). Thus, all these changes during puberty are associated with a sequence of events including gonadal maturation with an accompanying rise in serum testosterone, the appearance of hepatic androgen–receptor activity leading to the derepression of the α_{2u} gene, and the gradual accumulation of α_{2u}-mRNA with increased synthesis of hepatic α_{2u}-globulin. In this series of events, the appearance of a threshold level of hepatic androgen receptor activity seems to play a critical role in the induction of α_{2u}-globulin. Thus, prepubertal male rats that do not contain a threshold level of hepatic androgen receptor activity fail to show any induction of α_{2u}-mRNA even after androgen treatment. Similarly a mutant strain of androgen-insensitive (tfm) rats lacking hepatic androgen receptor activity failed to show α_{2u}-globulin induction after androgen administration (Roy 1973a; Roy et al. 1974, 1976b). Transcriptional regulation of α_{2u}-globulin by the androgens is also

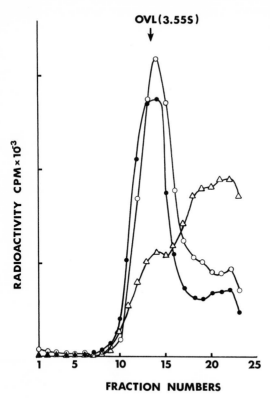

Fig. 7. Effect of unlabeled cyproterone acetate and unlabeled 5α-dihydrotestosterone on the uptake of [³H]DHT by the cytosol androgen-binding protein of rat liver. Sucrose density gradient pattern of cytosol incubated with (●) [³H]DHT (0.4 µmol/liter) (△) [³H]DHT (0.4 µmol/liter) + unlabeled DHT (12 µmol/liter); (○) [³H]DHT (0.4 µmol/liter) + cyproterone acetate (12 µmol/liter). Arrow shows the position of 3.55S marker protein (ovalbumin) and the last fraction represents the top of the gradient. (Roy AK (1976) J Endocrinol 71: 265)

indicated by the absence of α_{2u}-mRNA in the liver of female rats and the induction of α_{2u}-mRNA in spayed female rats after androgen treatment (Sippel et al. 1975). As mentioned earlier, estrogenic repression of α_{2u}-globulin synthesis results in the depletion of hepatic androgen receptor activity. Although this seems to be the principal pathway for estrogen action, additional regulatory factors in the estrogenic inhibition of α_{2u}-globulin synthesis may also be involved. However, similar to androgen action, estrogenic influence on α_{2u}-globulin is mediated through changes in the hepatic concentration of the messenger RNA for this protein. Estrogen treatment of mature male rats results in the depletion and ultimate loss of hepatic mRNA for α_{2u}-globulin, which is correlated with the inhibition of α_{2u}-globulin synthesis (Roy et al. 1977). Unlike the estrogenic steroids, the synthetic androgen analog cyproterone acetate does not bind to the hepatic androgen receptor and seems to enhance the binding of DHT to the hepatic receptor (Fig. 7) (Roy 1976). These findings tend to support the concept of a synandrogenic effect of this androgen analog (Bullock et al., Chap. 13 in this volume). In the maturing male rats, cyproterone acetate also produces a significant increase in the hepatic synthesis of α_{2u}-globulin, which correlate with a corresponding increase in the hepatic concentration of α_{2u}-mRNA (A.K. Roy, unpublished data). These results sharply contradict those of Kurtz et al. (1976, 1978a), who have reported that cyproterone acetate inhibits hepatic synthesis of α_{2u}-globulin in the mature male rats, inhibits androgen binding to the hepatic androgen receptor, and causes reduction in the hepatic concentration of α_{2u}-mRNA.

VI. Multihormonal Modulation of the Androgen-dependent Synthesis of α_{2u}-mRNA

One of the most interesting aspects of the hormonal regulation of the α_{2u} gene is the multihormonal modulation of the primary androgen response. Thyroxine, glucocorticoids, growth hormone, and insulin are known to be required for the maximum expression of α_{2u} gene under androgenic stimulation (Irwin et al. 1971; Kumar et al. 1969; Roy 1973b; Roy and Leonard 1973). Investigation of the molecular mechanism of the above nonandrogenic modulators of α_{2u} gene has shown that these hormones also regulate α_{2u}-globulin synthesis primarily through transcriptional regulation. Thyroidectomy in the male rat results in drastic reduction (>90%) in the hepatic synthesis of α_{2u}-globulin (Roy 1973b). This reduction in α_{2u}-synthesis closely parallels the decrease in the hepatic concentration of the mRNA for α_{2u}-globulin. (Roy *et al.* 1976a). Androgen treatment of the thyroidectomized male rats does not cause any change in the hepatic α_{2u}-globulin and its mRNA (Roy et al. 1976b). The possibility of a direct effect of thyroxine on the synthesis of α_{2u}-globulin rather than through thyroidal modulation of growth hormone secretion is supported by the observed lack of effect of growth hormone treatment in the thyroidectomized animals. Treatment of thyroidectomized male rats with growth hormone does not cause any increase either in α_{2u}-globulin synthesis or in the hepatic concentration of α_{2u}-mRNA. Moreover, a high level of cytosol androgen receptor activity in the thyroidectomized male rats was also observed, suggesting that decreased α_{2u}-globulin synthesis in thyroidectomized male rats does not result from a loss of hepatic androgen sensitivity (Roy AK, unpublished)

Similar to thyroidectomy, decreased α_{2u}-globulin synthesis in adrenalectomized male rats is also associated with a parallel decrease in the hepatic concentration of α_{2u}-mRNA, which can be reversed with glucocorticoid supplementation (Sippel et al. 1975).

The mode of action of growth hormone in the hepatic synthesis of α_{2u}-globulin was investigated in the hypophysectomized rats treated with a hormone combination containing DHT, corticosterone, and thyroxine with and without growth hormone in the combination. The results show that inclusion of growth hormone in the three hormone combination associated with a correlative increase in the hepatic level of both α_{2u}-globulin and its mRNA (Roy and Dowbenko 1977). Kurtz et al. (1978b) have reported that growth hormone is not required for maintaining the hepatic concentration of

α_{2u}-mRNA. Based on their studies, the above investigators suggest that growth hormone is required for the conversion of the α_{2u}-mRNA containing polysome in the hepatic cells from a free to a membrane bound state.

An examination of the effect of diabetes and insulin supplementation to mature male rats shows that moderate diabetes causes more than a 50% reduction both in the rate of synthesis of α_{2u}-globulin, and in the hepatic concentration of α_{2u}-mRNA, and furthermore, insulin supplementation results in the elevation of the hepatic α_{2u}-mRNA concentration to approximately 75% of the normal control (Roy 1979). These results are also indicative of the role of insulin in the transcriptional regulation of α_{2u}-globulin synthesis.

VII. Superinduction and Possible Post-transcriptional Control of α_{2u}-mRNA

Besides all of the above examples of transcriptional regulation of α_{2u}-globulin synthesis by hormonal factors, we have recently obtained evidence for possible posttranscriptional regulation of α_{2u}-globulin synthesis as an additional mechanism for the regulation of α_{2u} gene expression. Many inducible proteins show an enhanced rate of induction in the presence of certain inhibitors of RNA synthesis such as actinomycin D, a phenomenon known as superinduction (Tomkins et al. 1969). Androgenic induction of α_{2u}-globulin in spayed female rats is also susceptible to superinduction by actinomycin D (Chatterjee et al. 1979). Figure 8 shows the effect of actinomycin D on the hepatic level of α_{2u}-globulin after initial induction of the protein with a single injection of DHT. Administration of actinomycin D at 1 and 3 h after androgen treatment resulted in approximately twofold enhancement of the androgen response. In spayed female rats with a threshold level of hepatic α_{2u}-globulin (possibly owing to the chronic action of adrenal androgens), actinomycin D alone was capable of causing about a fivefold increase in the hepatic concentration of α_{2u}-globulin. This increase in hepatic α_{2u}-globulin caused by actinomycin D can be inhibited by the translational inhibitor, cycloheximide. The relationship between the degree of superinduction and inhibition of the poly(A)$^+$ hepatic RNA synthesis by actinomycin D, at different doses of the drug, show that at the optimum superinducing dose (80 μg/100 g body wt), there is more than an 85% inhibition of 14S RNA (the size of α_{2u}-mRNA). In spite of drastic inhibition of hepatic RNA synthesis, superinduction of α_{2u}-globulin by actinomycin D caused a parallel increase in the hepatic concentration of the functional mRNA for this protein (Fig. 9).

Several hypotheses have been proposed to explain the paradoxical effect of actinomycin D in the superinduction of inducible proteins. Our results seem to favor the hypothesis of posttranscriptional regulation as the possible mechanism for the superinducing action of the drug. Examining the functional interferon mRNA after superinduction of this protein in the fibroblast by an inhibitor of RNA synthesis (5,6-dichloro-1-β-D-ribofuranosylbenzimidazole), Sehgal et al. (1977) have also reached a similar conclusion. Since

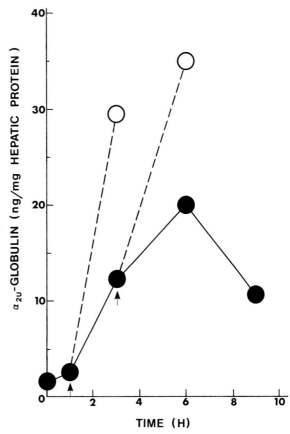

Fig. 8. Superinduction of α_{2u}-globulin by actinomycin D after initial induction by DHT (33 μg/100 g), and two groups of animals were subsequently treated with actinomycin D (80 μg/100 g) at 1 h and 3 h after DHT treatment (arrows). (●——●) Animals treated with DHT alone; (●——○) animals treated with DHT followed by actinomycin D. (Chatterjee B, Hopkins J, Dutchak D, Roy AK (1979) Proc Natl Acad Sci USA 76: 1833)

actinomycin D-mediated superinduction has been the basis of the posttranscriptional theory of hormone action (Tomkins et al. 1969), which was the center of extensive controversy for more than a decade, it may be worth recounting some of the principal alternative theories for the explanation of superinduction. Experimenting with actinomycin D-mediated superinduction of tyrosine aminotransferase (TAT) in rat hepatoma cells, Reel and Kenney (1968) have observed that the drug is able to decrease the rate of degradation of TAT in the deinduction phase (after hormone withdrawal) and have suggested that this may explain the mechanism of superinduction. Tomkins and his co-workers (1972) have argued that the decreased rate of TAT degradation at deinduction could only be observed under restrictive nutritional environment, the "step-down" cultural conditions, and cannot explain superinduction under normal conditions. In addition, Tomkins et al. (1972)

and Palmiter and Schimke (1973) have shown increased rate of TAT and ovalbumin synthesis under conditions of superinduction by actinomycin D. The relative ratio of the labeling time to the half-life of the protein in question may also have an important bearing on the resolutions of these discrepancies. However, Palmiter and Schimke (1973) have pointed out that two superinduced proteins of the chick oviduct, ovalbumin and conalbumin, both have long half-lives relative to the 1-h pulse used in their studies as a means of measuring the rate of synthesis of these proteins. Therefore, it seems unlikely that the increased incorporation of labeled amino acids into ovalbumin and conalbumin after superinduction results from the effect of

GEL FRACTIONS

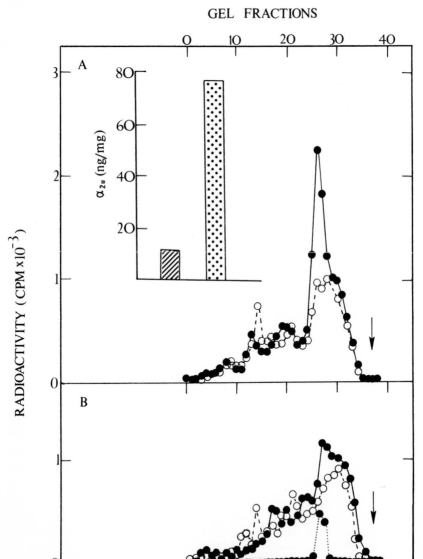

this drug on the inhibition of their degradation. In the case of α_{2u}-globulin, an additional consideration is the fact that it is a secretory protein, and, because of the short transit time (12–15 min) of the hepatic secretory proteins, it is unlikely that the rate of degradation (unless it is very rapid and faster than the rate of secretion) would play any significant role in the maintenance of its hepatic concentration.

Based on their studies on the effect of actinomycin D on the synthesis of ovalbumin in chick oviduct, Palmiter and Schimke (1973) have suggested that superinduction may involve an increased rate of translation of the mRNA for the inducible proteins. These investigators have proposed that under normal conditions, limiting concentrations of translational factors regulate the rate of translation of various mRNA species. However, after inhibition of RNA synthesis, mRNAs with short half-lives are quickly depleted so that the remaining stable mRNA species can be translated without any limitations for the translational factors. Contrary to this theory of "translational limitation," all inhibitors of RNA synthesis do not cause superinduction (Chatterjee et al. 1979; Sekeris et al. 1970), and the above theory fails to explain the superinduction of hepatic tyrosine aminotransferase and tryptophan oxygenase, the messengers for both of which possess short half-lives in the range of only few hours (Killewich et al. 1975; Steinberg et al. 1975).

Although the observation of the parallel increase in α_{2u}-globulin and its mRNA after superinduction supports the concept of posttranscriptional regulation of α_{2u}-mRNA in the liver cell, the results of the hormonal induction of this protein have consistently shown that both steroid and nonsteroid hor-

---

Fig. 9,A, B. Translatable mRNA for α_{2u}-globulin with the total hepatic mRNA in spayed female rats with and without actinomycin D treatment. Poly(A)-enriched hepatic RNA obtained from animals with and without actinomycin D treatment was found to be almost equally effective in promoting amino acid incorporation into trichloroacetic acid-insoluble proteins.
A Distribution of radioactivity in the $NaDodSO_4$–polyacrylamide gel after electrophoresis of the immunoprecipitated α_{2u}-globulin from in vitro translation product primed with hepatic mRNA obtained from animals treated with actinomycin D (●) or vehicle (○) for 3 h. Equal amounts of labeled released peptide chains (5.95 × 10⁶ cpm) were used for immunoprecipitation in the actinomycin D-treated and control samples. (*Inset*) Hepatic concentration of α_{2u}-globulin (ng/mg of hepatic protein) in the same group of animals. Hatched bar, vehicle-treated control; stippled bar, actinomycin D-treated rats. **B** (●) $NaDodSO_4$–polyacrylamide gel electrophoretic profile of the radioactivity of the second α_{2u}–anti-α_{2u} immunoprecipitate obtained after removal of the labeled α_{2u}-globulin from the released peptide chains synthesized in the presence of hepatic mRNA from the actinomycin D-treated animals by first immunoprecipitation (solid circles in A). The immunoprecipitation was 85% complete as judged by removal of ^{125}I-labeled α_{2u}-globulin. (○) $NaDodSO_4$ gel electrophoretic profile of the α_{2u}–anti-α_{2u} immunoprecipitate obtained from released peptide chains (5.96 × 10⁶ cpm) of the wheat germ cell-free translational product without any addition of RNA. (●) ^{125}I-labeled α_{2u}-globulin in the $NaDodSO_4$–polyacrylamide gel. Chatterjee B, Hopkins J, Dutchak D, Roy AK (1979) Proc Natl Acad Sci USA 76: 1833)

mones exert their influence by increasing the hepatic concentration of α_{2u}-mRNA. Moreover, convincing evidence for the transcriptional regulation of protein synthesis by various hormones in several other model systems have also been reported (Stein et al., Chapter 2 in this volume; Goldberger & Deeley, Chapter 3 in this volume; Rosen et al., Chapter 4 in this volume; Johnson et al., Chapter 10 in this volume; Garen & Lepesant, Chapter 16 in this volume). Therefore, it is reasonable to conclude that the hormonal regulation of α_{2u}-globulin and other hormone dependent proteins is principally achieved through transcriptional regulation of the specific mRNA concentrations. However, a minor degree of regulation at the level of post-transcriptional events still remains to be a likely possibility. (Supported by NIH grant AM-14744)

References

Berns A, Janssen P, Bloemendal H (1974) Biochem Biophys Res Commun 59: 1157
Chatterjee B, Hopkins J, Dutchak D, Roy AK (1979) Proc Natl Acad Sci USA 76: 1833
Deshpande AK, Chatterjee B, Roy AK (1979) J Biol Chem, 254, 8937
Gurdon JB, Lingrel JB, Marbaix G (1972) J Mol Biol 80: 539
Huez G, Marbaix G, Hubert E, LeClercq M, Nudel U, Soreq, H, Saloman R, Lebleu B, Revel M, Littauer UL (1974). Proc Nat Acad Sci USA 71: 3143
Huez G, Marbaix G, Burny A, Hubert E, LeClercq M, Cleuter Y, Chantrenne H, Soreq H, Littauer UL (1977) Nature 266: 473
Irwin JF, Lane SE, Neuhaus OW (1971) Biochim Biophys Acta 252: 328
Jung-Testas I, Bayard F, Baulieu EE (1976) Nature (London) 259: 136
Killewich K, Schutz G, Feigelson P (1975) Proc Natl Acad Sci USA 72: 4285
Kumar M, Roy AK, Axelrod AE (1969) Nature (London) 223: 399
Kurtz DT, Sippel AE, Ansah-Yiadom R, Feigelson P (1976) J Biol Chem 251: 3594
Kurtz DT, Chan K-M, Feigelson P (1978a) J Biol Chem 253: 7890
Kurtz DT, Chan K-M Feigelson P (1978b) Cell 15: 743
Lane SE, Neuhaus OW (1972) Biochim Biophys Acta 257: 461
Marbaix G, Huez G, Burny A, Cleuter Y, Hubert E, LeClercq M, Chantrenne H, Soreq H, Nudel U, Littauer UL (1975) Proc Natl Acad Sci USA 72: 3065
Milin B, Roy AK (1973) Nature (London), New Biol 242: 247
Morrison MR, Lingrel JB (1976) Biochim Biophys Acta 447: 104
Palmiter RD, Schimke RT (1973) J Biol Chem 248: 1502
Reel JR, Kenney FT (1968) Proc Natl Acad Sci USA 61: 200
Roy AK (1973a) Endocrinology 92: 957
Roy AK (1973b) J Endocrinol 56: 295
Roy AK (1976) J Endocrinol 70: 189
Roy AK (1977) Eur J Biochem 73: 537
Roy AK (1979) In: Litwack G (ed) Biochemical actions of hormones. Vol. 6, Academic Press, New York, p 481
Roy AK, Dowbenko DJ (1977) Biochemistry 16: 3918
Roy AK, Leonard S (1973) J Endocrinol 57: 327
Roy AK, Neuhaus OW (1966) Biochim Biophys Acta 127: 82
Roy AK, Neuhaus OW (1967) Nature (London) 214: 618
Roy AK, Raber DL (1972) J Histochem Cytochem 20: 618
Roy AK, Neuhaus OW, Harmison CR (1966) Biochim Biophys Acta 127: 72
Roy AK, Milin BS, McMinn DM (1974) Biochim Biophys Acta 354: 213
Roy AK, McMinn DM, Biswas NM (1975) Endocrinology 97: 1505

Roy AK, Schiop MJ, Dowbenko DJ (1976a) FEBS Lett 64: 396
Roy AK, Schiop MJ, Dowbenko DJ (1976b) FEBS Lett 70: 137
Roy AK, Dowbenko DJ, Schiop MJ (1977) Biochem J 164: 91
Ruh TS, Wassilak SG, Ruh MJ (1975) Steroids 25: 257
Schmidt WN, Sadler MA, Katzenellenbogen BS (1976) Endocrinology 98: 702
Sehgal PB, Dobberstein B, Tamm I (1977) Proc Natl Acad Sci USA 74: 3409
Sekeris CE, Niessing J, Seifart KH (1970) FEBS Lett 9: 103
Sippel AE, Feigelson P, Roy AK (1975) Biochemistry 14: 825
Steinberg RA, Levinson BB, Tomkins GM (1975) Cell 5: 29
Tomkins GM, Gelehrter TD, Granner D, Martin DT, Sammuels HH, Thompson EB (1969) Science 166: 1474
Tomkins GM, Levinson BB, Baxter JD, Dethlefsen L (1972) Nature (London), New Biol 234: 9
Van NT, Monahan JJ, Woo SLC, Means AR, O'Malley BW (1977) Biochemistry 16: 4090
Woo SLC, Rosen JM, Liarakos CD, Choi YC, Busch H, Means AR, O'Malley BW, Robberson DL (1976) J Biol Chem 250: 7027
Zava DT, McGuire WL (1978) Endocrinology 103: 624

Discussion of the Paper Presented by A. K. Roy

WILLIAMS-ASHMAN/ I'd like to ask two things. First, what do you think the function of this protein is in the male animal, and second, since as you pointed out so clearly that the size of the message seems to be somewhat larger than what is expected of this protein, is there any evidence for a pre-α_{2u} or preprotein, or something larger may be accumulating in tissues or being made in your translations?

ROY: We have a lead to look for the function of this protein. We have found that estrogen treatment not only stops α_{2u}-globulin synthesis; it also stops sperm maturation, and we were able to provide protection of the testes by supplementation of α_{2u}-globulin in the estrogen treated animals. About your second question, we are in the process of examining the various steps in the synthesis and secretion of α_{2u}-globulin and we soon will be able to make some definite conclusion regarding the nature of the signal sequence for this protein.

McGINNIS: Arun, I'm not sure what percentage of total protein is represented by α_{2u}. Is this something like 1%?

ROY: In the maximally induced animal it is about 1%.

McGINNIS: Does the sequence of addition of your hormone and/or effectors give you a different result?

ROY: We did such experiments on the hypophysectomized animals. For the recovery experiments, we used different sequences of hormone treatment, and it didn't make any difference.

JAFFE: Did you see if pharmacologic levels of thyroxine might inhibit production of this protein?

ROY: Yes, it does.

JAFFE: Do you have any idea or any evidence that the hepatic androgen receptor has two sites, one for the androgen and the other for estrogen? Do they compete for each other?

ROY: When the receptor was prelabeled with labeled DHT, it competed well with either cold DHT or cold estradiol. On the other hand, when we used labeled estradiol, unlike cold estradiol, cold DHT was only very weakly effective in competing out the labeled estradiol. This may suggest two independent sites, but I would say that at this point the results are not unequivocal.

GOLDBERGER: I don't remember whether you showed it, but what happens to the fall-off of the α_{2u}-globulin in the actinomycin experiments. I realize that you got the superinduction, and then did the duration of the response increase?

ROY: We did not check that for the superinduction experiments. We worked with animals after optimum induction with DHT.

GOLDBERGER: May I make a comment about Guy's (Williams-Ashman's) question concerning the size of the messenger RNA. I think just in this case, the messenger RNA is twice as big as it should be. But that just means 600 nucleotides or so, and 600 nucleotide is about the extra size one sees on secreted protein messenger RNAs in many cases. It just seems so dramatic that the protein is so small.

ROY: Yes, that is true. Besides, I think that in the case of ovalbumin messenger RNA, Dr. O'Malley's group and others have shown that the mRNA size is also quite large and really much larger than what could be accounted for by its coding function.

THOMPSON: Did the actinomycin effect the proportion of mRNA with short poly(A) tails to long poly(A) tails?

ROY: We have not looked into that. The superinduction experiments were done with total unfractionated messenger RNA.

THOMPSON: A historical comment. Because Gordon Tomkins evolved his model of posttranscriptional repression of hormone action on the basis of actinomycin D superinduction, many people have the misconception that he never believed in hormonal regulation of transcription. Before his untimely death, he published a paper in the Cold Spring Harbor symposium in which he said exactly what you and perhaps others may be finding, although I think that this is the first direct case of increased translatable message after superinduction. In that Cold Spring Harbor symposium he said that there was probably dual control, both at the transcriptional and translational levels. Just for the record I wanted to mention that.

ROY: Yes, I also think that the concept of dual control makes a lot of sense. And I may add that we are not first to show an increase in the messenger RNA after superinduction, as some of you may know that Igor Tamm and his colleagues have shown that the superinduction of interferon is also associated with corresponding rise in its translatable mRNA.

Discussants: R.F. GOLDBERGER, R. JAFFE, J. MCGINNIS, A.K. ROY, E.B. THOMPSON, and H.G. WILLIAMS-ASHMAN

Ecdysteroids

Chapter 15

Ecdysone: Introduction to the Molecular Mechanism of Steroid Action in Insects

F. M. BUTTERWORTH

Insect endocrinology is a recent science compared with that of vertebrates. At about 60 years of age insect endocrinology is only one-half as old as vertebrate endocrinology. When most of the major vertebrate hormones had been crystallized in 1933, the actual existence of the molting hormone, ecdysone, had just been confirmed. Although the knowledge of steroid hormone receptor proteins is well established in vertebrates, ecdysone-binding proteins were only first found six years ago (Butterworth and Berendes 1974; Emmerich 1972; Thamer and Karlson 1972; Yund and Fristrom 1975). In fact, 17 years after the discovery of Jensen and Jacobson (1962), the first truly conclusive report that insects do in fact have receptor proteins is described (O'Connor et al., Chap. 17 in this volume).

Insects form an important class of animals in terms of the overwhelming number of species, the decided medical and economic impact on humanity, and the interesting model systems they offer. However, in spite of the need to develop knowledge about the physiology and endocrinology of insects, their small size presented a formidable obstacle to experimentation. The classical tools of endocrinology: extirpation, ligation, and transplantation, to be employed on the tiny insect required great skill, determination, and patience from the early workers.

Although attempts to demonstrate gonadal hormones were unsuccessful in the late nineteenth century, insect endocrinology really began in 1922 when Kopec (1922) unequivocally demonstrated the existence of the growth and metamorphosis-controlling hormone, ecdysone. Through an intricate series of surgical and physiological experiments he was able to show that this hormone was secreted by the brain in the caterpillar *Lymantria*.

The work of Kopec was suspect, because at the time, it was thought that only gonads could produce hormones. However, in the early 1930s this "gonad only" concept was finally put to rest by various laboratories. For example, Wigglesworth (1934), using extirpation, ligation, and parabiotic pairs with the bug *Rhodnius*, Bodenstein (1933), using transplantation in the butterfly *Vanessa*, and Fraenkel (1935), using ligation on the fly *Calliphora*, confirmed Kopec's results. Indeed, insects produced a blood-borne substance that emanated from the brain region and controlled molting. This substance is ecdysone.

Although there are other hormones in insects that have been discovered subsequently, ecdysone will be the principal hormone for this discussion, since it is, as far as is known, the only steroid hormone in insects. The first relatively pure extract of ecdysone was produced about 20 years later in 1954 by Butenandt and Karlson (1954). From 1000 lb of silkworm pupae, they, by a monumental effort, were able to extract 25 mg of the steroid. With this material two active fractions, α and β, were demonstrated. Within the decade larger amounts were crystallized (Karlson et al. 1963), from which great strides were made in the elucidation of the steroidal structure. Two years later Karlson's colleagues Huber and Hoppe (1965) made the final structural determinations (Fig. 1). The α and β fractions turned out to be α-ecdysone and β-ecdysone or ecdysterone. There are two other forms of ecdysone that have been found in insects and possess hormone activity, inokosterone and ponasterone (Fig. 1). Also, two forms found to be synthesized by plants are active in insects. Collectively, the term "ecdysteroids" is employed.

The insect species that may well be the most useful in future studies on the mechanism of steroid hormone action is the fly *Drosophila melanogaster* be-

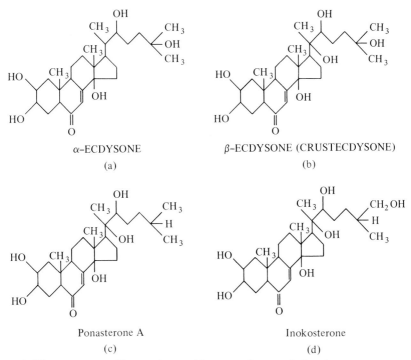

Fig. 1,a–d. The structure of four ecdysteroids: **a** α-ecdysone; **b** β-ecdysone; **c** ponasterone A; **d** inokosterone. (Turner CD, Bagnara JT (1976) General Endocrinology, 6th edition. W.B. Saunders Co., Philadelphia).

cause the species is perhaps the best known genetically of all higher organisms. Furthermore, its entire genome is already available in clones as recombinant DNA fragments. Ecdysteroid activity can be detected at all developmental stages, which include the embryo, the three larval instars, the pupa, and adult. Although it is unknown which ecdysteroid is the most prominent in *Drosophila*, transplantation studies (Bodenstein 1943) show that the lateral cells (or prothoracic gland) of the ring gland are the source of ecdysteroid activity in the lárva. Figure 2 illustrates the larval ring gland and its close proximity to the brain (King et al. 1966).

Ecdysteroids are commonly referred to as molting hormones, since they include molting or ecdysis to occur. Molting is a complex process involving a critical balance of ecdysone and another hormone, juvenile hormone, re-

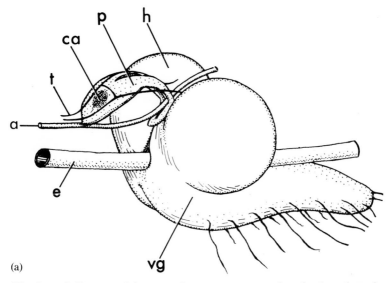

(a)

Fig. 2. **a** A diagram of the central nervous system, ring gland, and portions of the circulatory and digestive systems of a third-instar larva of *Drosophila melanogaster:* (a) aorta; (ca) corpus allatum portion of the ring gland; (e) esophagus; (h) brain hemisphere; (p) left limb of prothoracic gland; (t) trachea; (vg) ventral region.

b Diagram of an ideal transverse section through the ring gland of a third-instar larva. The corpus allatum is stippled to demonstrate its greater electron density. The solid filaments and short rods drawn in the cytoplasm of the corpus allatum and prothoracic gland cells represent mitochondria. The particles drawn in the cytoplasm of the corpus cardiacum cells and in the axoplasm of the swollen axonal terminations seen in the corpus allatum represent single and clustered neurosecretory spheres. The tracheal tubes are surrounded by an epithelium of tiny polygonal cells. The intima secreted by these cells contain spiral ridges or taenidia, which project toward the air-filled lumen. The path of the transverse trachea through the gland is shown by parallel dashed lines: (a) aorta; (ax) axon from corpus cardiacum to corpus allatum; (ca) corpus allatum cell; (cc) corpus cardiacum cell; (lt) left lateral trachea; (p) cell of prothoracic gland; (s) secretory deposit; (t) acellular tunica; (tt) transverse trachea. The arrow points to a pore connecting adjacent prothoracic gland cells. (King RC, Aggarwal SK, Bodenstein D (1966) Z Zellforsch 73: 272–285)

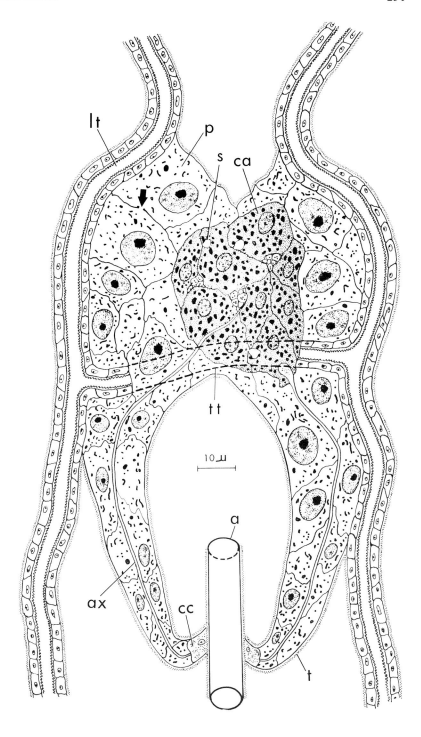

(b)

sulting in the production of a new underlying cuticle and the shedding of the outer cuticular layer. The most obvious receptor tissue in this process is the hypodermis, which is responsible for cuticle formation. However, all tissues of the insect are probably targets for ecdysteroids at some time during their development. Two examples that will be discussed in the following papers are the Kc cell line and the fat body.

The Kc cell line derived from *Drosophila* embryos (Echalier and Ohanessian 1970) responds to ecdysone by developing into cells that resemble axons and synthesize acetylcholinesterase (Cherbas et al. 1977; also Chap. 18 in this volume). Presumably these cells in the animal may become elements of the insect nervous system. An ecdysteroid receptor has been identified and characterized by O'Connor and co-workers (Chap. 17 in this volume) in these cells. As will be shown, these cells form three subclasses in response to ecdysteroid.

The fat body is a larval tissue, functionally likened to the vertebrate liver. It developes throughout the larval period and undergoes cell death in the young adult (Butterworth et al. 1965). Certain of the steps preparative to cell death require ecdysone (Thomasson and Mitchell 1972). Garen and Lepesant (Chapter 16 in this volume), using recombinant DNA technology have isolated a gene in the fat body that is induced by ecdysone. Since this gene is expressed just prior to the time of cell death, it is reasonable to speculate that the gene may play a role in this developmental process.

Another reason for choosing *Drosophila* is that many of its tissues divide endomitotically, producing giant, polytene chromosomes. Since the daughter strands (as many as 1024 in the salivary gland nuclei of the larva) synapse, characteristic banding patterns are produced, which have been correlated extensively with genetic maps of the chromosome. Gene activity can be visualized cytologically on these chromosomes in the light microscope by the puffing phenomenon. Puffs are the sites of highly unwound DNA and rapid RNA synthesis. Not only can one perceive gene activity, but also

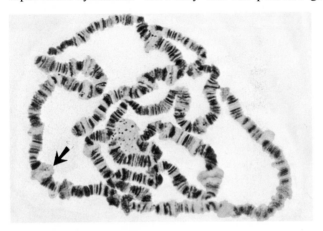

Fig. 3. Photomicrograph of the polytene chromosomes of the larval salivary gland of *Drosophila hydei*. Arrow indicates ecdysone-specific puff.

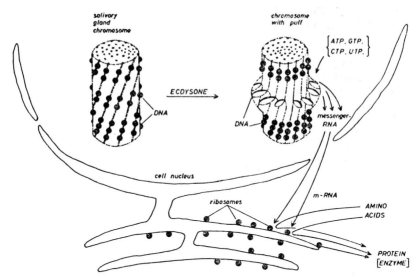

Fig. 4. Mechanism of action of ecdysone as originally proposed by Karlson. The hormone acts first on the DNA producing a puff, which is shown on the left as an unwound region. In the puff, RNA is synthesized from precursors. This RNA is believed to be transferred to the cytoplasm and attached to the ribosomes. As "messenger RNA," it carries the information about the amino acid sequence. According to this information, the specific protein is synthesized on the ribosome from activated amino acids. The whole chain of events explains the formation of specific proteins (e.g., certain enzymes) as the response of the target cell to a hormone. (Karlson P (1963) Perspec Biol Med 6: 203–214)

when, for how long, and how much judging by the time of occurrence, the length of time the puff is present, and the relative size the puff reaches, respectively. Ecdysone is known to induce specific regions of the chromosomes to form puffs (Fig. 3); for example, there are 125 such ecdysone-specific loci in *Drosophila melanogaster* (Ashburner 1972). Of course, the polytene chromosomes attain their maximal size toward the end of larval life; thus, this unusual view of gene activity is limited to a relatively short span during the animal's life cycle. However, it is this phenomenon that prompted Karlson (1963) to construct an hypothesis of hormonal control of gene activity (Fig. 4), a model that has been basic to the overall development of thinking in the field of the mechanism of steroid hormone action.

References

Ashburner M (1972) Chromosoma 38: 255–281
Bodenstein D (1933) Arch Entw Mech 128: 564–583
Bodenstein D (1943) Biol Bull 84: 34–58
Butenandt A, Karlson P (1954) Z Naturforsch 9b: 389–391
Butterworth FM, Berendes HD (1974) J Insect Physiol 20: 2195–2204
Butterworth FM, Bodenstein D, King RC (1965) J Exp Zool 158: 141–154
Cherbas P, Cherbas L, Williams CM (1977) Science 197: 275–277
Echalier G, Ohanessian A (1970) In Vitro 6: 162–172

Emmerich H (1972) Gen Comp Endocr 19: 543–551
Fraenkel G (1935) Proc Roy Soc B 118: 1–12
Huber F, Hoppe W (1965) Chem Berlin 98: 2403–2424
Jensen EV, Jacobson HI (1962) Rec Prog Horm Res 18: 387–414
Karlson P (1963) Perspec Biol Med 6: 203–214
Karlson P, Hoffmeister H, Hoppe W, Huber F (1963) Liebig Ann Chem 662: 1–20
King RC, Aggarwal SK, Bodenstein D (1966) Z Zellforsch 73: 272–285
Kopec S (1922) Biol Bull 42: 322–342
Thamer G, Karlson P (1972) Z Naturforsch 276: 1191–1195
Thomasson WA, Mitchell HK (1972) J Insect Physiol 18: 1885–1899
Turner CD, Bagnara JT (1976) General Endocrinology, 6th edn. W.B. Saunders Co., Philadelphia.
Wigglesworth VB (1934) Quart J Microsc Sci 77: 191–222
Yund MA, Fristrom JW (1975) Develop Biol 43: 287–298

Chapter 16

Hormonal Control of Gene Expression and Development by Ecdysone in *Drosophila*

ALAN GAREN AND JEAN-ANTOINE LEPESANT

It was shown 45 years ago that the molting of the bug *Rhodnius* is under hormonal control (Wigglesworth 1934), and later the induction of molting was used as a bioassay for the isolation of the molting hormone, called ecdysone, from the silkworm *Bombyx* (Butenandt and Karlson 1954). The purified hormone was identified as a pentahydroxysteroid (Huber and Hoppe 1965; Karlson et al. 1965), which is converted in vivo into a hexahydroxysteroid (King and Siddall 1969); the pentahydroxy form is designated α-ecdysone and the hexahydroxy form β-ecdysone or ecdysterone. Ecdysones have also been found in a variety of other organisms, all of which share in common a molting process during development (Morgan and Poole 1976). In addition to functioning as a molting hormone, ecdysones have also been implicated in other aspects of insect development, including larval growth, puffing of the polytene chromosomes in the salivary glands toward the end of the larval stage, pupariation, and the subsequent differentiation of the imaginal discs (Doane 1973; Wigglesworth 1964). Most of this information has been obtained either by surgical procedures involving ligation, extirpation, or transplantation, or by injecting or feeding larvae with an ecdysone, or by exposing isolated organs or cells to an ecdysone. The recent isolation of temperature-sensitive mutants of *Drosophila melanogaster*, which become deficient for ecdysone after a temperature shift from 20° to 29°C at various stages of development (Garen et al. 1977), provides for the first time a genetic approach to the problem of ecdysone action.

The ecdysone-deficient mutants, called *ecd*, were found in the course of examining our extensive collection of temperature-sensitive lethal mutants (Suzuki 1970) that can be propagated at the permissive temperature of 20°C but fail to produce adult progeny when the temperature is raised to 29°C during the periods of embryonic and first- and second-instar larval development. This collection was prepared by inducing mutations with ethylmethanesulfonate and generating lines homozygous for the mutagenized third chromosome. The *ecd* mutants have three distinguishing characteristics: (1) the temperature-sensitive lethal period does not begin until the second instar; (2) a temperature shift from 20° to 29°C early in the third instar blocks

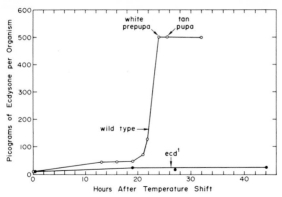

Fig. 1. Ecdysone titers in third-instar larvae at 29°C. The larvae were grown at 20°C for 4.5 days after hatching, and at the start of the measurements the temperature was shifted to 29°C. Both the *ecd¹* and wild-type larvae weighed 1.1 mg at the time of the shift, and both grew to the maximum larval weight of 2.1 mg; the wild-type pupariated 24 h after the shift, in contrast to *ecd¹*, which remained living larvae for the duration of the experiment. Further details are given in Garen et al. (1977).

pupariation and results in a marked prolongation of larval life for upwards of three weeks; (3) adults grown at 20°C become sterile soon after a temperature shift to 29°C (Garen et al. 1977). Measurements of ecdysone titers by radioimmunoassay show that second- and third-instar mutant larvae and adult mutant females become deficient for ecdysone after temperature shifts to 29°C (Fig. 1, Tables 1 and 2). This correlation between the occurrence of developmental abnormalities and ecdysone deficiencies suggests that a normal ecdysone titer is essential for normal development and function during all of these stages. The embryonic stage, in contrast to later stages, is not affected in the mutants by a temperature shift, although during this period the ecdysone titer increases markedly (Table 1). The insensitivity of the mutant embryos to a temperature shift might result from a maternal contribution of components required for ecdysone synthesis during early development until the zygotic system becomes established. The third-instar *ecd* larvae, which become deficient for ecdysone at 29°C, retain the capacity to respond to an ecdysone signal as indicated by the induction of pupariation when ecdysone is subsequently added to the larval food. Therefore, it is probably the biosynthetic or regulatory system for ecdysone, rather than the target system, that is affected in the mutants.

Table 1. Ecdysone Titers during Embryonic and Early Larval Development at 29°C[a]

	Protein per organism (μg)		Ecdysone per organism (pg)	
Stage	Wild type	*ecd¹*	Wild type	*ecd¹*
Blastula	0.84	Not done	0.036	Not done
Newly hatched larva	0.72	0.72	0.30	0.32
Early second-instar larva	6.1	5.7	1.1	0.41

[a] Further details are given in Garen et al. (1977).

Table 2. Ecdysone Titers in Adult Females[a]

	Ecdysone per fly (pg)	
	Wild type	*ecd*[1]
Newly eclosed at 20°C	17	15
After 4 days at 20°C	16	15
After 4 days at 29°C	14	2

[a] Further details are given in Garen et al. (1977).

A total of eight mutants expressing the *ecd* phenotype, comprising about 3% of the mutants examined, were isolated from the collection of temperature-sensitive lethals. In genetic complementation tests between pairwise combinations of the *ecd* mutants, all except two completely complemented for the developmental block at 29°C, indicating that mutations at numerous sites on the third chromosome, and probably throughout the genome, can produce the *ecd* phenotype.

In the course of normal larval development, cells of the imaginal discs proliferate extensively until about the time of pupariation when proliferation stops and imaginal differentiation begins. The switch from proliferation to differentiation of the imaginal cells appears to be initiated by the increase in ecdysone titer (Fig. 1), as indicated by the behavior of imaginal discs after transfer to adult hosts or to in vitro culture medium; in adult wild-type female hosts, which contain about as much ecdysone as do third-instar larvae during the period of imaginal cell proliferation (Table 2), imaginal cells continue to proliferate (Gehring 1968; Hadorn 1963) but can be induced to differentiate by an injection of a relatively large amount of ecdysone into the host (Postlethwait and Schneiderman 1970); analogous effects have also been obtained in vitro (Davis and Shearn 1977; Mandaron 1973). We examined the role of ecdysone in imaginal development using *ecd* larvae that were prevented from pupariating by a temperature shift to 29°C in the third instar. During five days of larval life past the time that pupariation should normally have occurred, the imaginal discs continued to grow by cell proliferation (Fig. 2); the imaginal cells in the enlarged discs were still capable of differentiating when tested by transfer to wild-type larval hosts, which subsequently metamorphosed. Thus, in the absence of the hormonal signal normally provided by an increase in ecdysone titer, the proliferative phase of imaginal development persists for prolonged periods, during which the imaginal cells retain the capacity to differentiate when subsequently exposed to an appropriate concentration of ecdysone.

The conditional ecdysone-deficient phenotype of *ecd* mutants provides a sensitive system for analyzing the control of specific developmental processes by a steroid hormone. We initially focused on the late third-instar larval stage of *Drosophila* development, when the ecdysone concentration increases dramatically, by a factor of about 20 within a few hours. This increase can be blocked in *ecd* larvae by a temperature shift to 29°C earlier in the third instar (Fig. 1), and consequently all of the late larval functions normally induced by ecdysone should also be blocked. Such functions might

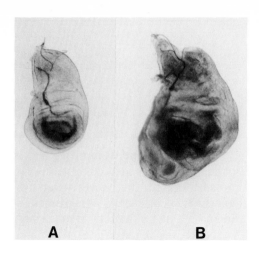

Fig. 2,A,B. Extended growth of an imaginal wing disc in *ecd*[1] larvae at 29°C. The larvae were grown as described for Fig. 1. The wing disc shown in **A** was dissected from a larva one day after the temperature shift, at about the time pupariation should normally have occurred, and for **B** the dissection was done 4 days later. The disc in **A** is the maximum size attained during normal larval development. Magnification, × 185.

be identified by an appropriate comparison between late third-instar *ecd* larvae that have an abnormally low concentration of ecdysone as a result of an earlier temperature shift and *ecd* larvae that have a higher concentration either because ecdysone was provided exogenously after the shift or the larvae were kept at 20°C until the ecdysone titer had increased in the course of normal development. The tissue chosen for this comparison was the larval fat bodies, in which major changes in protein content were detected after the increase in ecdysone concentration (Lepesant et al. 1978). Several new proteins appeared in the fat bodies as a result of two distinct responses induced by ecdysone; one response was an incorporation back into the fat bodies of proteins synthesized earlier in the same tissue and subsequently secreted into the circulating hemolymph, and the other response was de novo synthesis mainly of one new protein called *P1*.

The ecdysone-induced synthesis of the *P1* protein appears to be specific to the fat body tissue and to the late third-instar stage of development; synthesis could not be induced in four other tissues of late third-instar larvae examined, namely the salivary glands, imaginal discs, brain with the attached ganglion and female gonads, nor could synthesis be induced prematurely in fat bodies of early third-instar larvae (Lepesant et al. 1978). The reaction was further analyzed by in vitro translation of RNA isolated from fat bodies of low- and high-ecdysone samples of late third-instrar *ecd* larvae. The results indicate that ecdysone induces a major increase in the amount of translatable messenger RNA for the *P1* protein (Fig. 3A, B, C). The RNA was used as a probe for the isolation of a clone of λ bacteriophage carrying a segment of the *Drosophila* genome complementary to the *P1* messenger RNA, which was identified by specific inhibition of in vitro translation of the messenger RNA after hybridization with the cloned DNA (Fig. 3D, E, F). Measurements of the amount of poly(A)–RNA in the fat bodies hybridizeable to the cloned DNA showed about 50 times more hybridization occurring with the poly(A)–RNA from the high-ecdysone sample of larvae than the low-ec-

Fig. 3. In vitro translation of messenger RNA from larval fat bodies. The in vitro translation was done with a rabbit reticulocyte extract labeled with [^{35}S]methionine and cysteine, and the products were analyzed by one-dimensional gel electrophoresis and fluorography. Channels A and B show the gel patterns obtained using as templates total RNA from the fat bodies of low- and high-ecdysone samples, respectively, of late third-instar larvae prepared as described for Fig. 1; channel C shows the control without added RNA. The *P1* protein is in the band appearing in the high-ecdysone sample at the position corresponding to about 110 000 daltons. The templates used for channels D, E, and F were purified poly(A)–RNA from the high-ecdysone sample of larval fat bodies; for channel E, the poly(A)–RNA was hybridized before translation with DNA from a λ bacteriophage stock without *Drosophila* DNA inserts, and for channel F the hybridization was done with DNA from a λ-*Drosophila* hybrid clone. Further details are given in Lepesant et al. (1978).

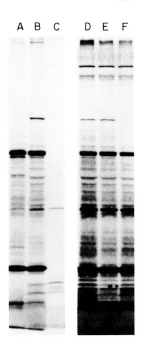

dysone sample (Table 3), in agreement with the in vitro translation results. The ecdysone-induced increase in *P1* messenger RNA could result from a higher rate of transcription of the *P1* gene, or a longer lifetime for the *P1* messenger RNA.

The organization of the *P1* gene in the cloned DNA was analyzed by cleaving the DNA with various restriction endonucleases, separating the fragments by gel electrophoresis, and testing for hybridization with a labeled high ecdysone sample of poly(A)–RNA. The resulting restriction map (Le-

Table 3. Hybridization of Poly(A)–RNA from Fat Bodies of Low and High Ecdysone Samples to Nitrocellulose Membranes Containing an Excess of the Cloned *Drosophila* DNA[a]

	Hybridized poly(A)–RNA			
	cpm		% of total	
DNA source	Low ecdysone	High ecdysone	Low ecdysone	High ecdysone
No DNA	220	230		
Control DNA (λ only)	230	240	0	0
Cloned DNA (λ + *Drosophila*)	590	18 300	0.017	0.87

[a] Further details are given in Lepesant et al. (1978).

pesant et al. 1979) shows a total of 16 kilobars (kb) of *Drosophila* DNA inserted into the λ genome, of which 3 kb hybridizes with the poly(A)–RNA probe. This 3-kb segment is about the size required to code for the *P1* protein, which has a molecular weight of 1.1×10^5 daltons, indicating that few, if any, noncoding sequences occur within the *P1* coding region.

The genetic map location of the gene coding for the *P1* protein, as determined by in situ hybridization with tritiated DNA from the complementary clone, is in region *70-D,E* of the third chromosome. We are currently isolating mutants that express an early pupal lethal phenotype in heterozygous combination with a third chromosome in which this region is deleted, on the assumption that the function controlled by the *P1* gene becomes essential for normal development at the time the gene begins to be expressed. The mutants will be examined for qualitative or quantitative abnormalities involving the synthesis of the *P1* protein by larval fat bodies, as evidence of defects in the coding or regulatory regions of the *P1* gene. The availability of mutants for a gene that responds to the hormonal signal is likely to be a key factor in further studies of the mechanism of gene regulation by a steroid hormone.

The other ecdysone-induced response of fat bodies in late third-instar larvae, which results in the incorporation of large quantities of circulating haemolymph proteins into the tissue (Lepesant et al. 1978; Thomasson and Mitchell 1972), might involve a direct effect of ecdysone on tissue permeability; alternatively, the incorporation of hemolymph proteins might be a secondary response initiated by the ecdysone-induced synthesis of one or more proteins in the fat bodies.

The remarkably varied effects induced by a steroid hormone during the development of an organism, as exemplified by ecdysone in *Drosophila* (Garen et al. 1977), depend on the coordinated occurrence of both quantitative changes in the amount of hormone and qualitative changes in the capacity of tissues to respond to the hormonal signal. Much of current research on steroid hormone action concerns the mechanism for eliciting the selective responses of certain tissues. The response mechanism appears to involve the regulation of gene transcription and is therefore at the center of molecular studies of development and eucaryotic cell function. With the availability of cloned DNA containing the coding and probably also regulatory sequences of genes that respond to a steroid hormone, rapid progress can be expected in this area. It will also be important to have available a system for isolating and analyzing mutants with coding and regulatory defects for such genes, and the prospects seem hopeful of accomplishing this with the *P1* gene of *Drosophila* as described above. There remains the problem of how appropriate levels of a steroid are established and maintained during development. Because genetic as well as biochemical methods will probably be needed to dissect the undoubtedly complex pathway of steroid biosynthesis and regulation, the possibility of obtaining conditional mutants with defects in this pathway (Garen et al. 1977) makes *Drosophila* an organism of choice for further studies of this problem.

Acknowledgments. This research was supported by grants from the American Cancer Society and the National Institute of General Medical Sciences of the U.S. Public Health Service.

References

Butenandt A, Karlson P (1954) Z Naturforsch B9: 389
Davis KT, Shearn A (1977) Science 196: 438
Doane WW (1973) In: Counce S, Wadding CH (eds) Developmental systems: Insects. Academic Press, New York, Vol 2, p 291
Garen A, Kauvar L, Lepesant JA (1977) Proc Nat Acad Sci USA 74: 5099
Gehring W (1968) In: Urspring H (ed) Results and problems in cell differentiation. Springer-Verlag, Berlin, Heidelberg, New York, Vol 1
Hadorn E (1963) Develop Biol 7: 617
Huber R, Hoppe W (1965) Chem Ber 98: 2403
Karlson P, Hoffmeister H, Hummel H, Hocks H, Spiteller G (1965) Chem Ber 98: 2394
King DS, Siddall JB (1969) Nature 221: 955
Lepesant JA, Lepesant J, Garen A (1978) Proc Natl Acad Sci USA 75: 5570
Lepesant JA, Garen A, Kejzlarova-Lepesant J, Maschat F, Rat L (1979) ICN Symp Mol Biol, in press
Mandaron P (1973) Develop Biol 31: 101
Morgan ED, Poole CF (1976) Adv Insect Physiol 12: 17
Postlethwait JH, Schneiderman HA (1970) Biol Bull 138: 536
Suzuki DT (1970) Science 170: 695
Thomasson WA, Mitchell HK (1972) J Insect Physiol 18: 1885
Wigglesworth VB (1934) Nature 133: 725
Wigglesworth VB (1964) Adv Insect Physiol 2: 247

Discussion of the Paper Presented by A. Garen

ROSEN: A question about your poly(A)–cDNA excess hybridization. I assume, since you didn't mention it, that this was a long-term labeling experiment since you got a 50-fold increase. If it was not a long-term labeling and it was a pulse, it would be more interesting because it would suggest a 50-fold increase in the rate of synthesis. Could you tell us a little bit more about that experiment, that critical detail that you omitted?

GAREN: The labeling was not done in vivo. Instead, the RNA samples were first isolated and labeled in vitro using $\gamma[^{32}P]ATP$ and polynucleotide kinase.

ROSEN: Have you tried to pulse label in vivo and look into the rate of synthesis and see if there is any increase in synthesis?

GAREN: That is an important point, but we have not done the experiment. Knowing the rate of appearance of the induced mRNA would help in determining whether ecdysone exerts a primary direct effect, or a secondary indirect effect, on the *Pl* gene.

THOMPSON: In the study you did regarding position 2 protein and its uptake could you tell whether the uptake you see represents a generalized uptake of serum proteins and is inextricably bound to the degradation of proteins, which would suggest a greater pinocytotic activity of these cells, or is it specific for these particular ones?

GAREN: No, we weren't able to tell if there were any other proteins involved in the uptake because those that you saw were virtually the only proteins detectable in the hemolymph.

THOMPSON: What about added proteins?

GAREN: Well, that was done some years ago by Thomasson and Mitchell. They injected a radioactive sample of total *Drosophila* proteins into the haemolymph of late third instar larvae and afterwards detected rapid incorporation of the label into fat bodies, suggesting that the uptake is not specific for the major haemolymph proteins.

O'MALLEY: I missed a couple details on how you did that cloning experiment, in which you found a pair of genes for the high molecular weight protein. You used a total *Drosophila* library, is that right? How was that made and how many genes were involved?

GAREN: We were generously provided with a *Drosophila* library, prepared in Tom Maniatis' laboratory, containing randomly sheared segments of genomic DNA incorporated into a lambda phage vector. The elegant procedure used to prepare this library is carefully described in a recent publication (Maniatis, Hardison, Lacy, Lauer, O'Connell, Quon, Sim and Estratiadis, Cell *15*, 687 [1978]). The library appears to include segments from the entire genome, averaging about 16 kilobases of DNA per segment.

O'MALLEY: How many total pieces of the *Drosphila* genomes were produced by that method.

GAREN: We tested about 40,000 cloned segments, or about 6×10^8 base pairs of DNA, which represents four copies of the entire genome.

O'MALLEY: Okay, then you took poly(A) from uninduced and induced and having only a few induced proteins you picked up those specifically. The proof of the feasibility of this method is the success of the experiment. Nevertheless, it appears quite difficult because you have so many clones and you are looking for only a few spots. You must have a rather sensitive detection method.

GAREN: Although the total poly(A)RNA from fat bodies is quite complex, we exposed the films for a relatively short time sufficient to detect only the clones that hybridize to the more prominent species of RNA.

O'MALLEY: But since the gene is a single copy species, it is only represented as 1/10,000 of the clones.

GAREN: Yes, but the use of a phage vector enables you to plate about 5000 clones on a petri dish, and therefore only a few dishes are required to test 40,000 clones.

O'MALLEY: No, it is very clever. Have you done translation-enhancement or translation-inhibition experiments yet to confirm the identity of the clones?

GAREN: No. (Note added in proof: The coding specificity of the clone for the *P1* gene has been confirmed by a hybrid-selection technique, in which the cloned DNA is immobilized on nitrocellulose, hybridized with the fat body poly(A)RNA, and the hybridized RNA is recovered from the nitrocellulose and translated in vitro.)

O'MALLEY: It should be remembered that the translational arrest sometimes creates some problems in that nonspecific inhibition is possible.

ROSEN: How do you kinase label without fragmenting the RNA or taking off the cap first?

GAREN: The RNA is first partially digested with alkali before labeling.

Discussants: A. GAREN, B.W. O'MALLEY, J.M. ROSEN, and E.B. THOMPSON

Chapter 17

Ecdysteroid Receptors in Cultured *Drosophila* Cells

JOHN D. O'CONNOR, PETER MAROY, CHRISTOPH BECKERS, ROGER DENNIS, CRISTINA M. ALVAREZ, AND BECKY A. SAGE

I. Introduction

The mode of action of ecdysteroids during the metamorphosis of insects or the molting of crustaceans has received increasing attention during the past few years. Indeed, the observation that β-ecdysone altered the pattern of puffs along a polytene chromosome first suggested the paradigm so in evidence here (Clever 1961; Karlson 1965) that steroid hormones modulate transcriptional activity. In the vertebrate systems studied thus far the effect of these steroids has been linked to the presence of a hormone receptor complex in the cytosol, which translocates to the nucleus, resulting in the accumulation of a specific mRNA species.

Although a similar paradigm was presumed to be operative in those systems effected by β-ecdysone, no direct evidence had been forthcoming that clearly demonstrated the existence of a specific receptor protein for ecdysteroids (Ashburner 1972; Yund and Fristrom 1975). The recent availability of both an established insect cell line that responds to β-ecdysone (see Cherbas et al., Chap. 18 in this volume) and high specific-activity ligand (Maroy et al. 1978; Yund et al. 1978) made possible the direct demonstration of a saturable, specific, high-affinity protein receptor for ecdysteroids. The observations of Fristrom and Yund (1976) have clearly indicated ponasterone A (25-deoxy-β-ecdysone) to be a more potent effector steroid in imaginal discs than β-ecdysone. It is certainly due in part to the latter observations that this study was undertaken. The materials and methods used in these experiments have been previously described (Maroy et al. 1978), and consequently will be detailed here only when pertinent to a specific conclusion.

II. Characterization of the Cytoplasmic Receptor

Illustrated in Fig. 1 is the rate at which both labeled ponasterone A (PNA) and β-ecdysone become associated with cultured *Drosophila* cells. At equilibrium the concentration of [³H]PNA associated with the cells compared with its concentration in an equivalent volume of medium is approximately 13:1 following incubation in 4×10^{-10} M ligand. This concentration ratio is reduced to approximately 3:1 in the presence of 100-fold excess of unlabeled PNA. The latter ratio probably represents nonspecific accumulation of the ligand since *Drosophila* cell lines in which no "receptor" activity can be demonstrated also appear to concentrate ponasterone A at a ratio of approximately 3:1. On the other hand, cells incubated with 3×10^{-8} M β-ecdysone effected a 1.5- to 2-fold concentration of ligand at equilibrium. Addition of 100-fold excess cold β-ecdysone or cold ponasterone A effectively inhibited any concentration of labeled [³H]β-ecdysone. The apparent discrepancy in the cells ability to concentrate β-ecdysone and ponasterone A ultimately reflects upon both the distinctive binding constants of the putative receptor for each of these ligands, which are demonstrated below, and the significantly different specific activities of the two ecdysteroids. Thus, a 13-fold concentration of [³H]PNA (126 Ci/mmol) incubated at 4×10^{-10} M rep-

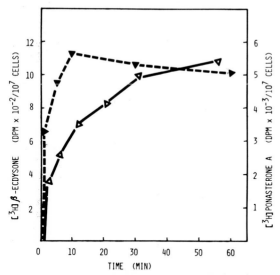

Fig. 1. The association of β-ecdysone (▼) and ponasterone A (▽) with cultured *Drosophila* cells. Kc cells were incubated with 10^{-7} M [³H]β-ecdysone (1.4 Ci/mmol) or 5×10^{-10} M [³H]ponasterone A (126 Ci/mmol) for the times indicated at 22°C in D-20 media (Echalier and Ohanessian 1970) without serum. [¹⁴C]Inulin was added to the incubation to correct for intracellular trapping of hormone. For each determination, 4×10^{7} cells were pelleted from the incubation medium by centrifugation (10 000*g*, 1 min), and subsequently transferred to a xylene–triton scintillation fluor for determination of associated radioactivity. Unless otherwise indicated the cells were harvested and the radioactivity determined in all subsequent experiments by this procedure.

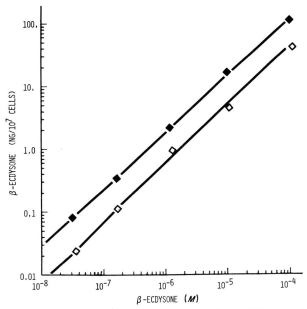

Fig. 2. The association of [³H]β-ecdysone (1.4 Ci/mmol) with Kc cells at 4°C (□) and 22°C (■). Cells were incubated with increasing concentrations of [³H]β-ecdysone for 5 min in D-20 media containing [¹⁴C]inulin. Incubation was terminated and the radioactivity determined as indicated in Fig. 1.

resents 1000 binding sites per cell. On the other hand, a two-fold concentration of 3×10^{-8} β-ecdysone (1.4 Ci/mmol) by a similar number of cells yields 2800 binding sites per cell.

Since the Kc cell line does not metabolize either β-ecdysone or ponasterone A nor does the association of these ligands with the cells appear to be carrier mediated (Fig. 2), the ability to concentrate steroid clearly suggested the existence of a binding moiety. In this regard the data obtained from the Kc cell line was reminiscent of that obtained earlier from imaginal discs (Fristrom and Yund 1976; Yund and Fristrom 1975).

The first direct evidence of an ecdysteroid-binding protein is presented in Fig. 3. Incubation of a post-100 000g supernatant with [³H]PNA and subsequent chromatography on Sephadex G-25 revealed that a significant portion of the label eluted in the void volume. The amount of bound hormone could be reduced dramatically by the addition of excess cold ponasterone A and to a lesser extent by excess unlabeled β-ecdysone (Fig. 3B and C, respectively). Scatchard (1949) analysis of the cytosol binding of ponasterone A revealed the concentration of binding sites to be 3.1×10^{-10} M/g of protein with a K_D of 3.8×10^{-10} M/g of protein (Fig. 4B). Expressed in a different fashion this represents approximately 1300 sites per cell.

It is of particular biological significance to document the dramatic temperature dependency which this binding exhibits. The data in Fig. 4A illustrate that cytosol incubated at 4°C with [³H]PNA binds less than 10% of the

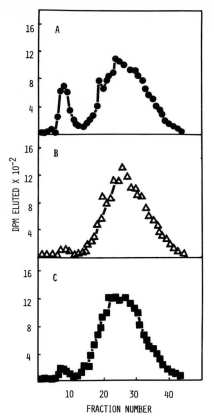

Fig. 3,A–C. Gel filtration of K_c cytosol–[³H]ponasterone A mixture on Sephadex G-25. Cells were harvested, washed in buffer (10 mM Tris, 5 mM MgCl₂, pH 6.9 at 22°C), and disrupted in a Potter–Elvejhem homogenizer. Cytosol represents the post 110 000g supernatant. In these experiments cytosol (100 μl, 1 mg total protein) was incubated with 7×10^{-10} M [³H]ponasterone A for 30 min at 22°C and then chromatographed at 4°C. The elution buffer was 10 mM Tris, 5 mM MgCl₂, 150 mM KCl, pH 7.4 at 4°C. **A** Cytosol from Kc cells; **B** cytosol plus 50-fold excess ponasterone A; **C** cytosol plus 200-fold excess β-or-ecdysone.

ligand even after 60 min. In contrast, if following a 4°C incubation the cytosol–ponasterone A mixture is elevated to the ambient temperature of 22°C, more than 35% of the ligand is bound. In view of this data it is no longer perplexing that even under endocrinological circumstances that induce elevated ecdysteroid titers in vivo, organisms are refractive to molting at low temperatures (Passano 1960; Roberts 1957).

The binding activity of the cytosol preparation appears to be reasonably stable when maintained at 4°C in the absence of ligand (Fig. 5). This observation is of significant consequence in future attempts at purifying the binding moiety. The dissociation rate of the cytosol receptor–hormone complex has a half life of greater than 4 h at 4°C and 26 min at 22°C as determined from the data in Fig. 6. These rate constants are consistent with the very rapid alteration in hormone concentrations in *Drosophila*. The peak titer of 240 pg of β-ecdysone per milligram of fresh weight occurs 12 h after pupariation in *Drosophila*. However, 18 h later this concentration has decreased to the baseline level of 30 pg/mg fresh weight (Hodgetts et al. 1977). Thus, within a rather short time span the organism has gone from a very high titer to a markedly lower baseline value. Were the dissociation rate constant not consistent with this rapid titer alteration, then bound ecdysteroid could be exerting a biological effect long after the organism attempts to clear itself of hormone.

The S value of the ecdysteroid binding moiety is very similar to those reported for other steroid binding proteins. In order to reduce the length of time necessary to sediment particles in the 4S range, a vertical reorienting rotor was employed and the cytosol–ponasterone A mixture was layered on a 10–40% sucrose gradient. The running time was 2.5 h at an average 370 000g. The profiles illustrated in Fig. 7 represent the results of the initial determination. The [³H]PNA–receptor complex migrates coincidently with a 4.2S hemoglobin marker. The bound label is effectively competed by a 100-fold excess of unlabeled ponasterone A (Fig. 7A). In addition, incubation of the cytosol with pronase completely inhibits binding (Fig. 7B), whereas similar incubations with either DNAse or RNAse were without effect (data not illustrated).

Thus far, then, the cytosol receptor for ecdysteroids appears to be a stable 4S particle present at a concentration of approximately 1300 molecules per cell. It exhibits high affinity and specificity along with a rapid rate of dissociation at 22°C.

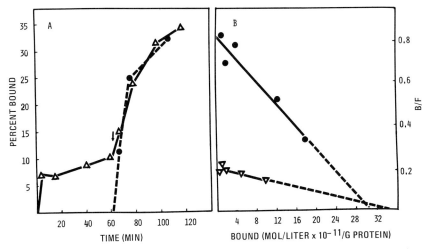

Fig. 4. **A** Temperature dependent cytosol binding of [³H]ponasterone A. A portion of a cytosol preparation was incubated with 5×10^{-10} M [³H]PNA at 4°C (△). After 60 min (arrow), the cytosol incubation was continued at 22°C for an additional 60 min. Another aliquot of the same cytosol preparation was maintained at 4°C for 60 min at which time [³H]ponasterone A was added to a final concentration of 5×10^{-10} M and the incubation was allowed to proceed at 22°C (●). Separation of bound from free hormone was performed using a dextran-coated charcoal (DCC) assay. (McGuire W (1975) Meth Enzymol 47: 469–478)

B Scatchard plot of cytosol binding of [³H]ponasterone A at 4°C (▽) and 22°C (●). Cytosol aliquots (100 μl, 1 mg total protein) were incubated for 30 min with increasing concentrations of [³H]PNA. Separation of bound from free was performed using the DCC assay.

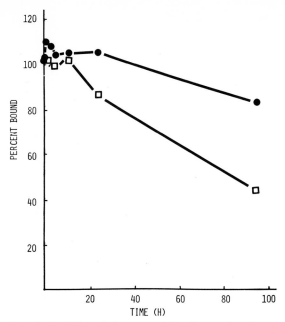

Fig. 5. Stability of the cytosol binding moiety in the presence or absence of ligand: (●) cytosol incubated with 5×10^{-10} M [³H]ponasterone A for 45 min at 22°C and then maintained at 4°C; (□) cytosol maintained at 4°C in the absence of radioligand, but at 45 min prior to the times indicated [³H]ponasterone A was added to a final concentration of 5×10^{-10} M and the mixture was incubated at 22°C. For both preparations, the separation of bound and free hormone was performed using the DCC assay. t_0 is designated as the time at which the initial cytosol incubation at 22°C was terminated.

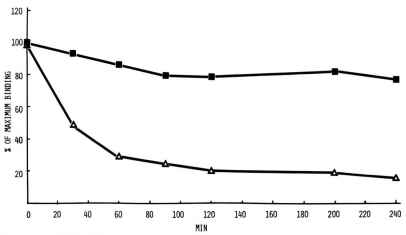

Fig. 6. The dissociation of Kc cytosol–ponasterone A binding at 4°C and 22°C. Kc cytosol was isolated as in Fig. 3 and incubated with 5×10^{-10} M [³H]PNA for 45 min at 22°C. Following the incubation, the cytosol sample was divided and unlabeled ponasterone A (10^{-7} M) was added to both samples at t_0. The samples were then maintained at 4°C (■) and 22°C (△). The dissociation of bound hormone was monitored using the dextran-coated charcoal assay.

Fig. 7,A,B. Velocity sedimentation of the ecdysteroid binding protein in Kc cytosol. Cytosol was isolated as in Fig. 3, and incubated with 5×10^{-10} M [³H]ponasterone A (45 min, 22°C). The cytosol–[³H]PNA mixtures were then sedimented on 10–40% sucrose gradients (10 mM Tris, 5 mM MgCl₂, pH 7.4, Sorvall TV-865 rotor at 65,000 rpm, 2.5 h, 4°C). **A** Kc cytosol-[³H]PNA mixture (×). Cytosol incubated with [³H]PNA plus 150-fold excess ponasterone A (○). **B** Kc cytosol incubated with [³H]PNA (○). Cytosol–[³H]PNA mixture incubated with pronase (10 mg/ml) for 30 min at 22°C (△). [¹⁴C]Globin was used as a 4.2S sedimentation marker.

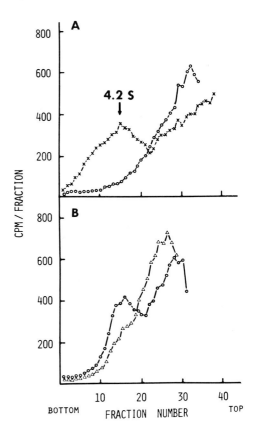

III. Translocation and Nuclear Binding

Evidence that the hormone receptor present in the cytosol complex undergoes a translocation to the nucleus is presented in Table 1 and Figs. 8–11. It is clear that, following a 30-min incubation of Kc cells in 0.3 nM [³H]PNA, almost 65% of the incorporated hormone is present in washed nuclei (Table 1). Increasing the concentration of [³H]PNA in the incubation medium resulted in an increase in the concentration of bound ligand in the nuclear preparations up to a concentration of approximately 4 nM free hormone

Table 1. Subcellular Distribution of [³H]Ponasterone A[a]

Labeling time (min)	Post-1700 × g supernatant (%)	Nuclear wash (%)	0.4 M KCl soluble (%)	0.4 M KCl insoluble (%)
5	55	6	23	3
30	27	5	47	17

[a] Cells were incubated for the indicated times in 3×10^{-10} M [³H]ponasterone A. After harvesting, cells were resuspended in buffer (10 mM Tris, 1.5 mM MgCl₂, 1.5 mM EDTA, 0.1% Triton X-100, pH 6.9 at 22°C), and homogenized by 15 passes in a Dounce homogenizer. The resulting nuclei were pelleted (10 min, 1700g) and washed before final extraction with 0.4 M KCl at 4°C for 10 min. Data are expressed as percentage of incorporated radioactivity.

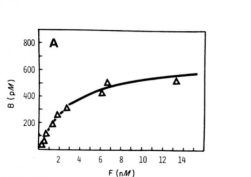

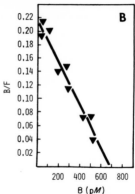

Fig. 8. **A** The concentration-dependent accumulation of [³H]ponasterone A by nuclei of Kc cells. Cells (10^8 cells/0.5 ml, approximately 4 mg total protein) were labeled with [³H]PNA (30 min, 22°C) and the nuclei were obtained using the method of Munck and Wira (1975). Briefly, labeled cells were diluted 50-fold with buffer (5 mM $MgCl_2$, 0.2% Triton X-100), and after 10 min at 4°C the resulting nuclei were harvested by centrifugation ($800g$, 10 min). The radioactivity present in both the nuclear pellets and their respective supernatants was then determined.

B Scatchard plot of the in vivo nuclear accumulation of [³H]ponasterone A in Kc cells.

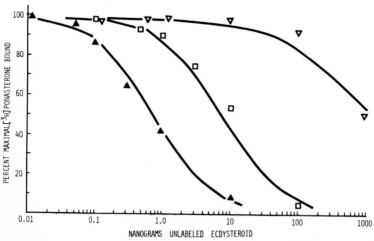

Fig. 9. Ecdysteroid inhibition of the in vivo nuclear accumulation of [³H]ponasterone A by Kc cells. Cells were incubated (30 min, 22°C) with increasing amounts of unlabeled ecdysteroids in the presence of $3 \times 10^{-10} M$ [³H]PNA. The nuclei were isolated using the technique of Munck and Wira (1975) as described in Fig. 8. The unlabeled ecdysteroids used as inhibitors were ponasterone A (▲), β-ecdysone (□), and α-ecdysone (▽).

(Fig. 8A).   The apparent saturation that occurs at approximately 6–8 nM free steroid is probably not a reflection of a limited number of nuclear "acceptor" sites but rather represents the saturability of translocatable cytoplasmic binding sites.   Replotting of the data on Scatchard coordinates (Fig. 8B) reveals the K_D to be $3 \times 10^{-9}\ M$ with a maximal binding of $7 \times 10^{-10}\ M$.   When corrected for protein concentration these values are very close to those that were previously illustrated for cytosol.   The similarity of K_D in both cytosol and nuclear preparations together with the observation that "naive" nuclei (that is, nuclei obtained from cells that have not been previously exposed to steroid) do not bind [³H]ponasterone A is consistent with a translocation paradigm.

It is both significant and reassuring to see that the K_D of the nuclear-binding component for various ecdysteroids parallels their biological activity.   Thus, the data in Fig. 9 reveal the relative binding efficiency of three ecdysteroids to be ponasterone A > β-ecdysone > α-ecdysone.   This order is identical to that obtained previously for the induction of imaginal disc evagination (Fristrom and Yund 1976).   In addition the difference between the

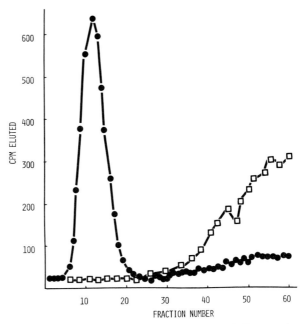

Fig. 10. Gel filtration of 0.4 M KCl extract of washed Kc cell nuclei.   Cells were incubated in $3 \times 10^{-10}\ M$ [³H]PNA at 22°C for 30 min and harvested.   The cells were then resuspended in buffer (10 mM Tris, 1.5 mM MgCl$_2$, 1.5 mM EDTA, 0.1% Triton X-100, pH 6.9 at 22°C) and homogenized by 15 passes in a Dounce homogenizer.   The resulting nuclei were pelleted (10 min, 1700g) and washed before the final extraction with 0.4 M KCl at 4°C for 10 min.   The KCl-soluble fraction (●) was then chromatographed on Sephadex G-25 at 4°C with an elution buffer of 10 mM Tris, 1.5 mM EDTA, 1.5 mM MgCl$_2$, 0.3 M KCl, pH 7.4.   The elution of [³H]ponasterone A (□) is indicated for comparison.

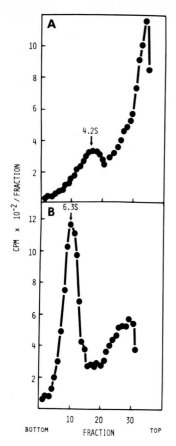

Fig. 11,A,B. Velocity sedimentation of ecdysteroid binding moieties in Kc cells. **A** Kc cytosol, isolated as in Fig. 3, was incubated with 5×10^{-10} M [^{3}H]ponasterone A for 45 min at 22°C. **B** KCl extract of washed Kc nuclei incubated in vivo with 5×10^{-10} M [^{3}H]PNA. Cells were homogenized and nuclei were isolated according to the procedure in Fig. 10. The washed nuclear pellet was extracted with an equal volume of 0.8 M KCl at 4°C for 10 min. The sedimentation parameters for both samples were similar to those specified in Fig. 7. Radiolabeled sedimentation markers (hen albumin, BSA, and aldolase) were prepared by a reductive alkylation of protein amino groups with [^{14}C]formaldehyde. (Means GE (1977) Meth Enzymol 47: 469–478)

concentrations of α- or β-ecdysone necessary to inhibit 50% of [^{3}H]ponasterone A binding is similar to concentrations of these two steroids observed by Ashburner (1971) to induce the puffing phenomenon in the polytenic chromosomes in *Drosophila* salivary glands. It should be pointed out that in order to conserve labeled ligand these displacement values were obtained at a concentration of [^{3}H]ponasterone A below the apparent K_D. Therefore, the K_D for α- or β-ecdysone could not be obtained directly from the figure. Rather these values were derived following the method of Rodbard (1973). The K_D for α-ecdysone was determined to be 5.5×10^{-6} M, whereas the K_D for β-ecdysone was calculated to be 3.3×10^{-8} M. In those insect and crustacean species in which careful determinations of α- and β-ecdysteroid concentrations have been performed, β-ecdysone levels are very close to 5×10^{-7} M at their peak near ecdysis (Chang et al. 1976; Hodgetts et al. 1977). In contrast α-ecdysone never reaches a concentration greater than 10^{-7} M and is generally well below this. These data are consistent with the suggestion that α-ecdysone acts as a precursor to the hormone β-ecdysone but does not act as a hormone itself, or alternatively if α-ecdysone does elicit specific hormonal responses, it does so without receptor protein mediation.

The data in Table 1 indicate that a single extraction of labeled nuclei with 0.4 M KCl removed 74% of the associated [^{3}H]ponasterone A. Chromatography of this salt extract revealed that the overwhelming majority of the label was associated with the macromolecular fraction (Fig. 10). Extraction of the labeled void volume with methanol and subsequent silica gel chromatography documented that only a single labeled component, which cochromatographed with authentic ponasterone A, was present (Maroy et al. 1978). In addition, if a KCl extract of in vivo labeled nuclei was applied to a 10–40% sucrose gradient, a peak of radioactivity was observed at 6.3S (Fig. 11B). This is in contrast to the 4.2S value obtained for the cytosol binding components under similar experimental conditions (Fig. 11A).

Preliminary data suggest the S value for neither the nuclear- nor cytosol-binding moieties to be dependent upon ionic strength, although the binding efficiency of the cytosol preparation decreases dramatically above 0.2 M NaCl. At the present time the reasons for the increased sedimentation velocity of the nuclear binding activity are not known although it is assumed to result either from either a dimerization of the cytosol protein, a

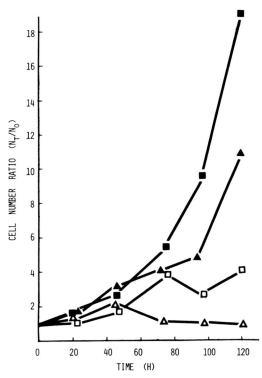

Fig. 12. Ecdysteroid inhibition of Kc cell division. (■) control; (▲) α-ecdysone; (□) β-ecdysone; (△) ponasterone A. Kc cells were maintained in suspension at 22°C in D-20 medium. At a cell density of 1.5×10^6 cells/ml (t_0), ecdysteroids were added to duplicate cultures at a final concentration of 10^{-8} M. Cell viability was determined by the criterion of trypan blue exclusion.

change of shape of the cytosol receptor, adventitious binding of a nuclear protein to the cytosol receptor, or all of the above. Appropriate experiments are currently underway to attempt to differentiate among the possibilities.

The presence of receptor activity in the Kc line of *Drosophila* cells was not altogether surprising since these cells show rather remarkable alterations in both morphological and biochemical behavior in the presence of hormone.

IV. Response of Cultured Cells to Ecdysteroids

Early reports indicated that ecdysteroids either killed or inhibited cell division in the Kc line (Courgeon 1975). Indeed, similar results have recently been obtained with the Kc line in this laboratory (Fig. 12). More importantly, however, there seem to be three distinct types of responses to ecdysteroids. Some cells are lysed (class 1), others remain viable (class II),

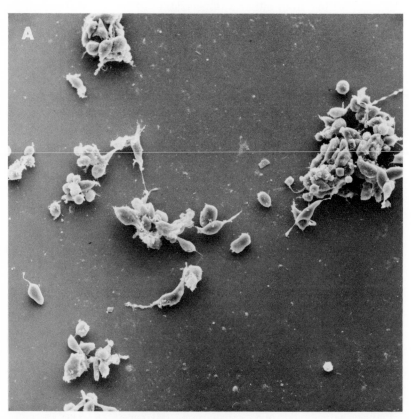

Fig. 13,A,B. Scanning electron micrographs of Kc cells. **A** Class II cells (Clone 2·E·5) grown in the presence of 5×10^{-7} M β-ecdysone for 48 h ($\times 2000$). **B** Class III cells grown in the continual presence of 10^{-7} M β-ecdysone ($\times 770$).

while a third class (III) continues to divide in the presence of hormone. Cloned cell lines representing the class I and class II types possess nearly identical receptor activity while class III cells possess no detectable binding activity. Illustrated in Fig. 13 are the class II and class III cells in the presence of β-ecdysone. Presumably the acetylcholine esterase inducible cells described by Cherbas et al. (1977) are of the class II type.

The existence of three types of behavior in the presence of hormone among the cells of the Kc line suggests that this line might well represent an excellent model analog for a developing insect. The lytic response of class I cells is very similar to many larval cells (fat body, epidermal, etc.) that die during metamorphosis to the adult. The cell death in the animal and the lysis of the cultured cells occur in the presence of β-ecdysone and the absence of juvenile hormone (JH). On the other hand, those cells of class II behave in a manner predicted for imaginal disc cells or neural elements, which are known to undergo differentiative events in response to β-ecdysone. The cells of class III respond in a fashion analogous to either germ cells, blood cells, or perhaps mutant tumor cells whose mitoses are not normally regulated in a detectable manner by β-ecdysone or have lost such a regulatory level during their time in culture.

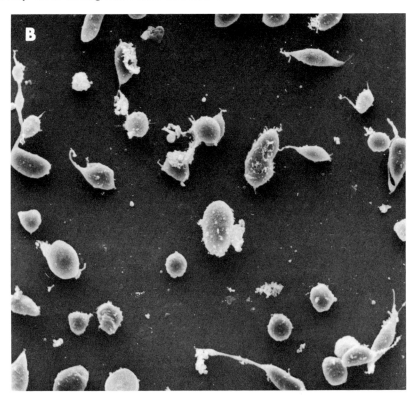

At the present time there is no compelling evidence that the cells of class I are indeed larval cells or that those of class II are imaginal. It is only that their response to ecdysteroids is similar. It is hoped, however, that an understanding of these responses will ultimately help to elucidate the mode of action of ecdysteroids in the developing organism.

Acknowledgments. We thank Drs. M.A. Yund and J.W. Fristrom for their gift of [³H]ponasterone A, and Dr. Ernest S. Chang for his purification of ponasterone A from the many products of the reductive tritiation of stachysterone C. Additional thanks are also expressed to DuPont Instruments for the use of a Sorvall TV-865 ultravertical rotor needed to perform the initial velocity sedimentation analyses. This work was generously supported by Grant NS 08990 from the National Institutes of Health and Grant PCM 75-20146 from the National Science Foundation. One of us (P. Maroy) was supported in part by Grant PCM 78-05471 from the National Science Foundation in conjunction with the Eastern European Cooperative Science Program.

References

Ashburner M (1971) Nature New Biol 230: 222–223
Ashburner M (1972) FEBS Lett 22: 265–269
Chang ES, Sage BA, O'Connor JD (1976) Gen Comp Endocrinol 30: 21–33
Cherbas P, Cherbas L, Williams CM (1977) Science 197: 275–277
Clever U (1961) Chromosoma 12: 607–675
Courgeon A-M (1975) Exp Cell Res 94: 283–291
Echalier G, Ohanessian A (1970) In Vitro 6: 162–172
Fristrom JW, Yund MA (1976) In: Maramorasch K (ed) Invertebrate tissue culture. Academic Press, New York, Vol 14, 161–178
Hodgetts RB, Sage BA, O'Connor JD (1977) Develop Biol 60: 310–317
Karlson P (1965) J Cell Comp Physiol Suppl 1: 66–69
Maroy P, Dennis R, Beckers C, Sage BA, O'Connor JD (1978) Proc Nat Acad Sci USA 75: 6035–6038
McGuire W (1975) Meth Enzymol 36: 248–254
Means GE (1977) Meth Enzymol 47: 469–478
Munck A, Wira C (1975) Meth Enzymol 36: 255–264
Passano LM (1960) In: Waterman TH (ed) Physiology of crustacea. Academic Press, New York, Vol 1, pp 473–536
Roberts JL (1957) Physiol Zool 30: 232–242
Rodbard D (1973) Adv Exp Med Biol 36: 289–326
Scatchard G (1949) Ann NY Acad Sci 51: 660–672
Yund MA, Fristrom JW (1975) Develop Biol 43: 287–298
Yund MA, King DS, Fristrom JW (1978) Proc Natl Acad Sci USA 75: 6039–6043

Discussion of the Paper Presented by J.D. O'Connor

SCHRADER: What is the lower limit of sensitivity of your assay? If the receptor level is one tenth of the wild type, would you still score that as positive?

O'CONNOR: We would probably score that as negative.

SCHRADER: So there could have been a fair amount of receptor there and yet you just might not have seen it.

O'CONNOR: Yes, that's a good point.

SCHRADER: Another question I wanted to ask was about the translocation to the nuclei. You were jumping back and forth from the cells to the isolated nuclei. This was done by giving the cells the ponasterone A, right?

O'CONNOR: The cells were given ponasterone A and then the nuclei extracted.

SCHRADER: As I understand from your presentation, it looks like there is a very sharp response curve to these compounds with a half maximal response at about 10^{-8} M for β-ecdysone. Does the equilibrium constant that you detect for this binding reaction bear any resemblance to that of the biologically predicted equilibrium constant, and if not, why not?

O'CONNOR: The binding constant of ponasterone to nuclear receptor is about 3×10^{-9} M, and I think that is very favorable with the kinds of concentrations that Jim Fristrom sees in the imaginal disc.

SCHRADER: Well, one of the speakers showed that there is no reaction to 10^{-9} M ecdysone.

O'CONNOR: That was with different ecdysteroids. With α- and β-ecdysone the binding constants are 5×10^{-6} M and 3×10^{-8} M.

CLARK: That was a very interesting talk. I would just like to make some general comments about the binding studies that you have done and discuss the whole concept of cytoplasmic to nuclear transfer, which many of us in this audience, including myself, have been prone to tout in various places, chapters, articles, and wherever we can put it. In animals that have high levels of hormone (10^{-8} to 10^{-6} M) it is not necessary to have special high-affinity receptors for cytoplasmic to nuclear translocation. Indeed, it is not necessary to have high-affinity receptors at all because they would be constantly occupied and hence not serve in a regulatory capacity.

O'CONNOR: May I make a remark?

CLARK: Yes.

O'CONNOR: In our hands at a level of receptor activity where you can show binding activity in cytosol you cannot find it in the isolated nuclei.

CLARK: In what cells are those?

O'CONNOR: In the Kc cell line. Nuclear binding and labeling in vivo is demonstrable in isolated virginal cells. I think that a distinction should be made, that is, they have not seen ecdysone before.

CLARK: Naive will be a better term than virgin. I prefer to save that other word for other things.

O'CONNOR: Where did you say you are from? In the in vivo experiment with whole animal it is very tough to get discs that have not been exposed to ecdysone previously.

CLARK: The rate studies though were not done in the presence of any competitive displacement hormone. So if there were fine differences, you still couldn't see it.

BUTTERWORTH: Dennis (O'Connor), I wanted to ask a question regarding some of your data. You had a picture of the RNase-treated systems and then the unextracted one, and the shape of these curves seemed to be quite different. The unextracted was quite broad, and the RNase treated one was very narrow. Is there any explanation for that?

O'CONNOR: No.

Discussants: F.M. BUTTERWORTH, J.H. CLARK, J.D. O'CONNOR, and W.T. SCHRADER.

Chapter 18

Ecdysteroid Hormone Effects on a Drosophila Cell Line

P. CHERBAS, L. CHERBAS, G. DEMETRI*,
M. MANTEUFFEL-CYMBOROWSKA**, C. SAVAKIS,
C. D. YONGER***, AND C. M. WILLIAMS

I. Introduction

We, like many others who study steroid hormone action, are starting out from an interest in the fundamental aspects of gene regulation in eukaryotic cells. From this perspective the steroid appears to be a powerful tool, potentially capable of piercing the maze of on-going control processes and leading us directly to the start of a well-defined transcriptional pathway. Thus, since the beginnings of the molecular study of prokaryotic inducible systems, steroid effects have appeared to be the likely eukaryotic analogs, and the study of vertebrate steroid hormones has flourished. That substantial progress has been made is well documented in this volume. We can now say with confidence what we could only surmise before, namely that steroid treatments lead to striking changes in the mRNA populations of target cells. Still we are ignorant of the details.

At this juncture there are very good reasons for those interested in the molecular basis of steroid action to consider the virtues of the fruitfly, *Drosophila melanogaster*. Preeminent among these reasons is the relative simplicity of the genome of *D. melanogaster*; its haploid DNA content is only 50 times that of *Eschericia coli* and in terms of kinetic complexity the difference is still less. We place special emphasis on this characteristic because we are persuaded that the purification of rare DNA sequences whose activities are directly controlled by hormone–receptor complexes is now the outstanding challenge and unifying theme of our field. One who chooses to work with *D. melanogaster* can benefit additionally from a half-century of accumulated knowledge about gene structure and its manipulation in this organism (see, for example, Garen and Lepesant, Chap. 16 in this volume). These virtues must be taken very seriously, especially when considered

---

* Present address: Stanford University, Stanford, California
** Present address: Polish Academy of Sciences, Nencki Institute of Experimental Medicine, Warsaw, Poland.
*** Present address: Department of Surgery, Yale Medical School, New Haven, Connecticut.

alongside the additional advantages of high-resolution in situ hybridization and the puffing response.

A final point concerns the nature of the hormonal responses available for study. It is now clear that insect tissues respond to the steroid molting hormones, the ecdysteroids,[1] with lengthy and impressive developmental changes, and that many tissues and organs retain their hormone responsiveness and their normal developmental potentials in short-term culture. This means that it is not only possible to study the earliest stages of the hormone response in detail, but that it is also possible to define with rigor the hormonal requirements at the target tissue level (Ashburner 1971; Chihara et al. 1972; Fristrom and Yund 1976). Thus, we shall have substantial recourse to this background of organ culture experience in what follows.

It is curious that, although the ecdysteroid-induced puffing seen in the giant polytene chromosomes of flies has often been cited as a vivid and virtually paradigmatic instance of a steroid hormone regulating transcription, in fact, the biochemical study of ecdysteroid action has lagged far behind that of the vertebrate steroids. It is now evident, however, that ecdysteroid-responsive cells contain receptors and that these receptors do interact with the nucleus (Maroy et al. 1978; Yund et al. 1978).

With all of these considerations in mind, we determined to study an ecdysteroid-responsive cell line from *D. melanogaster*. The virtues of cell lines are obvious: (1) they can provide large quantities of homogeneous cells; and (2) because the population in culture is growing, one can sometimes devise methods for selecting interesting mutant cells. Set against these virtues is the problem of finding a cell line that exhibits a palpably authentic and interesting hormone response. Fortunately, the pioneering efforts of others have led to the appearance of many *D. melanogaster* cell lines (Echalier and Ohanessian 1969; Kakpakov et al. 1969; Schneider 1972). In 1972, Courgeon (1972a,b) reported that the cells of one line, the Kc line of Echalier and Ohanessian, respond to ecdysteroids by ceasing to multiply and by changes in morphology. We have been studying a line derived from Kc cells. In what follows we will summarize our initial efforts, which have been devoted to the demonstration that these cells do exhibit an interesting ecdysteroid response, one that can, in some senses at least, be called authentic, and one that is certainly robust and amenable to biochemical study. We will go on to indicate why we expect such study to be worthwhile.

II. Background: Drosophila Cell Lines

During the past decade a number of workers have reported procedures for obtaining cell lines starting from *D. melanogaster* embryos. To understand

---

[1] In the nomenclature proposed by Goodwin et al. (1978) and adopted here, ecdysteroid is a generic term akin to corticosteroid or glucocorticoid. Ecdysone is the compound formerly called α-ecdysone. 20-Hydroxyecdysone is the compound formerly known variously as β-ecdysone, ecdysterone, or crustecdysone.

the results it is important to be aware that a fly embryo really harbors two essentially distinct populations of cells. Very early in embryogenesis virtually every cell is assigned to one of two possible fates. Either the cell and its progeny will participate in larval organs or they will participate later in adult (imaginal) life. Cells destined to form larval organs differentiate during embryogenesis and, thereafter, as the larva grows, so do they, becoming progressively more polyploid or polytene. Meanwhile those cells destined to form the adult organs are set aside as small nests of undifferentiated, diploid cells, of which the best known are the imaginal discs. The discs grow by cell multiplication during larval life. Then at metamorphosis, the appearance of ecdysteroids provokes both the degeneration of larval tissues and the differentiation of the imaginal ones to take their places (Demerec 1965).

The methods for starting cell lines have been reviewed by Schneider and Blumenthal (1978). Different workers have chosen embryos at different stages of development ranging from very early (just post-gastrulation) to very late (essentially complete larvae, within 1–2 h of hatching). The embryos are dissociated by enzymes or by homogenization and then cultured. In the primary cultures numerous larval cell types can be identified as they divide a limited number of times and then differentiate nearly on schedule (Seecof and Unanue 1968; Shields et al. 1975). Coexisting with them are colonies of diploid cells, which exhibit vigorous cell division. Schneider (1972) and Dübendorfer (1976; Dübendorfer et al. 1975) have provided strong evidence that under appropriate circumstances these cells can metamorphose into typical adult structures. Thus, the diploid colonies may safely be identified as imaginal disc cells. In such a primary culture the larval cell types eventually degenerate, and the diploid colonies frequently enter a stage of mitotic quiescence. Eventually, however, in many cases a line of diploid or near-diploid cells emerges and overgrows the culture. The line may then be indefinitely subcultured.

Undeniably some selective processes work upon the cells of a primary culture, eventually rendering these cells capable of unlimited division. A priori, one might imagine that some groups of cells might be more susceptible than others to such selection. For example, Schneider and Blumenthal (1978) have noted that only imaginal disc cells, brain cells, germ cells, and blood cells are known to be capable of further division in the advanced embryos sometimes being used as starting material. Certainly one would like to know, first, whether the cell lines arise from a limited range of cell types, and second, which cell types these are. This information might predict the kinds of ecdysteroid responsiveness to look for. Nevertheless, no unequivocal answers to these questions can yet be given. Since the results to be reported here will bear at least tangentially on these questions, we defer further consideration of the origins of the lines to the discussion.

Among the first of the *Drosophila* cell lines were the two designated K and C by Echalier and Ohanessian (1970), isolated in 1968. The sublines Ka, Kb, and Kc were derived from subpopulations of K cells by growth from low density (Debec, 1974). Kc cells were mostly female and haplo IV in karyotype (Dolfini 1971). Further sublines and clones have been obtained from Kc cells in Echalier's laboratory and in ours. Figure 1 provides a rough pedigree for the clones of interest. The clone K52(84) was grown for about two years in the laboratory of Prof. C. Thomas (Harvard Medical School) without further cloning. When he gave us the cells, we found them to be almost entirely (>90%) XO, haplo IV, the remaining small fraction of cells containing precisely twice this chromosome complement (Fig. 2). In view of the altered karyotype and of the fact that these cells resembled but were apparently not identical to Kc cells in growth properties and ecdysteroid response, we refer to them as the Harvard subline, or simply Kc-H cells.

Kc-H cells grow readily in suspension, and although some clones will attach to surfaces, their growth is never attachment dependent. For technical reasons we have chosen to work almost exclusively with suspension-growing cells. We use Echalier's medium D22 (Echalier 1976) which, although it contains undefined components, is inexpensive and easy to prepare, and supports vigorous cell multiplication. The medium is supplemented with 10% heat-inactivated fetal calf serum. Cells are grown at 25°C in plastic (bacteriological type) petri dishes, or for larger volumes, in spinner flasks. The gas phase is air. Most clones grow exponentially with a doubling time of 20–24 h between the concentrations of 5×10^5 cells/ml and $1–2 \times 10^7$ cells/ml. There is no evidence for density-dependent growth control; that is, when the cells stop dividing, the medium is unable to support further growth [see also Nakajima and Miyake (1976)].

We use a cloning technique that is identical in essentials to those described by Bernhard and Gehring (1975) and Richard-Molard and Ohanes-

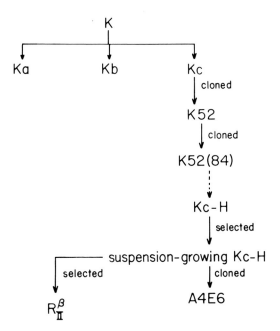

Fig. 1. Family tree of Kc-H cells. For details, see text.

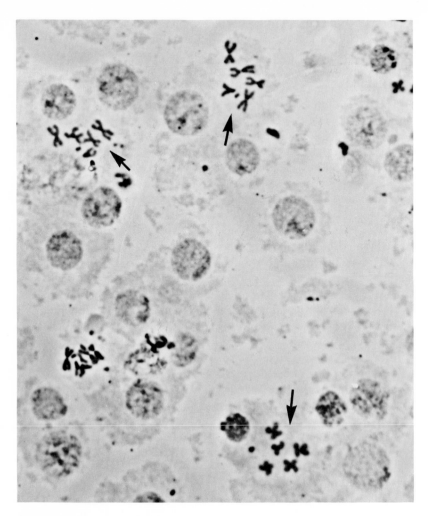

Fig. 2. Karyotype of Kc-H cells.   Arrows indicate three cells whose chromosomes are clearly spread.   In each case, four metacentric autosomes, one acrocentric X chromosome, and one dotlike chromosome IV are visible.

sian (1977).   One flask of cells is X-irradiated for use as a feeder layer (Puck and Marcus 1955).   As feeder layer we routinely use an ecdysteroid-resistant subline designated R_{II}^{β} derived from Kc-H cells.   A mixture is prepared containing feeder cells at 10^6 cells/ml and viable unirradiated cells at an appropriate concentration.   The suspension is then distributed amongst the wells of a 96-well microtiter dish at 100 μl per well.   The absolute cloning efficiency varies from clone to clone, but is always high (e.g., 85% for cells of the subline Kc-H, 50% for cells of the clone A4E6).

III. Ecdysteroid Effects

A. The Morphological Response

When Kc-H cells are treated with physiological concentrations of ecdy-steroid hormones, they undergo a dramatic morphological transformation, illustrated in Fig. 3. The untreated cells are small (7–8 μm in diameter) and undistinguished in appearance. When grown in suspension, these control cells are usually perfectly round (Fig. 3e), while their surface-growing counterparts generally bear short (<5 μm) processes (Fig. 3a). During 2–3 days of exposure to an ecdysteroid at saturating concentration, the cells elaborate distinctive long processes (Fig. 3b, c, d, f). Virtually every cell comes to bear one, two, three, or many such primary processes, which often extend to 50 μm or more in length. The primary processes themselves are always decorated with arbors of finer secondary processes (Fig. 3c and d), and the primary processes also bear inclusions (dark blebs in the photo-micrographs, Fig. 3b, c, d). A few minutes' examination of living cells reveals that the inclusions migrate along the processes.

The morphological transformation represents an exaggerated version of the ecdysteroid response of the parent Kc line (Courgeon 1972a) and of the Schneider's line 3 cells (Berger et al. 1978). The polarization of the cells is carried to an extreme, and the promiscuous elaboration of processes gen-erates a variety of cell shapes only hinted at in Fig. 3. The overall effect is most reminiscent of differentiating vertebrate neuroblastoma cell cultures as shown, for example, by Ross et al. (1975), Solomon (1979), or Spiegelman et al. (1979). Not surprisingly then, "differentiated" Kc-H cells also bear a striking resemblance to normal neural elements in culture [compare, for ex-ample, the study of chick embryonic optic lobe cells by Adler et al. (1979)].

We have observed the morphological progression of Kc-H cells in time-lapse films, using cells fastened to the plastic substrate by an intercalary layer of polylysine (Sanders et al. 1975). The films show that elongation is accompanied by a dramatic increase in the motility of the cell bodies. Typi-cally, the ends of long processes attach to the substrate and the cell body migrates back and forth between them. The appearance of this motility can be quantified (Fig. 4). In addition to this local motion, ecdysteroid-treated cells also aggregate. We find the rate and eventual extent of aggregation to be unpredictable, however, and to be extremely sensitive to minor altera-tions in the culture conditions.

It has seemed important to us to be able to assay independently each ec-dysteroid effect on these cells. In this vein we sought a way to quantify the morphological response. This turned out to be easy for, despite the variety of cell shapes represented, mean "cell length" defined in a particular way is a simple function of ecdysteroid concentration. Let λ be equal to the long-est possible path that can be drawn through one cell, starting at the visible tip of one process and extending through the center of the cell to the visible tip of a second process. We found that the mean value of λ increases with

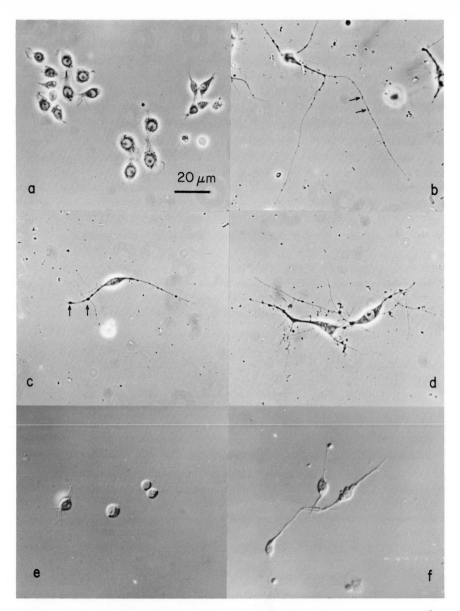

Fig. 3, a–f. The morphological response.   Kc-H cells (**a–d**) were grown on a glass surface; living cells were photographed in phase contrast optics.   A4E6 cells (**e, f**) were grown in suspension; living cells were photographed in Nomarski optics.   Cells were treated with 20-hydroxyecdysone as follows: **a** No treatment; **b** 10^{-5} M, 4 days; **c** 10^{-5} M, 3 days; **d** 10^{-5} M, 3 days; **e** no treatment; **f** 10^{-6} M, 3 days.   Arrows indicate inclusions in the processes, of the type discussed in the text.

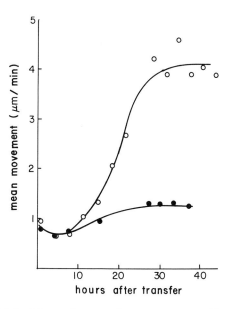

Fig. 4. Cell motion as a function of ecdysteroid treatment. Kc-H cells were transferred to fresh medium with (○) or without (●) 10^{-6} M 20-hydroxyecdysone, and observed by time-lapse cinematography. Motion was measured by plotting the center of an individual cell in 10 successive frames, photographed at 1-min intervals; the distances moved in 9 1-min intervals were averaged. Each point represents the mean of values from 13 to 23 cells.

ecdysteroid concentration in a sigmoid fashion. In practice, we use a still simpler procedure in which we determine the fraction of long cells (F_L), that is the fraction for which $\lambda > 17.2$ μm (a division on our ocular grid). With saturating ecdysteroid, F_L increases detectably by 8–10 h, continues to increase for 2–3 days, and then remains constant for another 1–2 days at least (L. Cherbas, C.D. Yonge, P. Cherbas, and C.M. Williams, unpublished manuscript). Figure 5 shows that even among uncloned Kc-H cells there is no significant fraction of nonresponding cells.

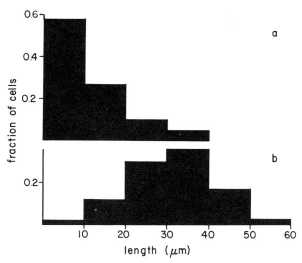

Fig. 5,a,b. Distribution of cell lengths as a function of ecdysteroid treatment. Kc-H cells were grown on a glass surface in the presence (**b**) or absence (**a**) of 10^{-7} M 20-hydroxyecdysone for 3 days, and then fixed and measured. Each histogram represents measurements of approximately 160 cells.

Using this simple and quantitative procedure to measure the morphological response we have been able to persuade ourselves that we are studying
an extremely sensitive and specific response to natural ecdysteroid hormones and their active analogs. The evidence for these assertions will be
published elsewhere; it can be summarized as follows:

1) Half-maximal induction of the morphological response in Kc-H cells
requires about 2×10^{-8} *M* 20-hydroxyecdysone (Fig. 6). This may be compared with published results showing that in organ culture the same steroid
(a) gives half-maximal puffing at 5×10^{-9} *M* to 2×10^{-7} *M*, depending on
the particular puff site (Ashburner, 1973), and (b) induces half-maximal
imaginal disc eversion at 3×10^{-8} *M* to 1×10^{-7} *M* (Chihara et al. 1972;
Fristrom and Yund 1976; Milner and Sang 1974). At metamorphosis,
D. melanogaster tissues are apparently exposed to concentrations of
20-hydroxyecdysone in the range $3-5 \times 10^{-7}$ *M* (Borst et al. 1974; Hodgetts
et al. 1977).

2) We have tested the activities of about 60 compounds that are structurally related to naturally occurring ecdysteroid hormones. Fourteen of
these compounds were active, and for each of these we determined a detailed dose–response curve. Every compound that exhibited activity in this
system was previously known to be active either in vivo or in organ culture.

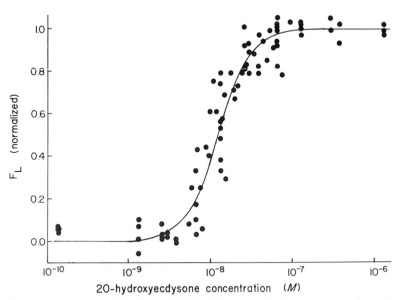

Fig. 6. Dose-response curve for the morphological response to 20-hydroxyecdy-
sone. A4E6 Cells were incubated for 2 days in various concentrations of 20-hydroxy-
ecdysone, and F_L measured for samples of 200 cells. Data were normalized by setting the mean value for 10^{-6} *M* 20-hydroxyecdysone at 1.0, and the mean value for
solvent-treated controls at 0. The line is fitted to a logit plot, using a least-squares
procedure. Details of the assay technique and statistical treatment of the data will be
published elsewhere.

Table 1. Relative Activities of Ecdysteroids in Cell and Organ Culture[a]

	Relative activity		
Compound	Kc-H cell morphology	Kc-H cell AChE induction	Imaginal disc eversion
Ponasterone A	8	9	20
Polypodine B	4	3	2
20-Hydroxyecdysone	[1]	[1]	[1]
Cyasterone	0.5	No data	1
Inokosterone	0.1	0.06	0.8
5-Hydroxy-22,25-deoxy-ecdysone	0.03	0.006	0.03
2-Deoxy-20-hydroxyecdysone	0.03	0.005	0.1
Ecdysone	0.006	0.004	0.002
22,25-Deoxyecdysone	0.003	0.004	Not detectable
2,22,25-Deoxyecdysone	<0.002	<0.002	Not detectable
14,15-Epoxy-22,25-deoxy-ecdysone	<0.002	<0.002	Not detectable
14,22-Deoxyecdysone	<0.002	<0.002	Not detectable

[a] Imaginal disc data from Fristrom and Yund (1976).

Furthermore, when we calculated relative activities among the active compounds, we found them to be in excellent agreement with the organ culture data. Table 1 shows one such comparison, with data reported by Fristrom and Yund (1976) for imaginal disc evagination in vitro.

3) We have been completely unsuccessful in attempts to elicit the morphological response without ecdysteroids. We have added vertebrate growth factors, insulin, fetuin, cholera toxin, cyclic nucleotides with and without phosphodiesterase inhibitors, dimethyl sulfoxide, butyric acid, and a long list of other potential agonists, all without effect. Similarly, we find that the response is neither induced nor inhibited by serum deprivation, or by prolonged maintenance in stationary phase. In short, we have been delighted to discover that ecdysteroid responsiveness in Kc-H cells is not only specific, but also exceptionally robust.

Our approach has been to exploit the morphological response as a convenient and accurate index of hormone responsiveness without pausing to investigate it as a phenomenon in its own right. Such investigation might prove rewarding; some of the directions it might take are indicated by previous studies of neuroblastoma cells. Presumably the elaboration of processes reflects or is accompanied by corresponding changes in the cytoskeleton. It is reasonable to expect, for example, that such changes might involve radical rearrangement of microtubule initiation sites (compare Spiegelman et al. 1979). Our results (see Sect. III, D) show that, whatever the basis of the transformation, it occurs without major changes in the pattern of protein synthesis. In a neuroblastoma line, Solomon (1979) has shown very clearly that daughter cells derived from division of a differentiated parent are morphologically similar even in detail. Our observations suggest that the same may be true in Kc-H cells. It is conceivable that the temporal and spatial regulation of the cytoskeleton might be particularly accessible experi-

mentally in a hormonally controlled system like Kc-H cells. Researchers at
at least one laboratory are now attempting to use Kc cells for just such studies
(Berger 1979).

B. The Effects on Proliferation

Ecdysteroids profoundly affect the rate at which Kc-H cells divide. The effects observed are illustrated in Fig. 7. 10^{-8} M 20-Hydroxyecdysone causes
a small and transient, but reproducible, increase in division. At 10^{-6} M this
stimulation is not observed; instead the cell number levels off after 24–48 h
and thereafter slowly declines. At the intermediate concentration of
10^{-7} M, the peak in cell number is followed by a very rapid declining phase.
Thus, there are three components of the response: stimulation, arrest, and
death.

Let us begin with the last, namely, the declining phase and cell death.
Time lapse films show that the rapid decline which occurs with 10^{-7} M but
not 10^{-6} M 20-hydroxyecdysone is accompanied by a high frequency of
abortive divisions, mitoses that yield only debris. These are very infrequent
at 10^{-6} M, at which concentration cells seldom enter mitosis at all. The
clone A4E6 derived from the Kc-H subline is significant because it undergoes an otherwise normal ecdysteroid reponse but does not die. When
treated with hormone these suspension-growing cells develop very long processes as in the case of the parent line, and their proliferation is arrested, but
they survive for at least 10 days healthy in appearance and with little debris
evident in the medium (Fig. 8). Therefore, proliferative arrest can effec-

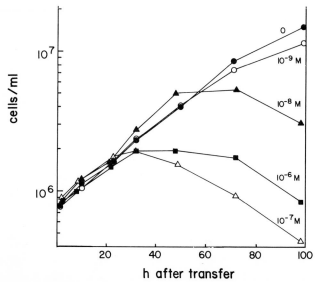

Fig. 7. Cell proliferation in the presence of 20-hydroxyecdysone. Suspension-growing Kc-H cells were transferred into fresh medium containing the indicated concentrations of 20-hydroxyecdysone. Cell density was determined by counting approximately 500 cells in a hemocytometer.

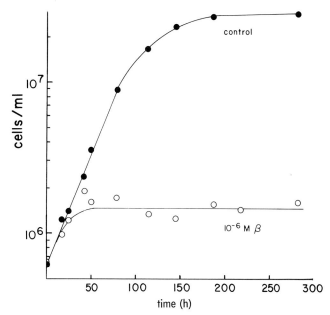

Fig. 8. Proliferation of A4E6 cells. Cells of the clone A4E6 were transferred to fresh medium with (○) or without (●) 10^{-6} *M* 20-hydroxyecdysone, and the cell density determined at intervals as in Fig. 7.

tively be separated from the complication of death. For this reason A4E6 has become our standard or wild-type clone. Occasionally even A4E6 cells die (cf. Fig. 9); this seems to depend on some uncontrolled variable in our culture conditions.

Our interpretation of the events in Fig. 7 is as follows. Ecdysteroids certainly cause a very early and small increase in cell numbers. By autoradiography we find that within 1 h of hormone addition there is a transient 10% increase in the fraction of cells labeled by a short [³H]thymidine pulse (C. Hsu, P. Cherbas, and L. Cherbas, unpublished observations, 1978). This is probably related to the comparably small increment in cell numbers that follows. We do not yet know the dose–response relationship for this stimulation, nor do we have a coherent explanation for the phenomenon. Stimulation is followed by an arrest of proliferation and, in some cases, by death. The time of onset of proliferative arrest depends on the hormone concentration (Figs. 7, 9). The resulting competition among stimulation, depression, and death results in a situation that is probably not fruitfully analyzed at this level.

The complexity of the response to 20-hydroxyecdysone does illuminate one long-standing question, however. Courgeon (1972b) reported that ecdysone and 20-hydroxyecdysone have qualitatively different effects on Kc cells. Her results were compatible with the idea that ecdysone stimulates morphological differentiation and division, while 20-hydroxyecdysone is relatively poor at inducing these and better at causing proliferative arrest. If 20-hydroxyecdysone alone can cause both stimulation and arrest, then there

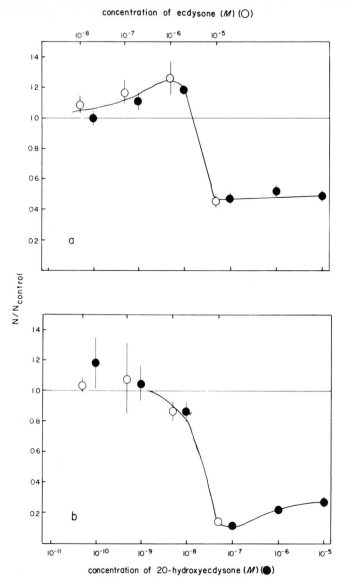

Fig. 9,a,b. Cell density as a function of 20-hydroxyecdysone or ecdysone concentration. A4E6 cells were transferred to fresh medium containing various concentrations of ecdysone (○) or 20-hydroxyecdysone (●) and the cell density determined at intervals thereafter. Dose-response curves are shown for 61 h (**a**) and 88 h (**b**) of culture. Note that the abscissa is not the same for the two compounds.

is the distinct possibility that what passes for a qualitative difference is really only a quantitative difference in activity. We tested this idea by exposing Kc-H cells to various concentrations of the two hormones and counting cells at various times thereafter. The results are summarized in Fig. 9, where the abscissa for ecdysone is shifted 200-fold with respect to that for 20-hydroxy-ecdysone to reflect their relative activities in our morphological assay (see Table 1). Evidently these two hormones are only quantitatively different. Wyss (1976) reported a similar result for Kc cells.

The ecdysteroid response in A4E6 cells resembles many developmental processes in that proliferative arrest is followed by morphological differentiation. We speculated that after hormone addition, each cell might undergo one final division prior to elongating. This would be consistent with Fig. 8, which shows that the A4E6 population approximately doubles after the ecdysteroid is added. Thereafter the cell number is constant, few cells die, and few incorporate [³H]thymidine. Thus, on the average an A4E6 cell divides once during continuous hormone treatment. However, time lapse films show that the careers of individual cells are quite variable. Many cells do divide precisely once during the first 50–60 h of treatment. But some cells do not divide at all, yet they elongate on schedule, while others divide twice or three times during this period irrespective of their lengths. Because ecdysteroid-treated cells aggregate, there is a substantial sampling bias that prevents us from analyzing the film data in a quantitative way. Still the records of individual cells are very striking. Occasionaly in cultures treated 3 days one sees a cell with giant processes that divides to yield daughters which immediately resume the elongate shape of the parent. Thus, we do not believe that there is any rigid linkage between division and differentiation in this system. The same conclusion is supported by observations of the pace of morphological differentiation in cells grown with limiting concentrations of serum. These serum-deprived cells grow extremely slowly, yet the kinetics of the ecdysteroid-induced morphological transformation and of acetylcholinesterase induction (see below) are unaltered (L. Cherbas, C.D. Yonge, P. Cherbas, and C.M. Williams, unpublished manuscript; M. Manteuffel-Cymborowska, L. Cherbas, P. Cherbas, and C.M. Williams, unpublished manuscript).

Cells treated with ecdysteroid eventually cease to proliferate. What is far more intriguing is that cells become committed to this behavior very early and very synchronously. We determine when a cell becomes irreversibly committed to reproductive death by suitable cloning experiments. The cells are exposed to hormone for various periods, washed, and then cloned in the absence of hormone. Kc-H cells that are simply washed and then cloned yield colonies with an efficiency (the absolute cloning efficiency) of 85%; for A4E6 cells the corresponding figure is 50%. These values are taken to represent 100% relative cloning efficiency for each set of cells. Figure 10 shows what happens to the relative cloning efficiency after pulses of 10^{-6} M 20-hydroxyecdysone (L. Cherbas, P. Cherbas, and C.M. Williams, unpublished manuscript). Addition of hormone is followed by a precipitous drop

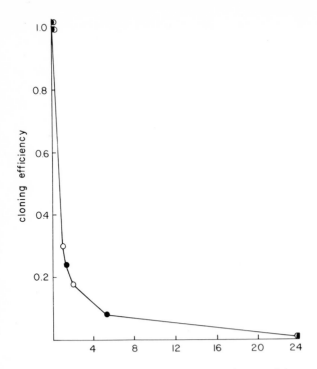

length of exposure to 20-hydroxyecdysone (h)

Fig. 10. Cloning efficiency after pulses of 20-hydroxyecdysone. Cells were treated as described in the text. Cloning efficiency for suspension-growing Kc-H cells (●) and A4E6 cells (○) is expressed as a fraction of the cloning efficiency of the corresponding untreated control cells.

in the cloning efficiency, so that by 1 h the value is about 30% and by 24 h much less than 1%. For comparison, note that about 1 cell in 2×10^5 (from these populations) can form clones in the continuous presence of ecdysteroid (see below).

Apparently commitment precedes any overt signs of declining proliferation by hours. For example, [^{3}H]thymidine pulses label a higher-than-control fraction of cells during the first 6 h of hormone treatment and only thereafter does the fraction fall below control.

Commitment probably requires protein synthesis because concomitant treatment with cycloheximide affords the cells virtually complete protection. In one such experiment, 5 h of $7 \times 10^{-5} M$ cycloheximide alone reduced the relative cloning efficiency to 63%. But cycloheximide plus 20-hydroxyecdysone for the same time gave 57%, that is, 90% of the cycloheximide value and fully 8 times the value obtained with hormone alone.

The rapidity of ecdysteroid-induced commitment in Kc-H cells may be instructively compared with the results in some other systems. Lymphoma cells, for example, are killed by glucocorticoids. Harris (1970) showed that in this system commitment occurs very slowly, over a span of several cell generations, and probably stochastically. Similarly, dimethyl sulfoxide treatment of Friend erythroleukemia cells leads to erythroid differentiation after a prolonged and apparently stochastic commitment period (Gusella et al. 1976). In each of these cases it is plausible to suppose that there is some probability that commitment will occur each time a cell passes through a particular period of the cell cycle. For Kc-H cells such a gate would have to include virtually the entire cell cycle, and would be meaningless. It is much more reasonable to suppose that commitment here is an early, programmed response to ecdysteroids akin to the induction of an early puff. Whatever the explanation, the synchrony of this early effect encouraged us to expect an early and synchronous biochemical response. We will return to this in Sect. III, D below.

C. The Induction[2] of Acetylcholinesterase Activity

The morphological differentiation of Kc-H cells is accompanied by the appearance of acetylcholinesterase (AChE) activity (Cherbas et al. 1977). *Drosophila* has one known AChE; it is coded for by the *Ace* locus, and in histochemical analyses it appears to be confined to the nervous system (Hall and Kankel 1976). The AChE found in ecdysteroid-treated Kc-H cells is identical in its substrate and inhibitor specificities and in its physical properties to the *Drosophila* enzyme, but its specific activity is about 50–100 times less than that in the *Drosophila* nervous system (Dewhurst et al. 1972). [The pseudocholinesterase of fetal calf serum is not a source of confusion in these experiments. Its properties are quite distinct from those of the *Drosophila* enzyme, and in addition, induction of AChE by 20-hydroxyecdysone occurs normally in cells incubated in the absence of serum (Manteuffel-Cymborowska et al., unpublished manuscript).]

The time course of AChE activity in cells treated with a saturating concentration of 20-hydroxyecdysone is illustrated in Fig. 11 (see also Cherbas et al. 1977). Acetylcholinesterase becomes detectable after about 24 h, and reaches a plateau after about 3 days. The dose–response curve is very steep, with induction increasing in the range of $2-5 \times 10^{-8}$ M 20-hydroxyecdysone (Fig. 12; Manteuffel-Cymborowska et al., unpublished manuscript). The hormone specificity of the induction of AChE activity is indistinguishable from that of the morphological response (Table 1; Manteuffel-Cymborowska et al., unpublished manuscript).

Knowing that cells are committed very rapidly to a later inhibition of cell division (see Fig. 10), we expected that short pulses of hormone would also commit cells to a later increase in AChE activity. This prediction was entirely wrong. When the hormone is withdrawn at any point during the induction process, not only do the cells fail to produce more AChE activity, but any enzyme activity already present disappears quite rapidly (Fig. 13;

---

[2] By "induction," we mean simply an elevation of enzyme activity, without implication of any particular molecular mechanism.

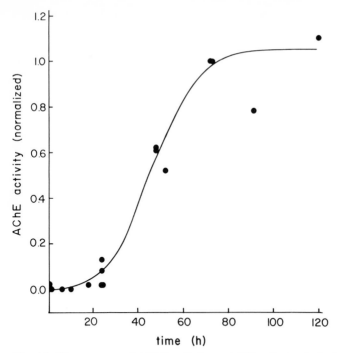

Fig. 11. Time course of AChE activity in A4E6 cells treated with 10^{-6} M 20-hydrox-yecdysone. Data from radiometric (Manteuffel-Cymborowska et al., unpublished manuscript) and colorimetric (L. Cherbas, and P. Cherbas, unpublished manu-script) assays are given; in both cases, the activity is normalized by setting the enzyme activity in extracts of cells treated for 3 days with 10^{-6} M 20-hydroxyecdy-sone as 1.0.

Manteuffel-Cymborowska et al., unpublished manuscript). Thus, whatever the early event is that makes cells unable to form clones, it does not commit them to a full hormone response. Instead, the ecdysteroid response seems to consist of a complex pathway, some parts of which—such as the inhibi-tion of proliferation—become independent of the hormone quite early, while other parts—such as the induction of AChE—require the continuous presence of the hormone.

Cells that have received a short pulse of ecdysone, and consequently do not have detectable AChE activity, are not, however, equivalent to un-treated cells with respect to their capacity to produce AChE. This point is illustrated by the following experiment. A4E6 cells are treated with 10^{-6} M 20-hydroxyecdysone for 24 h, incubated without hormone for a second 24-h period, and then treated again with 10^{-6} M 20-hydroxyecdysone. At the end of the first day, the cells have a barely detectable level of AChE activity (see Fig. 11). By the end of the second day, they do not contain detectable AChE. When the hormone is added a second time, AChE activity rises promptly and rapidly (Fig. 14). The response to the second hormone treat-

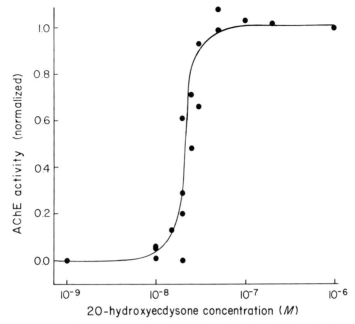

Fig. 12. Dose–response curve for the induction of AChE activity in A4E6 cells after three days of treatment with 20-hydroxyecdysone. Data were treated as in Fig. 11.

ment is much more rapid than the induction brought about by a single, continuous hormone treatment, though the final enzyme level is approximately the same in both regimens.

The time course of the subliminal "primary" effect was determined by the following experiment. Cells were treated with 10^{-6} M 20-hydroxyecdysone for various periods of time, and the effect of each treatment was measured by incubating the cells for 24 h without hormone, followed by a 24-h test period with hormone, and assaying the resultant AChE activity. The results of this experiment are shown in Fig. 15. Cells exposed to hormone for 1 h 24 h before the test period produced substantially more AChE activity than did naive cells; a 10-h initial pulse was sufficient to achieve a maximal effect.

Thus, while the actual induction and maintenance of AChE activity require the continuous presence of an ecdysteroid hormone, a portion of the induction process rapidly becomes independent of the presence of the hormone.

This sort of response to an interrupted hormone treatment is reminiscent of the primary and secondary stimulation of chick oviduct by estrogens (Towle et al. 1976). It is also possible that it simulates the response of tissues to the two pulses of ecdysteroid hormone associated with pupation in holometabolous insects. In the tobacco hornworm, *Manduca sexta,* pupation is preceded by two distinct pulses of ecdysteroid hormone, easily detectable by radioimmunoassay (Bollenbacher et al. 1975). The first hor-

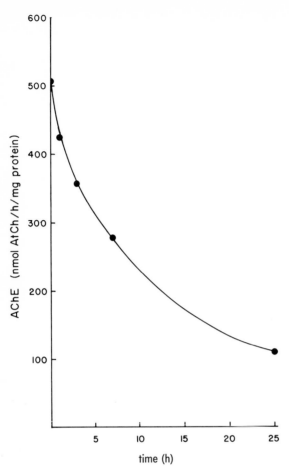

Fig. 13. The effect of ecdysteroid withdrawal upon AChE activity in A4E6 cells. Cells were treated for 3 days with 10^{-6} M 20-hydroxyecdysone, and then pelleted and resuspended in medium without hormone.   Cells were extracted at intervals, and AChE activity determined colorimetrically.

mone pulse leads to wandering behavior in *Manduca* larvae, and causes a juvenile hormone-sensitive commitment to metamorphosis in explanted larval epidermis (Riddiford 1976).   The second pulse induces the molt itself. Although analyses of *Drosophila* by radioimmunoassay have so far failed to demonstrate more than a single pulse of hormone before pupation (Hodgetts et al. 1977), two hormone treatments, separated by at least 3 h in hormone-free medium, are required to induce in isolated salivary glands the full series of chromosomal puffs associated with the pupal molt (Ashburner and Richards 1976).   Furthermore, in analogy with the *Manduca* system, the puffing changes induced by the second ecdysteroid treatment are at least partially sensitive to the presence of juvenile hormone during and after the first pulse (Richards 1978).

The Kc cell ecdysteroid response is partially sensitive to juvenile hormone

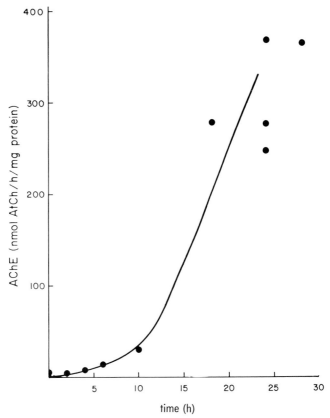

Fig. 14. The effect of a second ecdysteroid treatment upon AChE activity. A4E6 cells were treated with 10^{-6} *M* 20-hydroxyecdysone for 24 h, and incubated without hormone for 24 h. At time 0, 10^{-6} *M* 20-hydroxyecdysone was again added, and AChE determined at intervals thereafter, as in the experiment of Fig. 13.

(Courgeon 1975; Wyss 1976); we have obtained similar results with Kc-H cells. However, the period of juvenile hormone sensitivity is not restricted to the first pulse of ecdysteroid and the wash-out period as it is with salivary glands (L. Cherbas, unpublished observations, 1979). Hence, it is doubtful that the response of Kc-H cells to an interrupted ecdysteroid treatment can serve as a valid model for the response of tissues in vivo to a pair of ecdysteroid pulses. We should emphasize that this doubt does not detract from the usefulness of these experiments in dissecting the early steps of ecdysteroid action in Kc-H cells.

D. Early Changes in Protein Synthesis

Among the responses of Kc-H cells to ecdysteroid hormones are rapid changes in the capacity of the cells for growth (see Fig. 10) and in their ability to produce AChE activity when challenged with a second ecdysteroid treatment (see Fig. 15). The first effect, at least, appears to require protein

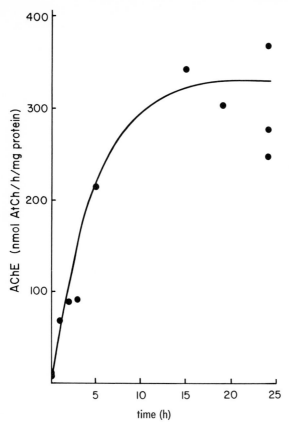

Fig. 15. The effect of ecdysteroid pulses upon the later induction of AChE. See text for details of the experiment. The abscissa indicates the length of the initial ecdysteroid treatment. AChE was determined as in the experiment of Fig. 13.

synthesis, since the decline in cloning efficiency is largely prevented by the presence of cycloheximide during the ecdysteroid treatment. One would expect, therefore, that ecdysteroid hormones cause rapid changes in the pattern of protein synthesis, and that these changes in protein synthesis are responsible for at least some of the hormone effects we have described.

In order to detect such changes in the pattern of protein synthesis, we incubated cells with or without 20-hydroxyecdysone, and then labeled them for 20 min with [^{3}H]leucine. The total cell protein was extracted and displayed by SDS–polyacrylamide gel electrophoresis, and the newly synthesized peptides visualized by autofluorography. The autofluorograms reveal no major changes in the pattern of protein synthesis during the first 3 days of ecdysteroid treatment, despite the enormous change in the appearance and division rate of the cells. But close examination of the pattern of radioactivity reveals a number of significant changes after hormone treatment. In particular, three clearly visible bands appear in the autofluorographic pattern within a few hours of the addition of hormone (Fig. 16; C. Savakis, G. Deme-

A B

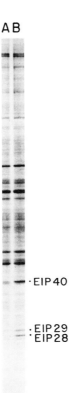

·EIP 40

·EIP 29
·EIP 28

Fig. 16. Pattern of protein synthesis before and after ecdysteroid treatment. A4E6 cells were labeled and their newly synthesized proteins visualized as described in the text. The photograph shows two channels from an autofluorogram. Channel A contains protein from untreated cells; channel B, from cells treated for 4 h with 10^{-6} M 20-hydroxyecdysone. The positions of 3 ecdysteroid-induced peptides are indicated.

tri, and P. Cherbas, unpublished manuscript). These bands correspond to peptides of molecular weight 40 000, 29 000, and 28 000 daltons, and are designated EIP40, EIP29, and EIP28, respectively. All three can be detected after 30 min of hormone treatment. The rate of synthesis of EIP40 increases for 4 h and declines thereafter; the synthesis of EIP29 and EIP28 increases for approximately the first 8 h of hormone treatment. All three peptides are induced only by active ecdysteroid hormones at appropriate concentrations, and only in ecdysteroid-sensitive clones (see below). In aggregate, they represent approximately 1% of the total protein synthesis of the cells after 4 h of hormone treatment.

We have used a mRNA-dependent cell-free protein-synthesizing system from rabbit reticulocytes (Pelham and Jackson 1976) to determine the titers of mRNA coding for the ecdysteroid-induced peptides in extracts of cells (C. Savakis and P. Cherbas, unpublished manuscript). The three induced peptides are translated from poly (A)-containing RNA species which are present at much higher concentrations in hormone-treated cells than in untreated control cells. The induction of EIP40, EIP29, and EIP28 is therefore brought about by an increase in the level of specific mRNA species.

IV. Ecdysteroid-Resistant Cells

Courgeon (1972a) reported that when Kc cells were maintained in the continuous presence of 20-hydroxyecdysone, the cultures were eventually taken over by round, dividing cells. Such "resistant cells" can also be selected easily from Kc-H cells; the subline R_{II}^{β}, which we use for feeder layers, was obtained in this way. In an effort to collect diverse types of resistant cells, we cloned A4E6 cells in the presence of 20-hydroxyecdysone (L. Cherbas, P. Cherbas, and C.M. Williams, unpublished manuscript). Approximately 1 in 2×10^5 cells can form a clone in the presence of the hormone. Twenty-five such clones have been collected and characterized to varying extents. So far, all of the resistant clones appear to be identical in their properties. Of those tested, none responds to 20-hydroxyecdysone or ecdysone with detectable elongation, induction of AChE activity, inhibition of proliferation, or synthesis of the ecdysteroid-induced peptides. In short, they all behave as if they were defective in the ecdysteroid receptor system.

V. Discussion

A. What Are Kc-H Cells?

We do not know what cells in the fruitfly Kc-H cells may be considered to represent. As we shall point out below, the usefulness of the cell line for studying steroid hormone action does not require that one know their origin or developmental state. Nonetheless, we have found it helpful to keep these questions in mind and to attempt occasionally to view our experiments in the perspective of the cells' history.

The Kc line is derived from a mixed culture of embryonic cells. Although the cell type(s) that gave rise to the line remains unknown, our observations on the ecdysteroid response may provide some clues. The two populations of cells in the *Drosophila* embryo, described in Sect. I, may be distinguished by the nature of their response to ecdysteroid hormones, both in vitro (Dübendorfer, personal communication, 1979) and in vivo. Differentiated larval cells die when treated with ecdysteroid hormones. Imaginal cells respond to ecdysteroid hormones by differentiating. The Kc-H cell response has elements of both death and differentiation. We would argue that differentiation—that is, the elaborate morphological transformation and the induction of AChE activity—is central to the response, while death—which does not necessarily occur—is an aberration attributable to defects either in the conditions of culture or in the cells themselves. We are suggesting, then, that Kc-H cells are imaginal, and their response to ecdysteroids is an abortive attempt at differentiation. This idea is consistent with a number of other observations. Imaginal disc cells remain diploid and divide extensively in the primary cultures from which the lines are derived. The serological properties of Kc cells are similar to those of imaginal discs and salivary glands (Moir and Roberts 1976). The spectrum of enzymatic activities in Kc cells is similar to that of imaginal discs and nervous tissue (Alahiotis and Berger 1977; Debec 1974, 1976).

If Kc-H cells are imaginal cells, then what is the cell type into which they attempt to differentiate? Both the presence of AChE activity and the appearance of the hormone-treated cells suggest some neural cell type, but here, too, the evidence is far from conclusive. The AChE specific activity of hormone-treated cells is low by neural standards, and the notorious morphological plasticity of cells in culture weakens any identification based upon the appearance of the cells. It is worth noting that AChE activity, either constitutive or inducible, has now been reported for Schneider's line 3 (Berger et al. 1978) and the C line of Echalier and Ohanessian (Best-Belpomme and Courgeon 1977), and is rumored to be found in a number of other independently derived cell lines. Perhaps AChE (at these low levels) is not confined to the nervous sytem. Or alternatively, neural precursors may have a strong selective advantage in the primary cultures.

B. Authenticity of the Response

Our major purpose is to use the Kc-H cell line for studying the mechanism of hormone action, and in particular the interaction of a hormone–receptor complex with hormone-sensitive genes. In such systems as the chick oviduct or the *Drosophila* imaginal disc, one is assured that the steroid hormone is acting in a normal manner, because it elicits a well-established normal response. But we do not know what the ecdysteroid response of Kc-H cells ought to be, and therefore we require some other form of evidence that the steroid response we are studying is similar in its basic mechanism to normal steroid responses in intact tissues.

Ecdysteroid hormones produce rapid changes in the patterns of transcription in target tissues (Ashburner and Richards 1976; Bonner and Pardue 1976, 1977). The experiments we have presented here suggest strongly that in Kc-H cells, as well, transcription at a number of genes is regulated by ecdysteroids. If gene regulation in Kc-H cells is mediated by a normal type of ecdysteroid receptor, that is, if 20-hydroxyecdysone is acting as an ecdysteroid hormone, then the regulation of transcription by the receptor–hormone complex in Kc-H cells can be used as a model for the regulation of transcription by steroid hormones.

Some of the properties of the ecdysteroid receptor system in Kc-H cells can be inferred from our data on the hormone specificity of the response. It is clear that the same compounds, at approximately the same concentrations, induce the ecdysteroid responses of Kc-H cells and cause imaginal disc eversion or initiate the puparial puffing pattern in salivary glands. This fact argues for a strong similarity between the ecdysteroid receptor systems in Kc-H cells and in intact *Drosophila* tissues. To the extent to which data are available, this conclusion is confirmed by studies of the binding of radiolabeled ecdysteroids (Maroy et al., 1978; Yund et al. 1978; O'Connor et al., Chap. 17 in this volume).

We would conclude, therefore, that the ecdysteroid hormones work by means of a normal receptor system in Kc-H cells, and therefore that the early effects of ecdysteroid hormones on transcription may be considered as a useful model for gene regulation by a steroid hormone.

C. The Number of Receptor Species

Our studies of hormone specificity have led us to conclude that there is a single receptor species—or at least a single species of hormone-binding site —that is responsible for all of the ecdysteroid effects we have studied, and is probably identical to the ecdysteroid-binding protein reported for Kc cells (Maroy et al. 1978; O'Connor, et al., Chap. 17 in this volume). This conclusion is based on the following generalizations:

1) All active ecdysteroids have parallel dose–response curves for a given hormone response, within the limits of certainty of our measurements. This has been shown for a large number of analogs in the case of the morphological response (L. Cherbas, C.D. Yonge, P. Cherbas, and C.M. Williams, unpublished manuscript) and the induction of AChE activity (M. Manteuffel-Cymborowska, L. Cherbas, P. Cherbas, and C.M. Williams unpublished manuscript), for 20-hydroxyecdysone and ecdysone for the effects on cell proliferation (Cherbas L., P. Cherbas, and C.M. Williams, unpublished manuscript), and the various effects of hormone withdrawal and re-addition on AChE activity (L. Cherbas and P. Cherbas, unpublished manuscript).

2) The ratio of activities of any two ecdysteroids is identical for all aspects of the response (Table I; Cherbas L., P. Cherbas, and C.M. Williams unpublished manuscript; Cherbas, L., and P. Cherbas, unpublished manuscript). Again, the morphological response and the induction of AChE activity have been compared for a number of compounds; for other responses only 20-hydroxyecdysone and ecdysone have been used.

3) Clones selected for the absence of one aspect of the response—the inhibition of proliferation—are lacking in all other aspects of the response. This holds true not only for unmutagenized cells (Cherbas, L., P. Cherbas, and C.M. Williams, unpublished manuscript; see above), but for mutagenized cells as well (L. Cherbas, unpublished observations, 1979).

4) The relative affinities of ecdysteroids for the ecdysteroid-binding protein of Kc cells (Maroy et al. 1978) is identical to the relative activities of the same compounds for inducing the various Kc-H cell responses.

All of these generalizations are readily compatible with the following simple model. A single ecdysteroid receptor species is found in the cells. The various hormone responses are determined by the concentration of ecdysteroid–receptor complex, regardless of the ecdysteroid bound. The activity of an ecdysteroid is therefore determined simply by its affinity for the receptor.

D. Prospects for Future Work

The experiments we have described in this contribution constitute a beginning in the development of a new system for the analysis of hormone action at the molecular level. The Kc-H cells provide a remarkably accessible system for studying the early steps in ecdysteroid action. The cells are easily grown in suspension and are much more stable in their properties than are most mammalian cell lines. The hormone response is dramatic, specific, and robust. Most important, the hormone appears to act by binding to a typical steroid receptor and causing a rapid induction of transcription at a small number of specific sites. Thus, we are now in a position to use the system to study some of the really interesting problems in molecular endocrinology.

The small size of the *Drosophila* genome, combined with the availability of large quantities of tissue culture material, makes the preparation of DNA clones containing rapidly induced genes relatively easy; our laboratory is now actively engaged in an attempt to isolate such clones. At the same time, the purification of an ecdysteroid receptor from Kc cells is well underway in another laboratory (O'Connor et al., Chapter 17 in this volume).

One particularly intriguing problem which has arisen in the course of these experiments is the nature of the early subliminal effect on AChE inducibility. Do ecdysteroids cause a change in the configuration of the *Ace* gene, converting it to a form whose transcription is controlled directly by the hormone? Or is there ecdysteroid control of translation or activation of AChE? Some of these questions may soon be approachable, since a DNA clone containing sequences of the *Ace* gene is now being isolated by an elegant combination of genetic, cytogenetic, and recombinant DNA techniques (Spierer, Bender, and Hogness, personal communication, 1979).

These directions for future work provide some examples of the ways in which the special properties of *Drosophila* may be used in studying the mechanism of steroid hormone action. Biochemical studies of the action of vertebrate steroids have met great success in the past two decades. It now appears that ecdysteroids act as typical steroid hormones, and that their action can be studied in a permanent cell line. The biochemical advantages of a cell line, combined with the unique technologies available for *Drosophila*, make the cell line an attractive experimental material in which to continue and complement the advances of the more traditional vertebrate systems.

Acknowledgments. The work done in our laboratory has been supported by grants from the National Institutes of Health to C. M. W., from the National Science Foundation to C. M. W. and to P. C. and L. C., from the American Cancer Society to P. C. and L.C., and from the Milton Fund of Harvard University to P. C. and to C. M. W. We thank Ms. Macy Koehler for her expert photographic assistance.

References

Adler R, Manthorpe M, Varon S (1979) Develop Biol 69: 424–435
Alahiotis S, Berger E (1977) Biochem Genetics 15: 877–883
Ashburner M (1971) Nature New Biol 230: 222–223

Ashburner M (1973) Develop Biol 35: 47–61

Ashburner M, Richards G (1976) In: Lawrence PA (ed) Insect development. Wiley, New York, pp 203–25

Berger, E. (1979). In Vitro 15: 211

Berger E, Ringler R, Alahiotis S, Frank M (1978) Develop Biol 62: 498–511

Bernhard HP, Gehring WI (1975) Experientia 31: 734

Best-Belpomme M, Courgeon A-M (1977) FEBS Lett 82: 345–347

Bollenbacher WE, Vedeckis WV, Gilbert LI, O'Connor JD (1975). Develop Biol 44: 46–53

Bonner JJ, Pardue ML (1976) Chromosoma 58: 87–99

Bonner JJ, Pardue ML (1977) Cell 12: 219–225

Borst DW, Bollenbacher WE, O'Connor JD, King DS, and Fristrom JW (1974) Develop Biol 39: 308–316

Cherbas P, Cherbas L, Williams CM. (1977) Science 197: 275–277

Chihara CJ, Petri WH, Fristrom JW, King DS (1972) J Insect Physiol 18: 1115–1123

Courgeon A-M (1972a) Exp Cell Res 74: 327–336

Courgeon A-M (1972b) Nature New Biol 238: 250–251

Courgeon A–M (1975) CR Acad Sci (Paris) 280: 2563–2565

Debec A (1974) Wilhelm Roux Arch 174: 1–19

Debec A (1976) Wilhelm Roux Arch 180: 107–119

Demerec M (ed) (1965) Biology of *Drosophila*. Hafner, New York

Dewhurst SA, Croker SG, Ikeda K, McCaman RE (1972) Comp Biochem Physiol 43B: 975–981

Dolfini S (1971) Chromosoma 33: 196–208

Dübendorfer A (1976) In: Kurstak E and Maramorosch K (eds) Invertebrate tissue culture. Academic Press, New York, pp 151–159

Dübendorfer A, Shields G, Sang JH (1975) J Embryol Exp Morphol 33: 487–498

Echalier, G. (1976) In: Kurstak E. and Maramorosch K. (eds.), Invertebrate tissue culture. Academic Press, New York, pp. 131–150

Echalier G, Ohanessian A (1969) CR Acad Sci (Paris) 268: 1771–1773

Echalier G, Ohanessian A (1970) In Vitro 6: 162–172

Fristrom JW, Yund MA (1976) In: Maramorosch K (ed) Invertebrate tissue culture. Academic Press, New York, pp 161–178

Goodwin TW, Horn DHS, Karlson P, Koolman J, Nakanishi K, Robbins WE, Siddall JB, Takemoto T (1978) Nature 272: 122

Gusella J, Geller R, Clarke B, Weeks V, and Housman D (1976) Cell 9: 221–229

Hall JC, Kankel DR (1976) Genetics 83: 517–535

Harris AW (1970) Exp Cell Res 60: 341–353

Hodgetts RB, Sage B, O'Connor JD (1977) Develop Biol 60, 310–317

Kakpakov VT, Gvozdev VA, Platova RP, Polokarova LG (1969) Genetika (USSR) 5: 67–75

Maroy P, Denis R, Beckers C, Sage BA, O'Connor JD (1978) Proc Natl Acad Sci USA 75: 6035–6038

Milner MJ, Sang JH (1974) Cell 3: 141–143

Moir A, Roberts DB (1976) J Insect Physiol 22: 299–307

Nakajima S, Miyake T (1976) In: Kurstak E, Maramorosch K (eds) Invertebrate tissue culture. Academic Press, New York, pp 279–287

Pelham HRB, Jackson RJ (1976) Eur J Biochem 67: 247–256

Puck TT, Marcus PI (1955) Proc Natl. Acad Sci USA 42: 432–437

Richard-Molard C, Ohanessian A (1977) Wilhelm Roux Arch 181: 135–149

Richards G (1978) Develop Biol 66: 32–42

Riddiford LM (1976) Nature 259: 115–117

Ross J, Olmsted JB, Rosenbaum JL (1975) Tissue Cell 7: 107–136

Sanders SK, Alexander EL, Braylan RC (1975) J Cell Biol 67: 476–480

Schneider I (1972) J Embryol Exp Morphol 27: 353–365

Schneider I, Blumenthal AB (1978) In: Ashburner M, Wright TRF (eds) The genetics and biology of *Drosophila*. Academic Press, London, Vol 2a, pp 266–316

Seecof RL, and Unanue RL (1968) Exp Cell Res 50: 654–660

Shields G, Dübendorfer A, Sang JH (1975) J Embryol Exp Morphol 33: 159–175

Solomon F (1979) Cell 16: 165–169

Spiegelman BM, Lopata MA, Kirschner MW (1979) Cell 16: 253–263

Towle HC, Tsai M-J, Hirose M, Tsai SY, Schwartz RJ, Parker MG, O'Malley BW (1976) In: Papaconstantinou J (ed) The molecular basis of hormone action. Academic Press, New York, pp 107–136

Wyss C (1976) Experientia 32: 1272–1274

Yund MA, King DS, Fristrom JW (1978) Proc Natl Acad Sci USA 75: 6039–6043

Discussion on the Paper Presented by P. Cherbas*

CLARK: You explained quite nicely that there was a dose–response curve for the early puffs, and the later puffs were essentially "on–off" type responses. Now, does that imply that, if you give a very low dose of hormone and create an early puff, you will see all other puffs or only some other puffs?

CHERBAS: The answer is simple. There is a threshold for the late puff. If you expose to a dose lower than that threshold, you get the early puffs. You simply never get the late.

CLARK: But what if you use just above the threshold, one-tenth of maximum, and get a puff, do you get later puffs?

CHERBAS: Yes, they all go on; they're only in some fraction of the cells.

CLARK: So then you have a case involving spare receptors. You have extra sites that don't necessarily have to be activated to observe a response.

CHERBAS: Well yes, if that has anything to do with receptors at all; perhaps in fact it's simply a question of a cooporation of early events which is what seems more likely. The model could have nothing to do with spare receptors, but it's simply a question of a mathematically, formally, higher order of response, where, for example, the induction of a late protein is dependent on a third or fourth power of some early effects. I think that's the simplest model.

STEVENS: In your presentation you made the point that one of the effects of ecdysone was that it works extremely rapidly on the cells you were studying and that washing out for about 45 min permitted a fair degree of expression of the ecdysone effects, and by analogy with the S49 cells it was suggested that in these lymphocytes the steroid effects caused by dexamethasone would be very, very slow in comparison to the insect system. Since I work with lymphocytes, I would like to throw in some experiments that have been done initially in Allan Munck's lab on the effect of dexamethasone and cortisol on inhibition of glucose uptake by thymocytes in which Dr. Munck and his co-workers were able to show that this effect required only 5 min exposure of the cells to the steroid for subsequent inhibition of glucose uptake to appear. In my own laboratory we have been able to show with the P1798 lymphoma cells that simple exposure of these cells for 30 min to cortisol and then washing out the steroid enabled subsequent appearance in the inhibition of uridine uptake to appear. Now admittedly neither the inhibition of glucose uptake nor the inhibition of uridine uptake can be correlated with something as dramatic as cell death, but I submit that other steroids and other systems also can bring about phenotypic effects very, very rapidly, and I wonder whether there is anyone in this audience, perhaps Dr. Thompson, who has tried very short exposure of this cloned lymphoma cells to see how long they have to see the steroid in order to obtain a response.

THOMPSON: Eighteen hours.

* All responses in this discussion are made by P. Cherbas.

CHERBAS: My point was a simple one. I wasn't trying to say there was anything unique about the early effects of steroid. I'm sure there are lots of early effects of steroids that have to deal specifically with this commitment question. As far as I can tell from the literature, there is no evidence of such a rapid commitment demonstrated before, something that can be shown to be sort of determinate. In fact, something that interests me more is the fact that it seems not to be stochastic. If other people have any contradictory data, I would like to hear them.

MUELLER: I think you showed that if you exposed your cloned cells to given levels of the ecdysone for different given periods, you dropped your clonability dramatically. Now, what I was wondering is whether you get the same kind of result again whether or not the survivors come through that situation, if you put them through a second cycle?

CHERBAS: We get the same result.

MUELLER: So in effect, you actually have a group of cells that for some reason or other do not respond to the ecdysone. There must be another factor present in these cultures that is conditioning some ecdysone sensitivity.

CHERBAS: Yes, there are variations that do not appear to be genetic within the population. I don't know the basis of that. Simply 90% of the cells are committed within 4 or 5 h.

GELEHRTER: I am a little confused about the ecdysone resistance. If I understood you correctly the Kc line was isolated because it wasn't killed by ecdysone and yet it obviously has ecdysone responses, morphologically, and biochemically, in effect, growth inhibition. Now from that line of isolated ecdysone resistant cells, how stable is that resistance and how much resistance is there? That also gets tied up with Dr. Mueller's question about resistance. Is it a random event in the population and not a mutation?

CHERBAS: I am not sure that I understand the beginning of the question.

GELEHRTER: Would you say that when you expose the cells to ecdysone, that they changed their morphology.

CHERBAS: The initial report was that most cells eventually died in the presence of ecdysone. We find that it is very easy to get clones of cells which do not die, but simply survive, in the presence of ecdysone, but they do not divide. They are functionally ecdysone sensitive. I can't say anything quantitative about the frequency of these clones. It is also possible with the frequency of about 1 in 10^6 to get cells that continue dividing at a normal rate in the presence of ecdysone, so far as we can tell, and are entirely stable during growth in the absence of ecdysone. I have been growing both types of cells for two or three years.

MOUDGIL: I compliment you on a good talk. The first question is you showed that cholinesterase was induced. I was wondering if you looked at the activity of cholineacetylase. Whether the ratios between the two, that is, acetylcholine formation and degradation, has anything to do with induction because the data shown do not really justify the term induction. It could just result from lower degradation or some other factor.

CHERBAS: I did not intend by the statement "induction" to say anything about the mechanism. I know nothing about whether this is protein synthesis. I have my biases, but we have no data on this question. There is enzymatic activity that we can measure. In mixing experiments between what we call induced cells and noninduced cells the total activity is the sum of the two extracts and there is no inhibition.

MOUDGIL: The second question is related to the specificity of induction. Have you examined any other steroid besides ecdysone for induction?

CHERBAS: Yes, as I said, we actually tested some 60 compounds, both in terms of morphology and cholinesterase. These include biosynthetic precursors of ecdysone, the natural ecdysones, and many compounds of related structure, and so far as we can tell, the vast majority of these are inactive and all of the standard mammalian steroids are inactive.

MOUDGIL: To be specific, what I mentioned was something like estradiol. It is known that in diversified tissues like lung, brain, and some other tissues, estradiol induces acetylcholinesterase. I was wondering if you tried any sex steroids, like estradiol?

CHERBAS: Well, in terms of morphology, we tried them all. But to be quite honest, I don't know whether we tried them in terms of cholinesterase. The general pattern is so specific that I would be very surprised if it worked.

DESHPANDE: Is differentiation essential for the survival of these cells?

CHERBAS: The ecdysone resistant cells survive very well without differentiating at all. I am not sure that I understand the question.

DESHPANDE: You get the cells differentiated into something like neural tissue, that is the cells tend to get elongated on continued culture and start producing acetylcholinesterase. Have you looked for any neurosecretory granules or something like that in those cells?

CHERBAS: I don't know whether these cells are neural. I somehow doubt it. Our attempt to find neurotransmitters in these cells has been uniformly negative. The cells do not have choline acetyl transferase. In what John Hildebrand calls the "hot zap" experiment, where you add lots of hot precursors for a variety of known invertebrate neurotransmitters, we can't find any of them in these cells.

DESHPANDE: How can you say that these differentiated tissues are from the larval type and not from the adult type?

CHERBAS: No, I didn't say that. What I said is that these cells, in response to ecdysone, undergo morphological changes. They also develop cholinesterase activity. They may well be related to neurons. That doesn't mean that they are larva or adult. I don't think that there is any very pressing reason to think that they are related to neurons; however, the possibility exists. I am not sure why you think that makes them larval.

DESHPANDE: Because if they are larval, then they must die.

CHERBAS: The dogmatic description of the nervous system in insects holds that larvae contain neurons which die during metamorphosis, and neuron mother cells that give rise to adult neurons. The description that I have gotten from the insect neurobiologists is that the situation is entirely in flux and that almost any pattern of response to ecdysone may be typical of the nervous system. No one is willing to give me an exact answer. Frankly, I think the question, at this time, is entirely unanswerable.

DESHPANDE: Do you think that the neural tissue that is differentiated from your cells might be producing some juvenile hormone.

CHERBAS: I have no evidence on the subject.

DESHPANDE: Only one hormone may not have the final effect on these systems. An interaction between both of these hormones seems to be necessary, but I have not heard you refer to the other hormone (the juvenile hormone).

CHERBAS: The reason why you haven't heard very much about juvenile hormone is that not very much is known about juvenile hormone in *Drosophila*. In the case of some of the simpler systems like the salivary glands, which have been more extensively studied, one can in fact mimic most of the in vivo response with ecdysone, and so somebody will have to discover something about juvenile hormones before we can say much about it.

O'CONNOR: In the Kc cells at least we know that the juvenile hormone is metabolized very rapidly. Within about 5 min the epoxide group is opened up to a diol and the methyl ester and carboxyl end is taken off, so instead of juvenile hormone, you end up with an acid diol floating around in your cells.

FRIEDMAN: I think the last questioner made an assumption that was in error, that is, that all larval tissues are histolyzed during metamorphosis. Malphigian tubules, brain, ovaries and some other tissues, like fat bodies, are carried over intact to adults.

MOUDGIL: One point I wanted to bring to notice was that the presence of acetylcholinesterase does not imply that the tissue or the cell is nervous in origin because actylcholinesterase has also been found in sperm and some other nonneural tissues.

CHERBAS: This may be relevant in the case of vertebrates. However, in the case of flies there is a dogmatic statement that one can find all over the literature. That is, there is only one cholinesterase in insects and there is no evidence for any pseudo-cholinesterase. If one runs a histochemical test on fresh frozen section, one finds it in one tissue and one tissue only, that is, the nervous system. What I am saying is that we must be more conservative and I don't know whether cholinesterase is at $1/100$th of the brain activity is a specific marker. We simply have to look and see. However, I want to emphasize that I know of no evidence of acetylcholinesterase in fly or any other insects in any tissue other than the nervous tissue.

Discussants: P. CHERBAS, J.H. CLARK, A.K. DESHPANDE, T.B. FRIEDMAN, T.D. GELEHRTER, G.C. MUELLER, J.D. O'CONNOR, J. STEVENS, and E.B. THOMPSON.

Subject Index